Biology of Methylotrophs

BIOTECHNOLOGY

JULIAN E. DAVIES, Editor
PASTEUR INSTITUTE
PARIS, FRANCE

Editorial Board

L. Bogorad	Harvard University, Cambridge, USA
J. Brenchley	Pennsylvania State University, University Park, USA
P. Broda	University of Manchester Institute of Science and Technology, Manchester, United Kingdom
A.L. Demain	Massachusetts Institute of Technology, Cambridge, USA
D.E. Eveleigh	Rutgers University, New Brunswick, USA
D.H. Gelfand	Cetus Corporation, Emeryville, California, USA
D.A. Hopwood	John Innes Institute, Norwich, United Kingdom
S.-D. Kung	University of Maryland, College Park, USA
J.-F. Martin	University of Leon, Leon, Spain
C. Nash	Schering-Plough Corporation, Bloomfield, New Jersey, USA
T. Noguchi	Suntory, Ltd., Tokyo, Japan
W. Reznikoff	University of Wisconsin, Madison, USA
R.L. Rodriguez	University of California, Davis, USA
A.H. Rose	University of Bath, Bath, United Kingdom
P. Valenzuela	Chiron, Inc., Emeryville, California, USA
D. Wang	Massachusetts Institute of Technology, Cambridge, USA

BIOTECHNOLOGY SERIES

1. R. Saliwanchik — *Legal Protection for Microbiological and Genetic Engineering Inventions*

2. L. Vining (editor) — *Biochemistry and Genetic Regulation of Commercially Important Antibiotics*

3. K. Herrmann and R. Somerville (editors) — *Amino Acids: Biosynthesis and Genetic Regulation*

4. D. Wise (editor) — *Organic Chemicals from Biomass*

5. A. Laskin (editor) — *Enzymes and Immobilized Cells in Biotechnology*

6. A. Demain and N. Solomon (editors) — *Biology of Industrial Microorganisms*

7. Z. Vaněk and Z. Hošťálek (editors) — *Overproduction of Microbial Metabolites: Strain Improvement and Process Control Strategies*

8. W. Reznikoff and L. Gold (editors) — *Maximizing Gene Expression*

9. W. Thilly (editor) — *Mammalian Cell Technology*

10. R. Rodriguez and D. Denhardt (editors) — *Vectors: A Survey of Molecular Cloning Vectors and Their Uses*

11. S.-D. Kung and C. Arntzen (editors) — *Plant Biotechnology*

12. D. Wise (editor) — *Applied Biosensors*

13. P. Barr, A. Brake, and P. Valenzuela (editors) — *Yeast Genetic Engineering*

14. S. Narang (editor) *Protein Engineering: Approaches to the Manipulation of Protein Folding*

15. L. Ginzburg (editor) *Assessing Ecological Risks of Biotechnology*

16. N. First and F. Haseltine (editors) *Transgenic Animals*

17. C. Ho and D. Wang (editors) *Animal Cell Bioreactors*

18. I. Goldberg and J.S. Rokem (editors) *Biology of Methylotrophs*

Biology of Methylotrophs

Edited by

Israel Goldberg and J. Stefan Rokem
Department of Applied Microbiology
The Hebrew University
Jerusalem, Israel

Butterworth–Heinemann
Boston   London   Oxford   Singapore   Sydney   Toronto   Wellington

Copyright © 1991 by Butterworth–Heinemann, a division of Reed Publishing (USA) Inc.
All rights reserved.

No part of this publication may be reproduced, stored in a retrieval system, or transmitted, in any form or by any means, electronic, mechanical, photocopying, recording, or otherwise, without the prior written permission of the publisher.

Recognizing the importance of preserving what has been written, it is the policy of Butterworth–Heinemann to have the books it publishes printed on acid-free paper, and we exert our best efforts to that end.

Editorial and production supervision by Science Tech Publishers, Madison, WI 53705.

Library of Congress Cataloging-in-Publication Data
Goldberg, Israel, 1943–
   Biology of methylotrophs / edited by Israel Goldberg and J. Stefan Rokem.
     p. cm. — (Biotechnology series)
   Includes bibliographical references and index.
   ISBN 0-7506-9188-3
   1. Methylotrophic microorganisms.   I. Rokem, J. Stefan.
   II. Title.   III. Series.
QR88.4.G65   1991
589.9′01—dc20
                                                               90-22483
                                                                   CIP

British Library Cataloguing in Publication Data
Biology of methylotrophs.
  1. Methylotrophy. Applications of biotechnology.
  I. Goldberg, Israel II. Rokem, J. Stefan III. Series
  576.162

ISBN 0-7506-9188-3

Butterworth–Heinemann
80 Montvale Avenue
Stoneham, MA 02180

10 9 8 7 6 5 4 3 2 1

Printed in the United States of America

To Frida and Ann-Marie

CONTRIBUTORS

Christopher Anthony
Department of Biochemistry
University of Southampton
Southampton, England

William A. Corpe
Department of Biological Sciences
Barnard College and Columbia University
New York, New York

James M. Cregg
Department of Chemical and Biological Sciences
Oregon Graduate Institute
Beaverton, Oregon

Lubbert Dijkhuizen
Department of Microbiology
University of Groningen
Haren, The Netherlands

Johannis A. Duine
Department of Microbiology and Enzymology
Kluyver Laboratory of Biotechnology
Delft, The Netherlands

Thomas Egli
Department of Technical Biology
Swiss Federal Institute for Water Resources and Water Pollution Control
Swiss Federal Institute of Technology
Zurich, Switzerland

Lars Ekeroth
Department of Biotechnology
Technical University of Denmark
Lyngby, Denmark

Larry E. Erickson
Department of Chemical Engineering
Kansas State University
Manhattan, Kansas

Israel Goldberg
Department of Applied Microbiology
The Hebrew University
Jerusalem, Israel

Richard S. Hanson
Gray Freshwater Biological Institute
University of Minnesota
Navarre, Minnesota

Thomas E. Jensen
Department of Biological Sciences
Lehman College C.U.N.Y.
West Bronx, New York

Kazuo Komagata
NODAI Research Institute
Tokyo University of Agriculture
Tokyo, Japan

Mary E. Lidstrom
Department of Environmental Engineering Science
California Institute of Technology
Pasadena, California

Colin A. Mason
Department of Technical Biology
Swiss Federal Institute for Water
  Resources and Water Pollution
  Control
Swiss Federal Institute of
  Technology
Zurich, Switzerland

J. Rodney Quayle
Vice-Chancellor
University of Bath
Bath, England

J. Stefan Rokem
Department of Applied
  Microbiology
The Hebrew University
Jerusalem, Israel

Victoria A. Romanovskaya
Department of Biology of Gas-
  Oxidizing Microorganisms
Institute of Microbiology and
  Virology
Kiev, USSR

Ivan G. Sokolov
Institute of Microbiology and
  Virology
UkrSSR Academy of Sciences
Kiev, USSR

Yoshiki Tani
Research Center for Cell Tissue
  Culture
Faculty of Agriculture
Kyoto University
Kyoto, Japan

Juerg F. Tschopp
Telios Pharmaceuticals Inc.
San Diego, California

Yuri D. Tsygankov
Institute for Genetics and
  Selection of Industrial
  Microorganisms
Moscow, USSR

Pravate Tuitemwong
Department of Chemical
  Engineering
Kansas State University
Manhattan, Kansas

Johannes P. van Dijken
Department of Microbiology and
  Enzymology
Kluyver Laboratory of
  Biotechnology
Delft, The Netherlands

John Villadsen
Department of Biotechnology
Technical University of Denmark
Lyngby, Denmark

Elizabeth V. Wattenberg
Gray Freshwater Biological
  Institute
University of Minnesota
Navarre, Minnesota

CONTENTS

Preface	*xv*
Introduction by J.R. Quayle	*xix*

PART I. TAXONOMY AND MORPHOLOGY

1. Taxonomy of Methylotrophic Bacteria — 3
V.A. Romanovskaya
- 1.1 The Methane-Oxidizing Bacteria — 4
- 1.2 Obligate and Restricted Facultative Methanol-Utilizing Bacteria — 13
- 1.3 Facultative Methylotrophic Bacteria — 16
- 1.4 Relationship of Methylotrophic Bacteria Taxa with Other Prokaryotic Taxa — 17
- 1.5 A New Classification Scheme for the Methylotrophs — 20
- References — 21

2. Systematics of Methylotrophic Yeasts — 25
K. Komagata
- 2.1 General Characteristics of Methanol-Utilizing Yeasts — 26
- 2.2 Methane-Utilizing Yeasts — 34
- 2.3 Applications of Methanol-Utilizing Yeasts — 34

xii   Contents

	2.4	Conclusion	35
		References	35
3.	**Ultrastructure of Methylotrophic Microorganisms**		39
	Thomas E. Jensen and William A. Corpe		
	3.1	Prokaryotic Cells	40
	3.2	Eukaryotic Microbes	69
		References	71

PART II. GROWTH AND METABOLISM

4.	**Assimilation of Carbon by Methylotrophs**		79
	C. Anthony		
	4.1	The Ribulose Bisphosphate (RuBP) Pathway	80
	4.2	The Ribulose Monophosphate (RuMP) Pathway	87
	4.3	The Dihydroxyacetone (DHA) Cycle of Formaldehyde Assimilation in Yeasts	98
	4.4	The Serine Pathway of Formaldehyde Assimilation	101
	4.5	Distribution and Occurrence of the Assimilation Pathways	107
		References	109
5.	**Oxidation Pathways in Methylotrophs**		111
	J. Stefan Rokem and Israel Goldberg		
	5.1	Bacterial Oxidation of C_1 Compounds	111
	5.2	Oxidation of C_1 Compounds in Yeasts	122
	5.3	Conclusions and Biotechnological Implications	124
		References	124
6.	**Regulation of Oxidation and Assimilation of One-Carbon Compounds in Methylotrophic Bacteria**		127
	L. Dijkhuizen and I.G. Sokolov		
	6.1	Regulation of the Oxidation of C_1 Compounds in Bacteria	128
	6.2	Regulation of the Assimilation of C_1 Compounds in Bacteria	139
		References	145
7.	**Growth Yields, Productivities, and Maintenance Energies of Methylotrophs**		149
	L.E. Erickson and P. Tuitemwong		
	7.1	Basic Bioenergetic Concepts	150
	7.2	Aerobic Growth and Product Formation	157
	7.3	Anaerobic Growth and Product Formation	167

	7.4	Conclusions	169
		References	169
8.	**Mixed Substrates and Mixed Cultures**		173
	T. Egli and C.A. Mason		
	8.1	Introduction	173
	8.2	Methylotrophic Bacteria	175
	8.3	Methylotrophic Yeasts	183
		References	198

PART III. USES OF METHYLOTROPHS

9.	**Single Cell Protein Production from C_1 Compounds**		205
	L. Ekeroth and J. Villadsen		
	9.1	Overview of Single Cell Protein Use	205
	9.2	Industrial SCP Production Using C_1 Compounds	210
	9.3	Is There a Future for SCP Production?	224
		References	230
10.	**Enzymes of Industrial Potential from Methylotrophs**		233
	J.A. Duine and J.P. van Dijken		
	10.1	Hydrolases and Lyases	236
	10.2	Oxidoreductases	238
	10.3	Perspectives	247
		References	248
11.	**Production of Useful Chemicals by Methylotrophs**		253
	Yoshiki Tani		
	11.1	Production of Chemicals by C_1-Oxidative Enzymes	254
	11.2	Production of Respiratory Coenzymes	257
	11.3	Glycerol Production in Yeast	261
	11.4	Polyol Production in Yeast	263
	11.5	Amino Acid Production	264
	11.6	Poly-β-Hydroxybutyric Acid Production	265
	11.7	Production of Other Metabolites	266
	11.8	Conclusions	266
		References	267

PART IV. GENETICS

12.	**Molecular Genetics of Methylotrophic Bacteria**		273
	Mary E. Lidstrom and Yuri D. Tsygankov		
	12.1	General Molecular Genetics	274
	12.2	Mutagenesis	281
	12.3	Surrogate and Classical Genetics	283

	12.4	Gene Cloning in the Methylotrophs	289
	12.5	Regulation and Genetics of Amino Acid Biosynthesis	289
	12.6	Genes Involved in C_1 Oxidation and Assimilation	294
		References	301

13. Heterologous Gene Expression in Methylotrophic Yeast — 305
Juerg F. Tschopp and James M. Cregg

13.1	Vectors and Hosts Used for Heterologous Gene Expression	307
13.2	Expression of Heterologous Proteins in the Cytoplasm	312
13.3	Expression and Secretion of Heterologous Proteins	316
13.4	Intracellular Transport and Post-Translational Modifications	319
13.5	Discussion	320
	References	320

PART V. ECOLOGY

14. Ecology of Methylotrophic Bacteria — 325
R.S. Hanson and E.V. Wattenberg

14.1	Role of Methane in the Biosphere	326
14.2	Ecological Importance of Soluble Methane Monooxygenase in Methanotrophic Bacteria	328
14.3	Methods for Estimation of Populations of Methane-Utilizing Bacteria in Soil and Water Samples	330
14.4	Methane Production and Oxidation in Aquatic Ecosystems	334
14.5	Distribution of Methane-Oxidizing Bacteria in Freshwater Environments	336
14.6	Methanotrophic Bacteria in Seawater	338
14.7	Methane Utilization in Soils	339
14.8	Are Soils and Aquatic Environments Sinks for Atmospheric Methane?	339
14.9	Fortuitous Metabolism of Methanotrophs and Its Environmental Significance	340
14.10	Anaerobic Methane Oxidation	341
14.11	Ecology and Diversity of Bacteria That Grow on C_1 Substrates Other Than Methane	342
	References	344

Index — *349*

PREFACE

This book is intended for all those who feel in need of some orientation in the vast and fascinating field of the methylotrophs—a group of microorganisms that utilize carbon compounds lacking carbon-carbon bonds, i.e., one-carbon (C_1) compounds.

This book highlights the most important biological features of these biosynthetically versatile organisms and provides information and data on their taxonomy, physiology, biochemistry, nutrition, genetics, and ecology. We hope this volume will serve as a time-saving reference when information about the biology of a particular methylotrophic microorganism is needed before a process utilizing it can be developed.

The ultrastructure, physiology, biochemistry, and genetics of the methylotrophs differ in many aspects from that of the more abundant, heterotrophic microorganisms, primarily because the methylotrophs are able to build up cell biomass and produce energy during aerobic growth on reduced C_1 compounds, which the heterotrophs cannot do.

The C_1 compounds, such as methane, methanol, methylamine(s), formaldehyde, and formate, occur in abundance throughout nature, and therefore it is not surprising that a wide variety of microorganisms (mainly bacteria and yeasts) capable of utilizing such compounds have been isolated. The methylotrophs also play a significant role in the earth's carbon cycle (recycling methane back to carbon dioxide) and in the nitrogen cycle (oxidizing ammonia to nitrite or nitrate and fixing atmospheric nitrogen) in nature.

Interest in methylotrophs began in 1956 when M. Dworkin and J.W. Foster reisolated N.L. Söhngen's methane and methanol-utilizing strain called *Bacillus methanicus* (which he had isolated in 1906) and noted some of its physiological and biochemical peculiarities. Since then, the microbial utilization of C_1 compounds, especially methane and methanol, has come under intense study, motivated primarily by the potential of these carbon substrates to serve as cheap and convenient feedstocks for industrial fermentations, such as the production of microbial biomass (referred to as single cell protein), as a nonagricultural means of producing foods and feeds, and the production of useful metabolites, for example, amino and organic acids, polysaccharides, vitamins, and coenzymes. This research has unraveled many of the basic biochemical problems of the metabolism of C_1 compounds and has resulted in discoveries of new assimilatory and dissimilatory processes and new coenzymes, as well as unusual ultrastructural and genetic properties.

The ongoing accumulation of data on the physiology, biochemistry, and, recently, the molecular biology of methylotrophs has enabled workers in the field to achieve cellular yields and productivities that are very close to those thought to be the theoretical maxima. These data were necessary for the development of the highly efficient processes for the production of microbial biomass that resulted in the widespread acceptance of the potential of noncarbohydrate carbon and energy substrates for fermentation processes. These processes also demonstrated the use of some novel principles in fermentation technology, such as 1) large-scale, stable, chemostatic, continuous-flow, aerobic cultures, 2) fermentor designs to improve mass transfer, 3) implementation of microprocessors for controlling sterilization and fermentation processes, and 4) process water recycling.

The text of this book is divided into the following six sections: First, there is a very brief introductory guide to the methylotrophs written by J.R. Quayle, F.R.S., who is undoubtedly the "father" of methylotrophy. Chapters 1, 2, and 3 provide an overview of the taxonomy and morphology of methylotrophic bacteria and yeasts, including new developments in the classification of these organisms. Chapters 4 to 8 cover many aspects of the growth of these organisms. The oxidation and assimilation pathways are described first, followed by a detailed discussion of the cellular regulation of these pathways. Single or mixed cultures on C_1 and non-C_1 compounds and mixtures of these carbon sources are described. Chapters 9, 10, and 11 deal with the commercial exploitation of the methylotrophs for the production of useful metabolites, enzymes, and single cell protein. Chapters 12 and 13 describe our current understanding of the molecular genetics of methylotrophic bacteria and yeasts, including the mechanism of heterologous gene expression in these organisms. Finally, Chapter 14 describes their ecological characteristics.

We wish to thank each of the distinguished contributors for their generous investment of effort and time in sharing their expertise in this volume. We gratefully acknowledge the help of the editorial staff of Butterworth-Heinemann, with special thanks to Gregory Franklin.

Israel Goldberg
J. Stefan Rokem

INTRODUCTION
J.R. Quayle

The study of methylotrophs is a fascinating example of the emergence of a new area of biological science from many scattered observations over a long period of time. In many such cases there comes a time when these scattered observations become integrated into a clearly focused field which brings together different research groups. Their combined efforts then widen the problem along paths which could not have been foreseen at the beginning.

The first of these scattered observations and findings, which were to lead half a century later to the luxuriant field of methylotrophic biology and biochemistry, was the isolation by Loew at the University of Munich in 1892 of an organism capable of growth on a variety of one-carbon (C_1) compounds (Loew 1892). He called the organism *Bacillus methylicus*, and it was clearly a very similar organism to the *Pseudomonas* AMI strain isolated by Peel and Quayle 70 years later (Peel and Quayle 1961). A somewhat similar case of discovery and rediscovery also happened when Söhngen isolated a bacterium (*Bacillus methanicus*) able to grow on methane as sole source of carbon and energy (Söhngen 1906). Söhngen was mainly interested in organisms which could utilize hydrocarbons, rather than methane per se, and the organism was not preserved for later study. It was reisolated 50 years later by another group of hydrocarbon microbiologists (Dworkin and

Foster 1956) and renamed *Pseudomonas methanica* (Söhngen). Other scattered references can be found in the literature to organisms capable of utilizing C_1 compounds but until the 1960s, they were regarded as little more than microbiological oddities.

At this point we need to turn to the rapid developments in intermediary metabolism which were being made by biochemists in the 1950s and 1960s. They unraveled the biosynthetic pathways resulting in the net biosynthesis of key intermediates of glycolysis and the tricarboxylic acid (TCA) cycle from C_3 compounds, such as pyruvate (Krebs 1954); from C_2 compounds, such as acetate (Kornberg and Krebs 1957); and from CO_2, in the case of photosynthetic tissue, by Calvin and his colleagues (Bassham et al. 1954). These studies, particularly those leading to the discovery of the ribulose bisphosphate cycle and the glyoxylate cycle, highlighted the challenging problems of intermediary metabolism raised by the growth of methylotrophs on C_1 compounds. Thus from the 1960s onwards, the methylotrophs attracted the attention of those microbial biochemists who concentrated on the elucidation of the biosynthetic pathways which enabled these organisms to synthesize in net fashion their cell constituents from C_1 compounds. A wealth of new metabolic cycles and linear sequences emerged, which are the subject of Chapter 4 by Anthony.

At the same time, investigation of the generation of energy from oxidation of C_1 compounds led to a surprisingly complex picture involving new electron transport chains, new prosthetic groups, and elegant cellular ultrastructures, which are described in Chapter 5 by Rokem and Goldberg and in Chapter 3 by Jensen and Corpe, respectively.

It is unlikely that this explosion of activity in the methylotrophic field would have taken place if the occurrence of methylotrophy had been confined to the few exotic organisms known to the microbiologists in the early 1960s. Yet the very ubiquity of C_1 compounds in nature, at every oxidation level from methane to carbon dioxide, argued against such microbial singularity. Rather, it appeared likely that the methods of selective culture used up to that time were not appropriate for isolation of these kinds of organisms. That this was indeed the case was convincingly shown by Whittenbury, Wilkinson, and their colleagues in 1970 by their systematic use of many different isolation methods, which resulted in the isolation of over 100 strains of methane utilizers alone (Whittenbury et al. 1970). From such studies as these, a wealth of new organisms, both prokaryotic and eukaryotic, have been isolated, and their taxonomic relationships have been studied. With the powerful genetic techniques now available, microbial taxonomy has become much more of a science than the art it used to be, and it is timely that the present state of knowledge has been brought together in Chapter 1 by Romanovskaya and Chapter 2 by Komagata.

One of the driving forces in the opening-up of the large-scale study of the methylotrophs was the recognition of the industrial potential of these organisms. The simplicity and cheapness of the growth substrates, particularly methane and methanol, and the simplicity of their oxidation products appealed to the biochemical engineers and led to the hope that high-quality single cell protein could be made more cheaply in the industrial fermentor than through conventional agriculture. The enormous versatility of the methane oxygenase enzyme system also suggested that it could be used to effect a whole array of specific organic transformations. Despite this optimism, the early promise of the 1960s has yet to be realized in terms of commercial profit. While the technology of large-scale growth of the methanol utilizer *Methylophilus methylotrophus* was brilliantly developed by Imperial Chemical Industries in the 1970s and a high-quality single cell protein "Pruteen" was produced from it, the economics of the process were not competitive. The future must surely lie in the production of high-cost, low-volume products, both by whole cell transformations carried out by methylotrophs, and by some of the unique enzymes isolated from them. It is of course no longer necessary to think only in terms of monocultures and single substrates, because continuous culture methods, backed up by detailed knowledge of energetics, can make use of the potential of consortia of organisms under carefully controlled conditions, as described in Chapter 8 by Egli and Mason.

The editors asked me to share my thoughts on the future of methylotrophs. As far as the study of metabolic pathways goes, I believe that the golden period of discovery of entirely new pathways of C_1 assimilation, rather than variants of existing ones, has almost come to an end. There still remains the tantalizing problem of the *icl*⁻ variant of the serine pathway, which has resisted solution for so long. I should be surprised if its final solution turns out to be a simple one. There is much interesting work to be done on metabolism of mixed substrates and the mixing of metabolic modes under different growth conditions. These studies have already illuminated new areas of metabolic regulation.

Complex enzymes such as methane monooxygenase, methanol dehydrogenase, and ribulose bisphosphate carboxylase still provide fascinating insights into the mechanism of enzyme action as more and more powerful genetic and physicochemical techniques are brought to bear on them. However, such detailed studies are passing out of the realm of methylotrophic biology.

As far as methylotrophic biology goes, the future must lie with the application of molecular genetics to a wider variety of methylotrophic organisms and to the long-term study of their ecology. Chapter 12 by Lidstrom and Tsygankov and Chapter 14 by Wattenberg and Hanson deal with these two topics, respectively.

A methylotrophic bacterium growing on methane or methanol as sole carbon source, which can have inserted in it the genetic capability of synthesizing or overproducing a high-cost metabolite, can provide a powerful tool for the biotechnologist. The fact that energy generation may come from a short linear oxidative sequence, divorced from central intermediary metabolism, can be a valuable simplification. The extension of genetic manipulation techniques into the methylotrophic yeasts, as described in Chapter 13 by Tschopp and Cregg, has opened the way to extending these advantages into studies of eucaryotic biology and biotechnology.

Finally, as in so many other fields of microbial biochemistry and physiology, it cannot be overemphasized how artificial the growth conditions of many laboratory studies are as compared with the natural habitat. With environmental considerations pressing in upon so many aspects of everyday life, the ecological study of the methylotrophs will become of increasing importance.

REFERENCES

Bassham, J.A., Benson, A.A., Kay, L.D., et al. (1954) *J. Am. Chem. Soc.* 76, 1760–1770.
Dworkin, M., and Foster, J.W. (1956) *J. Bacteriol.* 72, 646–659.
Kornberg, H.L., and Krebs, H.A. (1957) *Nature* 179, 988–991.
Krebs, H.A. (1954) in *Chemical Pathways of Metabolism* (Greenberg, D.M., ed.), pp. 109–171, *Academic Press*, New York.
Loew, O. (1892) *Z. Bakteriol.* 12, 462–465.
Peel, D., and Quayle, J.R. (1961) *Biochem. J.* 81, 465–469.
Söhngen, N.L. (1906) *Z. Bakteriol.* Parasitenk Abt. II, 15, 513–517.
Whittenbury, R., Phillips, K.C., and Wilkinson, J.F. (1970) *J. Gen. Microbiol.* 61, 205–218.

PART I

Taxonomy and Morphology

CHAPTER 1

Taxonomy of Methylotrophic Bacteria

V.A. Romanovskaya

Methylotrophic bacteria (i.e., microorganisms utilizing reduced C_1 compounds such as methane and methanol) have been the subject of deep and thorough study in many laboratories throughout the world in the last two decades. The use of methylotrophic microorganisms as producers of protein, polysaccharide, and other biologically active materials is usually linked with the necessity to patent industrially valuable strains, methods of synthesis with these strains, etc. These applications of methylotrophs require a more precise definition of the taxonomic status of industrially valuable cultures than is currently available, especially for solving controversial patent questions. Unfortunately, the taxonomy of the methylotrophic bacteria has not yet been discussed in a generalized form in scientific literature. Most of the available information on their classification and nomenclature is uncertain and sometimes contradictory. Moreover, general and accepted criteria for the classification of methylotrophs are missing; this has led, unfortunately, to chaos in their nomenclature.

The author thanks R. Whittenbury, J. Heyer, and V. Galchenko for strains of methane oxidizers, S. Stolyar for cellular protein electrophoresis, A. Netrussov for 5S rRNA analysis, L. Andreev for fatty acid analysis, and V. Ponomarev for the English translation and the preparation of the manuscript for this chapter.

4   Taxonomy of Methylotrophic Bacteria

The situation with one classical species of methane-oxidizing bacteria can serve as an example: this bacterium was originally described in 1906 by Söhngen as "*Bacillus methanicus*"*; later it was renamed several times: "*Methanomonas methanica*," "*Pseudomonas methanica*," and finally *Methylomonas methanica*. Surprisingly, all these names are still used for this species. The species currently named "*Pseudomonas methanica*," which assimilates methanol and other organic substrates but does not oxidize methane, has also been described. Another example is a representative of the facultative methylotrophs now called *Methylobacterium extorquens*; it was previously called "*Bacillus extorquens*," "*Vibrio extorquens*," "*Pseudomonas extorquens*," "*Flavobacterium extorquens*," and "*Protomonas extorquens*."

Nomenclature complications were also caused by a proposal to allot some facultative methylotrophic bacteria to the genus "*Protomonas*" (instead of *Methylobacterium*), which has repeatedly been reassigned to *Methylobacterium*. An ambiguous situation exists for representatives of the genus *Methylomonas*; because for many years both methane-utilizing and methanol-utilizing bacteria were assigned to this genus. Considering the overall situation of the taxonomy of the methylotrophic bacteria, it should be noted that much of the data is controversial and, very few species are covered. Thus, the latest edition of *Bergey's Manual* (1984) includes descriptions only of two species of methane-oxidizing bacteria, while a book on taxonomy by Galchenko et al. (1986) presents a classification based primarily on their own isolates. Our own work on the classification of methane-utilizing bacteria (Romanovskaya et al. 1978) is now also unsatisfactory. The classification of the obligate methanol utilizers is somewhat better, but these taxa are not included in any of the existing manuals on bacterial systematics, and data are available in the form of separate articles only.

Today, methylotrophic bacteria are conventionally divided into the three following classes: obligate—utilizing C_1 compounds (methane, methanol, or methylamines) only; restricted facultative—utilizing, besides C_1 compounds, a limited range of simple organic compounds (e.g., glucose or fructose); and facultative—utilizing both C_1 compounds and a variety of simple and complex organic substrates.

1.1   THE METHANE-OXIDIZING BACTERIA

The history of the taxonomy of the methane-oxidizing bacteria can be revealed by an analysis of the changing family name of these bacteria: 1909, Oxydobacteria (Orla-Jensen 1909); 1949, Pseudomonadaceae (Krassilnikov 1949); 1957, "Methanomonadaceae" (*Bergey's Manual* 1957); 1974, "Methylomonadaceae" (*Bergey's Manual* 1974); and 1984, Methylococcaceae (*Ber-*

---

* Unapproved taxon names are given in quotation marks.

gey's Manual 1984). The only described species of methane-oxidizer, "*Bacillus methanicus*" (Söhngen 1906), has been included in two different families, Oxydobacteriaceae and Pseudomonadaceae. The family "Methanomonadaceae" was based on the description of a monotypic genus "Methanomonas" represented by the same single species. Today this species is called *Methylomonas methanica*.

It should be noted that while obligate bacteria utilizing methane or methanol as a carbon source are included in the family "Methylomonadaceae," obligate bacteria able to utilize methane only as a sole carbon and energy source, as well as "facultative methane-oxidizers," are included in the family Methylococcaceae. Bacteria unable to oxidize methane are excluded from the family even if they can grow on other C_1 compounds. The name Methylococcaceae has been validated (Validation 1984).

Sixty years after the initial description of "*Bacillus methanicus*" (*Methylomonas methanica*), another species, *Methylococcus capsulatus*, was isolated and studied (Foster and Davis 1966). In 1970, Whittenbury et al. (1970) described a number of diverse methane-oxidizing strains combined in 5 "groups" and 15 "subgroups," without any attempt at a formal description of species. In a subsequent review on methylotrophy, Whittenbury et al. (1975) reduced the range of microorganisms under study to three genera having the following type species: (1) *Methylomonas*—*M. methanica* or "*M. alba*"; (2) *Methylococcus*—*M. capsulatus*; (3) "*Methylosinus*"—"*M. trichosporium*" or "*M. sporium*." Later, however, the authors did not validate the names proposed (except for *M. methanica* and *M. capsulatus*; Whittenbury and Krieg 1984), though other researchers began to use these names, equating generic status to "group" and specific to "subgroup." Thus, an equivocal situation arose.

On the one hand, the names of the methane oxidizers were used in much scientific literature; on the other hand, they did not possess nomenclature status according to the existing international rules, and type strains were not defined. Therefore, we suggested using principles proposed by Whittenbury et al. (1970), the formation of "groups" and "subgroups" of methane oxidizers. In addition, general criteria for intragroup classification of obligate methane oxidizers, descriptions of their taxa (both for new species we isolated and for "subgroups" that we received from Whittenbury and examined in our laboratory), and diagnostic keys were also compiled (Maleshenko et al. 1978; Romanovskaya et al. 1978). In these studies we noted that the genera *Methylococcus* and *Methylomonas* were heterogenous. We studied chemotaxonomic and other characters, and also the data available in the literature, in order to determine precise positions for the species of these two genera. Based on the results obtained, we suggested the establishment of a new genus of methane oxidizers called "*Methylovarius*" with the type species "*Methylovarius ucrainicus*" (Romanovskaya 1984). The known taxa (genera and species) are listed in Section 1.5.

Our attempt to validate descriptions of taxa of methane oxidizers in 1975 by publication in the *International Journal of Systematic Bacteriology* (IJSB) was a failure. The IJSB editorial staff did not find it possible to validate the genus "*Methylomonas*" (and the family "Methylomonadaceae," correspondingly), due to the priority of the name "*Methanomonas*." Somewhat later (Validation 1981), only specific names of the genus *Methylococcus* (*M. luteus*, *M. whittenburyi*, *M. vinelandii*) were validated on the basis of the descriptions proposed earlier (Romanovskaya et al. 1978).

In recent years the data on methane utilizers have been supplemented by determinations of their fatty acid and phospholipid contents (Galchenko et al. 1986), the GC content of their DNA (Meyer et al. 1986), and their ubiquinone content (Urakami and Komagata 1986b). Such information undoubtedly widens our knowledge on the properties of methane oxidizers. In addition, the data support the classification scheme proposed by Whittenbury et al. (1970), although the scheme has been slightly revised (Whittenbury and Krieg 1984). The interpretation of all the studies mentioned above is hindered by the facts that these studies were mainly conducted with the author's own isolates or with limited number of known strains, and the authors had not included the results of other investigations. New species of methane oxidizers (validated names), *Methylococcus thermophilus* (Malashenko 1976), *Methylococcus mobilis* (Hazeu et al. 1980), and *Methylomonas pelagica* (Sieburth et al. 1987) have recently been described.

1.1.1 Descriptions of the Methane-Oxidizing Bacteria

The colony color and vegetative cell structures of methane oxidizers are variable. Colonies are often pigmented: ochre-pink, red, crimson (red-violet), yellow, brown, etc. Carotenoid pigments primarily containing β-carotene were found in *Methylomonas methanica* and "*M. rubra*" (which forms pinkish-red colonies); crimson strains possess pigment of the prodiginine type; brown and yellow strains (*Methylococcus thermophilus* and *M. luteus*) contain melanine. The cell shape and size of methane oxidizers varies over a broad range: rods, cocci, curved, and pear-shaped cells are observed. A high degree of polymorphism in some species of methane-oxidizing cells should be noted. Several different resting forms of cells are seen: exospores, cysts, and "immature" cysts.

The methane oxidizers are divided into two types on the basis of their cytoplasmic membrane (CM) structure (Davies and Whittenbury 1970). CM type I strain contain tightly packed and flattened vesicles that run parallel to each other (Figure 1–1A). This rod-coccoid group of bacteria includes the genera *Methylomonas*, *Methylococcus*, and "*Methylovarius*." Vibrioid organisms of CM type II ("*Methylosinus*" and "*Methylocystis*") have a system of double membranes located at the periphery of the cell, parallel to the cytoplasmic membrane (Figure 1–1B). In type I bacteria, CM is usually in close-packed bundles; however, "*M. rubra*" and *M. pelagica* strains are

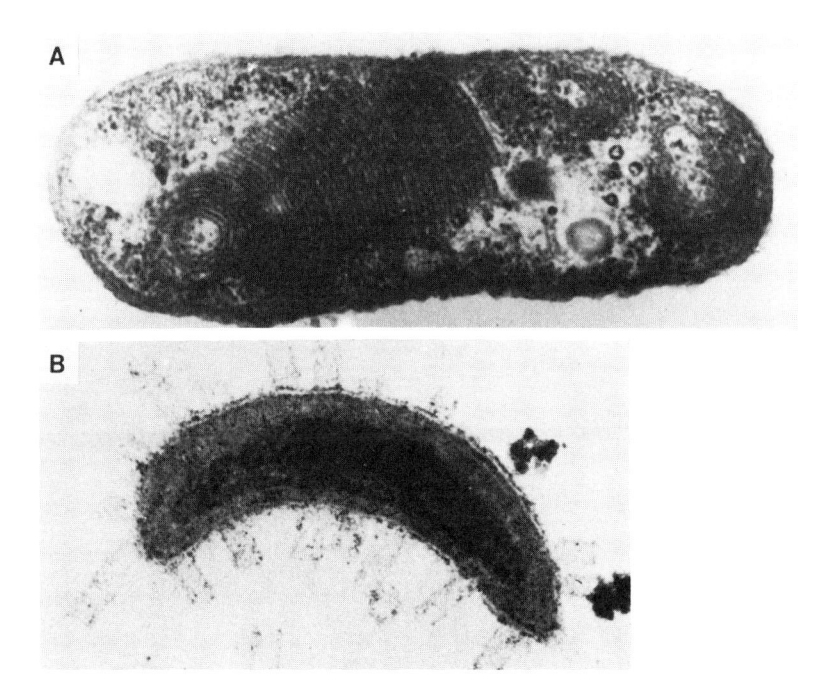

FIGURE 1-1 Electron micrographs of thin layers. (A) "*Methylomonas rubra*," CM type I; several initial sites of CM formation are visible, and the CM location is radial. (B) "*Methylocystis echinoides*," CM type II; the CM forms concentric circles at the periphery of cells. Figure 1-1B courtesy of J. Heyer.

exceptions, as their CM structures have a radial orientation and several sites where membrane formation is initiated (Figure 1-1A). Lawrence and Quayle (1970) suggested that the CM types of the methane oxidizers were correlated with their pathways of C_1-compound assimilation, i.e., organisms with type I CM assimilate methane via the ribulose-monophosphate pathway, and that of type II via the serine pathway. Later, however, it was found that many species of CM type I possess the key enzymes of both metabolic pathways (Malashenko 1976; Whittenbury et al. 1975). It seems that the membrane type should be strictly linked to the assimilation pathway type in obligate methylotrophs. For example, obligate and facultative methylotrophs having no CM structures and utilizing no methane, but assimilating other C_1 compounds via the serine or hexulose phosphate pathways are known to exist, however, the appearance of CM structures is dependent on the growth conditions (see Chapter 3).

The cellular fatty acid content is a useful chemotaxonomic criterion for the classification of organisms. The characteristic chromatograms of the fatty acid content of 11 strains of methane oxidizers are shown in Figure 1-2. Fatty acids with long carbon chains, C_{16}, prevail in rods and cocci; C_{18}

8 Taxonomy of Methylotrophic Bacteria

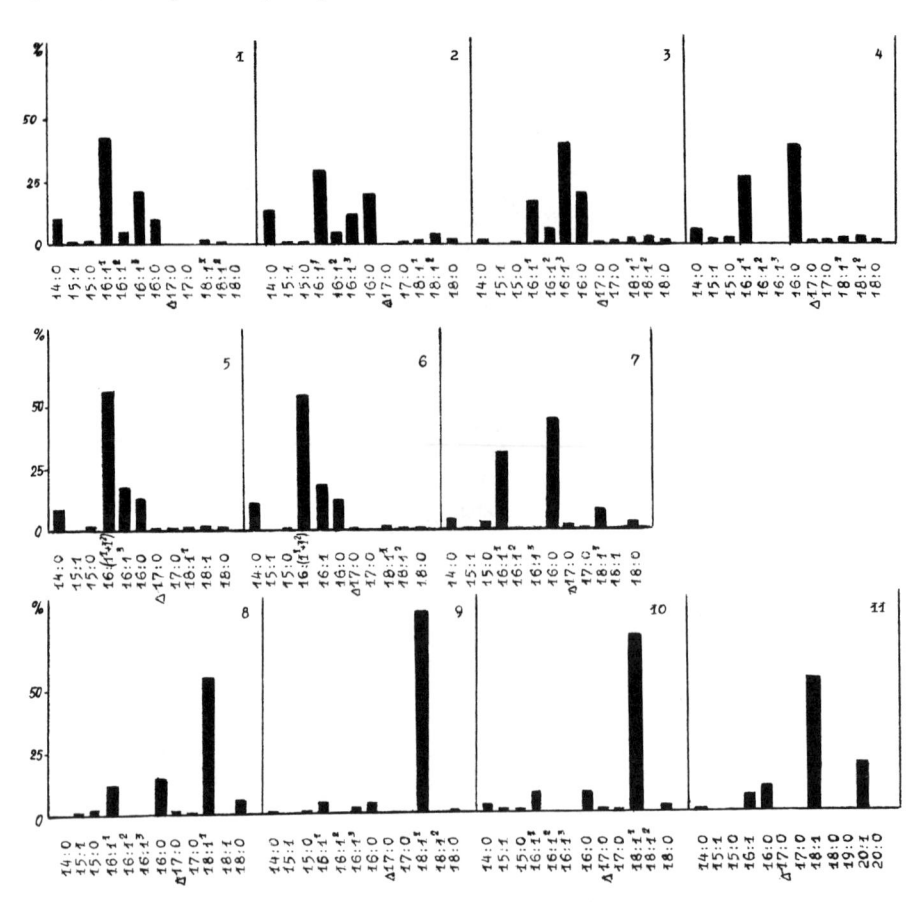

FIGURE 1-2 Chromatograms of the fatty acid content of 11 strains of methane-oxidizing bacteria. The strains shown are: (1) "*Methylomonas rubra*" 15sh; (2) *M. methanica* 15k; (3) "*M. alba*" BG8; (4) "*Methylococcus gracilis*" 14l; (5) "*Methylovarius ucrainicus*" 160; (6) "*Methylovarius luteus*" 53v; (7) *Methylococcus thermophilus* 2yu; (8) "*Methylosinus trichosporium*" OB3b; (9) "*Methylocystis parvus*" OBBP; (10) "*Methylosinus sporium*" 5w; and (11) *Methylobacterium organophilum* XX. Symbols 14:0–20:1 indicate the number of carbon atoms in the fatty acid: 14:0, tetradecane acid; 15:1, pentadecene acid; 15:0, pentadecane acid; 16:1, hexodecene acid; 16:0 hexodecane acid; 17:0, cis-9,10-methylene-hexodecane acid; 17:0, heptadecane acid; 18:1, octadecene acid; 18:0, octadecane acid; 1^1–1^3, isomers of corresponding acids; Δ, cyclopropane.

prevails in vibrioid organisms. The relative percentage of the fatty acid content in C_{16} or C_{18} in corresponding bacterial group ranges from 50–90%, i.e., it forms the major part of the total fatty acids in cells. Significant differences were found between the thermophilic and mesophilic forms of the methane oxidizers.

Phospholipids are present in large amounts in the membranes of methane oxidizers, at a concentration of 8-11% of cell dry weight (Makula 1978). The phospholipid content of methane-oxidizing cells has been reported by several workers (e.g., Galchenko et al. 1986; Makula 1978). The presence of phosphatidylglycerol (PG) and phosphatidylethanolamine (PE) or methylated derivatives of PE is characteristic of all studied representatives of these bacteria.

1.1.2 Two Tribes of Methane Oxidizers

Within the family Methylococcaceae, the methylotrophic genera are distinctly divided into two groups by a number of characters (obligate or facultative methylotrophs, the metabolic pathway of C_1 compounds, CM type, and fatty acid content), which coincide with their division by sensitivity to antibiotics and lytic enzymes (Romanovskaya 1984). Unfortunately, as if following the early traditions in the classification of obligate methylotrophs, formation of "groups" of organisms are mentioned again. What are these groups? Or more correctly, what range of taxa do they reflect? It should be remembered that if distinct differences in important differentiating characters exist between two groups, then within each group, differences among genera are less essential (Romanovskaya and el Sayed 1986). Therefore, these groups can be taxa of a higher level than that of genus. Therefore, we proposed dividing the family Methylococcaceae into two tribes, "Methylococceae" and "Methylosineae." The differentiation is based on a complex of characters as shown in Table 1-1. The numerical analysis of phenotypic features (Romanovskaya 1984) showed the formation of five phenons. The results obtained confirmed the existence of the new genus "*Methylovarius*." This genus includes four species of the "Methylobacter group" and three species formerly assigned to the genus *Methylococcus* (Romanovskaya 1984). Hence the genus "*Methylovarius*" is not identical to the "Methylobacter group." According to the Rules of the International Code of Nomenclature of Bacteria, it was not possible in 1984 to use the name "*Methylobacter*" for a new genus, since in 1976, the genus *Methylobacterium*, based on a description of bacteria utilizing no methane, was validated (Patt et al. 1976).

In recent years, the results of DNA-DNA hybridization studies and nucleotide sequence analyses of 5S ribosomal RNA were published. These allow us, to some extent, to evaluate evolutionary distances (similarity) between organisms. Unfortunately, these studies are still fragmentary, so their results are ambiguous. The results of the DNA-DNA hybridization of methane oxidizers (Lysenko et al. 1988) showed that methane oxidizers are a more diverse group phylogenetically than had been supposed earlier. Any discussion of these results should be done cautiously, since in that work the authors' own isolates were used to prepare reference DNA. The main results were as follows: four strains of *M. methanica* showed lower levels of ho-

TABLE 1-1 Properties Differentiating the Taxa of the Methane-Oxidizing Bacteria

Property	Tribe "Methylococceae"	Tribe "Methylosineae"
Pathway of C₁ compounds	Ribulose-monophosphate	Serine
Cytoplasmic membrane type	I	II
NAD⁺-dependent isocitrate dehydrogenase	+	−
TCA cycle	Incomplete	Complete
Major fatty acid[1]	16:1	18:1
Squalenes	+	−
Sensitivity to lysoenzymes	+	−

	Genus				
	Methylomonas	"*Methylovarius*"	*Methylococcus*	"*Methylosinus*"	"*Methylocystis*"
Cell shape	Rod	Oval, rod	Round, oval	Pear-shaped, curved rod	Vibrioid
Cell motility	+	− or +	− or +	+	−
Major fatty acid[1]	16:1¹; 16:1³	16:1¹; 16:1²; 16:1³	16:1¹; 16:0; Δ17:0	18:1	18:1
Polar lipids[2]	PG, PE	PG, PE, DPG	PG, PE, DPG, PC	PG, PE, PC	PG, PE, PC
Resting stage	Cyst	Azotobacter-type cyst	Azotobacter-type cyst	Spore	Cyst
Growth range (°C)	20–35	25–42	37–65	30–37	30–37
GC content of DNA (mol%)	49–54	47–50 (52.5)	56.6–62.7	60.7–65.7	59–64.1
Ribulose-diphosphate carboxylase	−	−	+	−	−

[1] Number of carbon atoms.
[2] Abbreviations: PG, phosphatidylglycerol; PE, phosphatidylethanolamine; DPG, diphosphatidylglycerol; PC, phosphatidylcholine.

mology (61–78%) than that of four species of "*Methylobacter*" (65–96%), while among three species of *Methylomonas*, the homology value was 12–32%. The analysis of the 5S rRNA nucleotide sequences (Bulygina et al. 1989) confirmed the distant positions of the phenotypically similar species, *M. methanica* and "*M. rubra*," and also the separation of the genus "*Methylovarius*" from the genus *Methylococcus*. The results also showed that *Methylococcus capsulatus* was closer to the representatives of the genera *Methylomonas* and "*Methylovarius*" than that of the tribe "Methylosineae" (genera "*Methylosinus*" and "*Methylocystis*").

Electrophoretic analysis of protein spectra has shown a similar resolution level as the DNA-DNA hybridization method, but it is faster and cheaper. The electrophoretic analysis of the cellular proteins of all the organisms available in our collection of methane-oxidizing bacteria (50 strains) were carried out. The main goal was to define more precisely the specific and generic status of these strains. Since the reproducibility of the results of protein electrophoresis is higher if they are run in one gel, various combinations of the strains under study were repeatedly analyzed. Usually, the same protein samples in various gels were electrophoretically identical, which simplified the interpretation of the results obtained. Therefore, the most typical electrophoretograms of cellular proteins of the methane-oxidizing strains studied are presented in Figure 1–3. They indicate that at least five different genera are involved.

1.1.3 Protein Spectra of the Methane-Oxidizing Bacteria

1.1.3.1 The Genus *Methylococcus*. Significant differences exist among *M. capsulatus* strains IMV B-3330 (= 9m) and *M. thermophilus* strains IMV B-3111 (= 111) and IMV B-3119 (= 119). All *M. thermophilus* strains possess identical protein electrophoretograms. "*M. gracilis*" IMV B-3012 (= 12r) is very similar to *M. thermophilus*, which confirmed the correctness of assigning "*M. gracilis*" to the genus *Methylococcus*.

1.1.3.2 The Genus *Methylomonas*. *M. methanica* strains ATCC 35067 (S_1) and IMV B-3016 (= 15k) and "*M. rubra*" strains IMV B-3076 (= 15m) and IMV B-3005 (= 2t) have differing cellular proteins spectra.

1.1.3.3 The Genus "*Methylovarius*". The following strains have similar protein spectra: "*M. ucrainicus*" IMV B-3159 (= 159) and IMV B-3160 (= 160), "*M. whittenburyi*" IMET 10596 (= Y), and "*M. vinelandii*" IMV B-3030 (= 30) and IMV B-3088 (= 87). "*M. luteus*" IMV B-3053 (= 53v) and IMV B-3098 (= 98) vary from the first group significantly.

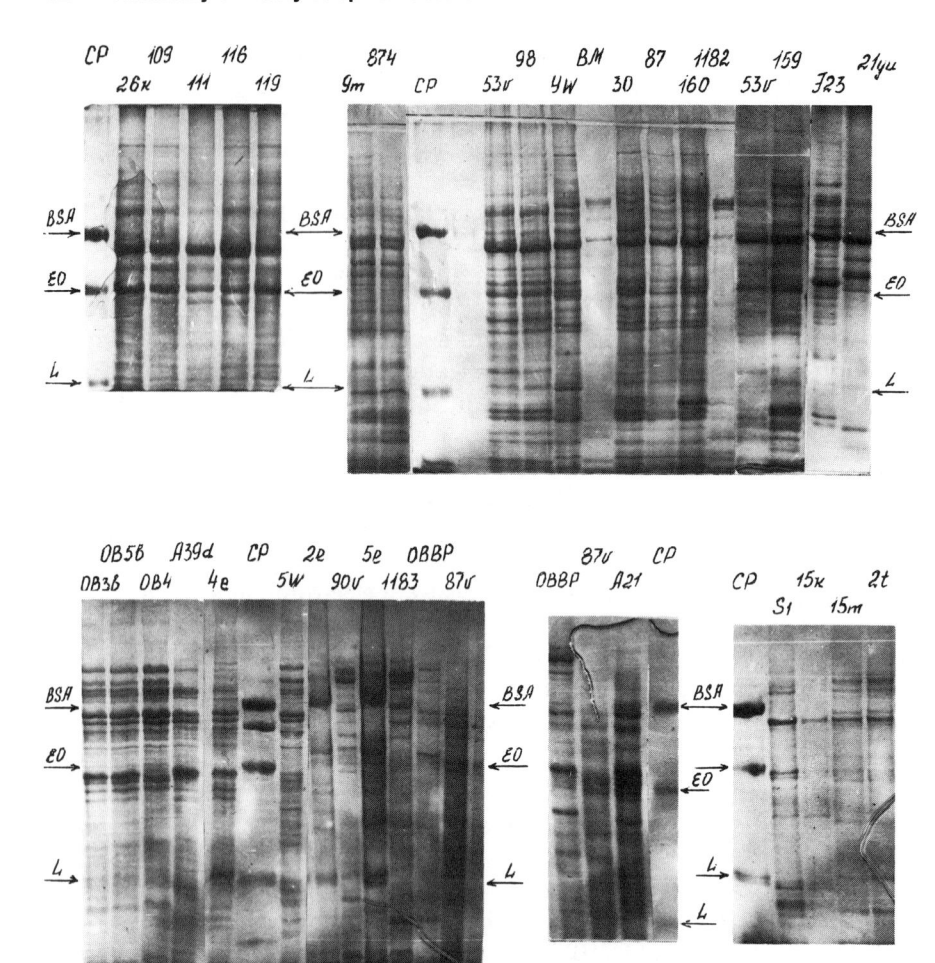

FIGURE 1-3 Electrophoretograms of cellular proteins from various methane-oxidizing strains obtained by electrophoresis in 10% homogeneous acrylamide gel with sodium dodecyl sulfate. Abbreviations: CP, calibration proteins of known molecular weight (BSA, bovine serum albumin; EO, egg ovalbumin; L, lysozyme). Identification of the strains: *Methylococcus thermophilus* 26K, 109, 111, 116, and 119; *Methylococcus capsulatus* 9m and 874; "*Methylovarius luteus*" 53v and 98; "*Methylovarius whittenburyi*" Yw; "*Methylovarius vinelandii*" 30 and 87; "*Methylovarius ucrainicus*" 160 and 159; "*Methylosinus trichosporium*" OB3b, OB5b, and OB4; "*Methylosinus sporium*" 5w; *Methylomonas methanica* S_1 and 15k; "*Methylomonas rubra*" 15m and 2t. The other strains were new isolates. (Strains OB3b, OB5b, OB4, Yw, and 5w courtesy of R. Whittenbury; F23, A39d, and A20d courtesy of J. Heyer; 30, 87, 98, and 100 courtesy of V. Galchenko.)

1.1.3.4 The Genus "*Methylosinus*".
Almost identical spectra of proteins were obtained for "*M. trichosporium*" strains NCIB 11131 (= OB3b), IMV B-3013 (= OB5b), IMV B-3400 (= OB4), IMV B-3039 (= A39d), and IMV B-3004 (= 4e). "*M. sporium*" NCIB 11126 (= 5w), IMV B-3020 (= A20d), and some other strains differ from "*M. trichosporium*," which confirms the existence of two species of "*Methylosinus.*"

1.1.3.5 The Genus "*Methylocystis*".
The protein spectra of "*M. parvus*" strains NCIB 11129 (= OBBP), IMV B-3087 (= 87v), IMV B-3492 (= 492), and "*M. echinoides*" IMET 10491 (= C493) differed significantly.

1.1.3.6 Strains of Uncertain Genus.
As shown in Figure 1-3, two separate groups were formed by strains IMV B-3102 (= 2e), IMV B-3025 (= 5e), and IMV B-3100 (= 100), and by strains IMV B-3090 (= 90v), IMV B-3182 (= 1182), IMV B-3183 (= 1183), and IMV B-3034 (= BM). It is impossible to assign these strains to any of the species studied. Unfortunately, *Methylococcus mobilis*, "*Methylomonas alba*," and *Methylomonas pelagica* species were not included in the analysis. It seems appropriate to continue these investigations, as the results obtained allowed us to differentiate phenotypically similar species of methane-oxidizing bacteria and also to conduct rapid identification of new isolates, since all strains of a species usually possess identical (or very similar) spectra of cellular proteins. Properties differentiating the genera of methane-oxidizing bacteria are listed in Table 1-1.

1.2 OBLIGATE AND RESTRICTED FACULTATIVE METHANOL-UTILIZING BACTERIA

A significant number of bacteria (obligate methylotrophs) assimilating methanol or methylamines and unable to grow on polycarbon compounds have already been described. Uncertainty about their taxonomy has impeded the use of the data on their physiological and biochemical properties, and also their potential application in biotechnology.

Historically, the first published genus of obligate methanol-utilizing bacteria was the monotypic genus *Methylobacillus*—type species, *Methylobacillus glycogenes*—(Yordy and Weaver 1977). Later, Urakami and Komagata (1986a) described the genus *Methylobacillus*. These authors suggested that all the other motile and nonmotile strains studied should be assigned to the species *Methylobacillus glycogenes*, including strains able to utilize D-fructose, in particular, "*Methylomonas espexii*," "*M. methylovora*," "*M. methanocatalalessica*," "*M. methanolica*," "*M. methanofructolica*," "*M. aminofaciens*," "*Methanomonas methylovora*," "*Pseudomonas methanolica*"

("*Methylomonas methanolovorans*"), "*Ps. insueta*," "*Ps. inaudita*," "*Ps. methanomonas*," "*Ps. methylonica*," "*Protaminobacter candidus*," "*Prot. thiaminophagus*," and also "*Achromobacter methanolophila*," *Methylophilus methylotrophus*, and "*Methylomonas clara*." Urakami and Komagata (1986a) suggested that, although the above-mentioned bacteria differ in their GC content, DNA-DNA homology, enzyme electrophoresis pattern, and polar lipid content, they all displayed very similar morphological and physiological properties, had identical fatty acid and ubiquinone contents, and therefore, "for practical purposes," they should all be assigned to one species. This however contradicts earlier results obtained by Jenkins et al. (1984) of the numerical analysis of phenotypic properties of methanol-utilizing bacteria. They found that the obligate and restricted facultative methanol-utilizers formed two distinct taxa of generic status: *Methylobacillus* and a comparable taxon containing organisms now related to *Methylophilus*. Jenkins et al. (1987) published, in the IJSB, the diagnoses of the genus *Methylophilus* and the species *Methylophilus methylotrophus* (type strain, NCIB 10515); *Methylomonas clara* ATCC 31226 was also assigned to this species. This proposal confirms the results that Byrom (1981) obtained with cellular protein electrophoresis and DNA-DNA hybridization. Properties differentiating *Methylobacillus* and *Methylophilus* are listed in Table 1–2.

The description of the genus *Methylophaga* (Table 1–2) was published (Janvier et al. 1985); these organisms which were isolated from seawater, grew on methanol, methylamine, and fructose, were vitamin B_{12}-dependent, and the GC content of their DNA was 38–46 mol%. The authors tried to determine whether methylotrophs isolated from seawater (i.e., salt water) differed from soil isolates. Hybridization results were negative. Chemical analysis and electron microscopy showed low content of peptidoglycanes in cell walls of marine isolates. Obligate methylotrophs isolated from seawater had been studied before 1985 and had been assigned to various taxa: "*Alteromonas thalasso-methanolica*," a strain utilizing fructose, along with C_1 compounds; "*Methylomonas thalassica*," a strain utilizing C_1 compounds only; and the genus "*Methylophilus*" (the GC content of the DNA is 45 mol%) (Strand and Lidstrom 1984). According to the data of Janvier et al. (1985), they should all be assigned to *Methylophaga thalassica*.

The taxa of obligate and restricted facultative methanol-utilizing bacteria are difficult to discern. However DNA-DNA hybridization confirmed the differentiation of the three genera *Methylobacillus*, *Methylophilus*, and *Methylophaga*, and also the two species *Methylophaga marina*, and *Methylophaga thalassica* (Janvier et al. 1985). Examination of 5S rRNA nucleotide sequences in some methylotrophs (Wolfrum and Stolp 1987) showed that the obligate methanol utilizers studied belonged to the beta subclass of the Proteobacteria (Stackebrandt et al. 1988), while methane oxidizers, i.e., *Methylomonas alba* BG8 belonged to the gamma subclass. It appears impossible to assign the genera of methanol-utilizing bacteria to any known family. Their isolated position was confirmed by the data from DNA-DNA

1.2 Obligate and Restricted Facultative Methanol-Utilizing Bacteria

TABLE 1-2 Properties Differentiating the Genera of the Methanol-Utilizing Bacteria[1]

Property	"Methylobacillaceae" Methylobacillus	Methylophilus	Methylophaga	Methylobacterium
Pathway of C$_1$ compounds	Ribulose-monophosphate	Ribulose-monophosphate	Ribulose-monophosphate	Serine
Major fatty acids	16:0; 16:1	16:0; 16:1	16:0; 16:1	18:1
Obligate and restricted facultative strains	+	+	+	−
Major ubiquinone	Q-8	Q-8	Q-8	Q-10
Cell shape	Rod	Straight or curved rod	Rod to coccus	Pleomorphic rod
Flagellation	Polar or −	Polar or −	Polar	Polar, lateral, or −
Growth range (°C)	32–42	30–37	10–40, optim. 30–37	25–30
GC content of DNA (mol%)	50.0–56.0	50.0–52.0	38.0–46.0	60–70
Halophilic	−	−	+	−
Colony pigmentation	White to pale yellow	Nonpigmented	White to pale yellow or pink	Pink to red
Polar lipids	PG, PE, DPG	PG, PE	n.d.	PG, PE, DPG, PC
Growth on methylamines	+ or −	+	+ or −	− or +
Growth on fructose	+ or −	+	+	+ or −
Growth on glucose	−	+ (weak)	−	+
Growth on ethanol	−	+ (weak)	−	+

[1] n.d., not determined. For other abbreviations, see Table 1-1.

hybridizations and 5S rRNA nucleotide sequences. Therefore, it seems appropriate to place them in the independent family called "Methylobacillaceae."

1.3 FACULTATIVE METHYLOTROPHIC BACTERIA

This bacterial group is composed of strains from a number of genera, such as nonpigmented strains of *Pseudomonas*; representatives of the genera *Flavobacterium* and *"Protaminobacter"*; Gram-variable strains of *Arthrobacter* and *Corynebacterium*; strains tentatively attributed to *Hyphomicrobium, Alcaligenes, Achromobacter,* and *Acinetobacter*; Gram-positive strains of the genera *Bacillus, Mycobacterium,* and *Brevibacterium*; and some facultative autotrophs or phototrophs capable of growth on methanol or formate. From the taxonomic point of view, these organisms are not difficult to identify, since most of them belong to well-known and well-studied bacterial genera. Their classification and nomenclature are traditional. In general, from the data presented, it seems that the ability to utilize C_1 compounds is widely spread among various microbial groups. Besides bacterial forms, a significant number of eukaryotic microorganisms able to grow on C_1 compounds are known.

Pink-pigmented bacterial strains growing on methanol and/or other C_1 compounds using the serine pathway compose a significant group. All of them form coenzyme Q-10, have a GC content of the DNA of 60–70 mol%, and their major fatty acid is $C_{18:1}$. They are combined in the genus *Methylobacterium* (Green and Bousfield 1983). DNA-rRNA hybridization data (Wolfrum et al. 1986) indicate that pink-pigmented strains of facultative methylotrophs, some of which had been attributed to the genus *Pseudomonas*, did not really belong to that homology group on the basis of rRNA similarity. Instead, these data suggest that the pink-pigmented facultative methylotrophs belong in the genus *Methylobacterium*, as proposed by Green and Bousfield (1983). This genus is phenotypically and genetically separated from obligate and restricted facultative methanol utilizers; therefore it seems incorrect to consider it to be a member of the family Methylococcaceae or "Methylobacillaceae," although historically *Methylobacterium* was considered to be an obligate methylotroph. However, molecular studies (in particular, 16S rRNA nucleotide sequencing) showed that this genus belongs to the alpha group of the Proteobacteria (Stackebrandt et al. 1988), while obligate methanol utilizers have been attributed to the beta group of the Proteobacteria.

The genus *Methylobacterium* was first proposed in 1976 (Patt et al. 1976). At that time, the newly isolated facultative methane oxidizer *Methylobacterium organophilum* was placed in this genus. "*Methylobacterium ethanolicum*" and "*M. hypolimneticum*," which were also able to utilize methane facultatively, were isolated. It was shown that *M. organophilum* and "*M. ethanolicum*" were syntrophic associations of two methylotrophic bacteria—an obligate methane oxidizer and a facultative methanol utilizer, respectively (Lidstrom-O'Connor et al. 1984). Therefore, at present, there are no facultative methane-utilizing bacteria among species of the genus *Methylobacterium* (Green and Bousfield 1983; Romanovskaya and el Sayed 1986; Hood et al. 1987).

Less than six months after additional data on the genus *Methylobacterium* were published (Green and Bousfield 1983), the IJSB published a proposal to combine the pink-pigmented facultative methylotrophs (i.e., representatives of the genus *Methylobacterium*) in the genus "*Protomonas*" (Urakami and Komagata 1984). Since the existence of double names contradicts to the international rules of nomenclature, somewhat later the IJSB reclassified the genus "*Protomonas*" into *Methylobacterium* (Bousfield and Green 1985). Thus, validly published names are: *Methylobacterium rhodium* (*Pseudomonas rhodos*), *Methylobacterium radiotolerans* (*Pseudomonas radiora*), *Methylobacterium mesophilicum* (*Pseudomonas mesophilicum*), and *Methylobacterium extorquens* ("*Protomonas extorquens*"). Green et al. (1988), based on the numerical analysis of phenotypic characters and the results of DNA-DNA hybridization (Hood et al. 1987), suggested the formation of three new species. Green et al. (1988) noted that these species, *M. rhodesianum*, *M. zatmanii*, and *M. fujisawaense*, were phenotypically very difficult to separate.

Recently the genus "*Methylomicrobium*" has been described (Govorukhina et al. 1989). Formation of this genus is not sufficiently grounded, as the ability to utilize methanol facultatively as a source of carbon supply is known for many bacterial species. It seems likely therefore that the isolates of these authors are representatives of classical bacterial genera. In similar cases, facultative methylotrophs should be identified as strains of already known species, e.g., *Bacillus brevis* (Dijkhuizen 1988), or described as new species of existing genera, e.g., *Acetobacter methanolica* (Uhlig et al. 1986). The ability to utilize methanol facultatively, in our opinion, cannot serve as a criterion for placing bacteria in a new genus.

1.4 RELATIONSHIP OF METHYLOTROPHIC BACTERIA TAXA WITH OTHER PROKARYOTIC TAXA

To study the relationship of methylotrophic bacteria with other bacteria, we carried out the numerical analysis of several phenotypic characters of taxa (genera) of methylotrophs and other Gram-negative, aerobic and facultatively anaerobic, organoheterotrophic, and hemolytotrophic bacteria. Figure 1-4 shows the dendrogram constructed using this data, when the taxonomic distance value (D_{jk}) is used for determination of taxa pair similarity.

$$D_{jk} = \sum_{i=1}^{n} \left[(x_{ij} - x_{ik})^2 \right]^{1/2}$$

18   Taxonomy of Methylotrophic Bacteria

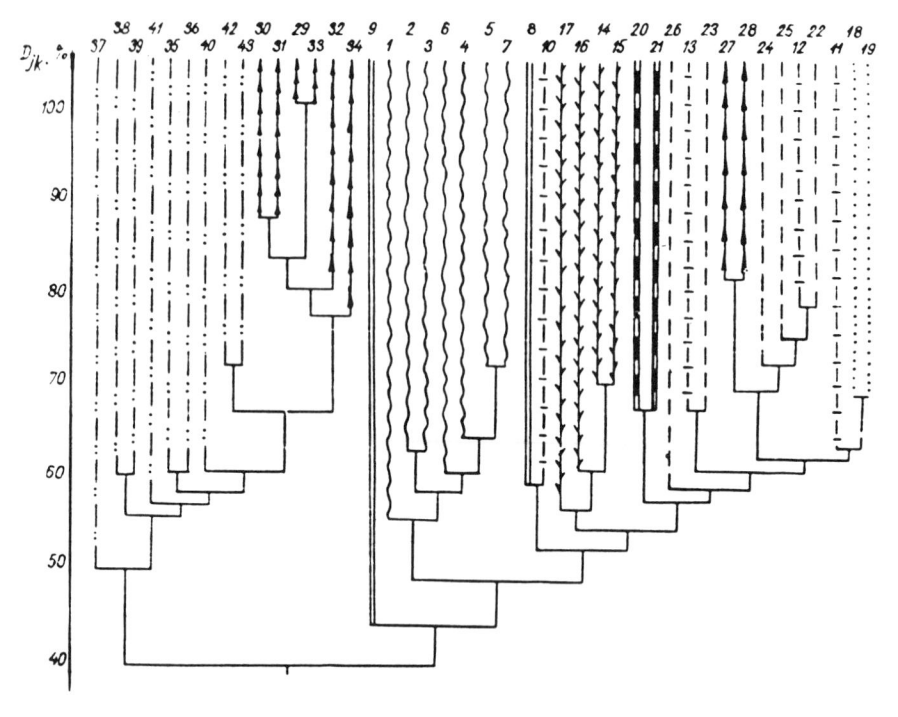

FIGURE 1-4 Dendrogram reflecting phenon formation when the taxonomic distance value (D_{jk}) is used for the determination of taxa pair similarity. Families (according to *Bergey's Manual* 1974) are indicated by a different kind of line. The numbers correspond to the following genera: 1, *Nitrobacter*; 2, *Nitrospira*; 3, *Nitrococcus*; 4, *Nitrosomonas*; 5, *Nitrosospira*; 6, *Nitrosococcus*; 7, *Nitrosolobus*; 8, *Thiobacillus*; 9, *Sulfolobus*; 10, *Pseudomonas*; 11, *Xanthomonas*; 12, *Zoogloea*; 13, *Gluconobacter*; 14, *Azotobacter*; 15, *Azomonas*; 16, *Beijerinckia*; 17, *Derxia*; 18, *Rhizobium*; 19, *Agrobacterium*; 20, *Halobacterium*; 21, *Halococcus*; 22, *Alcaligenes*; 23, *Acetobacter*; 24, *Brucella*; 25, *Bordetella*; 26, *Thermus*; 27, *Escherichia*; 28, *Edwardsiella*; 29, *Citrobacter*; 30, *Salmonella*; 31, *Shigella*; 32, *Klebsiella*; 33, *Enterobacter*; 34, *Proteus*; 35, "*Methylovarius*"; 36, *Methylomonas*; 37, *Methylococcus*; 38, "*Methylosinus*"; 39, "*Methylocystis*"; 40, *Methylophaga*; 41, *Methylobacterium*; 42, *Methylophilus*; 43, *Methylobacillus*.

The letters j and k are the compared taxa; x_{ij}, the numerical value of i descriptor for j taxon (i = 1,2,...,n); x_{ik}, numerical value of i descriptor for k taxon; and n, the total number of descriptors considered.

These results indicate that the phenons that were formed by the numerical analysis correspond to some changes in the classification of bacteria made in *Bergey's Manual* of 1984 compared with the edition of 1974. (Figure 1-4: different kind of lines indicate the genera that were grouped in the families in *Bergey's Manual* 1974). Thus, for example, *Gluconobacter* (line 13, Figure 1-4) and *Acetobacter* (line 23, Figure 1-4) formed a separate

1.4 Relationship of Methylotrophic Bacteria Taxa with Other Prokaryotic Taxa

phenon. In *Bergey's Manual* of 1984 *Gluconobacter* and *Acetobacter* are grouped in the family Acetobacteriaceae.

These results indicate that taxa from Section 4 (*Bergey's Manual* 1984), including the taxa of facultative methylotrophs, form a compact group related to the obligate methylotrophs by their low similarity value (Romanovskaya et al. 1986). It is likely that such characters as the ability to use C_1 compounds facultatively has no effect on phenon formation. According to the criteria used in numerical analysis, only a set of characters (but not single tests) can alter taxon relationships and, consequently, dendrogram structure. Obligate methylotrophs form several phenons (Figure 1-4). Facultative methanol utilizers of the genus *Methylobacterium* occupy a separate position among obligate and restricted facultative methylotrophs.

It should be kept in mind that the dendrogram is not identical to the phylogenetic tree and does not reflect relative (or evolutionary) relationships. At present, this phylogenetic interpretation of prokaryotic classification is based on the results of DNA-DNA and DNA-rRNA hybridization data and on the comparative analysis of rRNA nucleotide sequences. Although only a few methylotrophs were analyzed using these methods, meaningful results were obtained. Bulygina et al. (1989) showed, by an analysis of 5S rRNA nucleotide sequences of 30 methylotrophic strains, that one of the branches combined methane oxidizers (*Methylococcus*, "*Methylovarius*," "*Methylosinus*," and "*Methylocystis*") and another branch combined obligate and restricted facultative methanol utilizers.

Among two species of the genus *Methylophilus* (*M. methylotrophus* and "*M. methanolovorus*") more differences are observed by nucleotide sequences analysis than among "*M. methanolovorus*," "*Pseudomonas insueta*," and "*Protaminobacter candidus*." Since the latter two species are related, according to our current view, to the genus *Methylobacillus*, it can be assumed that "*M. methanolovorus*" is also related to the genus *Methylobacillus* (or is its representative). Bacteria of the genus *Methylobacterium* form a monolithic group. They occupy an isolated position relative to obligate and restricted facultative bacteria (Bulygina et al. 1989), which allows us to treat them as an independent taxon. The data obtained, however, do not allow us to completely determine the interrelations of methylotrophs with other prokaryotic taxa. Wolfrum and Stolp (1987) tried to solve this problem: According to their data, "*Methylomonas alba*" BG8 is a member of the gamma group of the Proteobacteria; "*Methylomonas clara*" (= *Methylophilus methylotrophus*) ATCC 31226, "*Methanolomonas glucoseoxidans*" DSM 3031, and "*Methanomonas methylovora*" ATCC 21369 (attributed to the genus *Methylobacillus*) are members of beta group of the Proteobacteria. The dissimilarity values, D_{AB}, obtained by 5S rRNA nucleotide sequencing for "*Methylomonas alba*" (methane oxidizers) and "*Methylomonas clara*," "*Methanomonas methylovora*" (methanol oxidizers) are 0.3273 and 0.3399, correspondingly. The relationship between "*Methylomonas alba*" and *Thiobacillus* M1 is closer ($D_{AB} = 0.2326$). These data also confirm isolated taxonomic position of methane-utilizing and methanol-utilizing bacteria. Representatives of the genus *Methylobacterium* are members of the alpha group of the Proteobacteria (Stackebrandt et al. 1988). Since the alpha, beta, and gamma groups were proposed to be considered at the level of subclasses (Stackebrandt et al. 1988), it is believed that the groups of methylotrophs examined (methane oxidizers, obligate methanol oxidizers, and *Methylobacterium*) are taxonomically and phylogenetically distant from each other.

Another grouping of methylotrophs was based on the 16S rRNA sequencing of 12 methylotrophic strains (Tsuji et al. 1989). The strains analyzed were attributed to the α and γ groups of purple eubacteria. It is difficult to discuss these data (Tsuji et al. 1989) as they were only published as an abstract of a symposium. However, the position of at least one strain, "*M. alba*" BG8, attributed to the alpha group by Tsuji et al. (1989) and to the gamma group by Wolfrum and Stolp (1987), is controversial. Obviously, more molecular-biological studies to better define the phylogeny of the methylotrophs are needed.

1.5 A NEW CLASSIFICATION SCHEME FOR THE METHYLOTROPHS

It is necessary to develop a new classification scheme based on all the available information on methylotrophs and their nomenclature data, in order to eliminate the existing confusion. For example, by 5S rRNA nucleotide sequencing (Wolfrum and Stolp 1987), "*Methylomonas alba*" was placed in subclass gamma; "*M. clara*" and "*Methanomonas methylovora*" were placed in subclass beta. (It should be noted that "*Methanomonas*" is identical to *Methylomonas*.) Thus, it seems as if representatives of the same genus are in different subclasses! But, this is not true: "*Methylomonas alba*" is an obligate methane oxidizer; "*M. clara*" is a restricted facultative methylotroph; "*Methanomonas methylovora*" is an obligate methanol-utilizer now attributed to the genus *Methylobacillus*. More precise identification of these strains were later made on the basis of their phenotypic characters and confirmed by DNA-DNA hybridization. Consequently, it now seems quite logical to attribute the genus *Methylomonas* (methane oxidizers), and the genera *Methylophilus* and *Methylobacillus* (obligate and restricted facultative methanol utilizers)—based on 5S rRNA nucleotide sequencing—to different subclasses. Therefore, we propose the following classification scheme without awaiting final data on their phylogeny:

1.5.1 Obligate Methane-Utilizing Bacteria
Family Methylococcaceae
  Tribe I. "Methylococceae"
    Genus I. *Methylococcus*: *M. capsulatus*, *M. thermophilus*, *M. mobilis*, "*M. gracilis*" ("*Methylomonas gracilis*").

Genus II. *Methylomonas*: M. methanica ("*Methanomonas methanica*," "*Pseudomonas methanica*"), "*M. rubra*," "*M. alba*," M. pelagica.
Genus III. "*Methylovarius*": "*M. ucrainicus*" ("*Methylococcus ucrainicus*"), "*M. luteus*" (*Methylococcus luteus*), "*M. whittenburyi*" (*Methylococcus whittenburyi*, "*Methylobacter capsulatus*"), "*M. vinelandii*" (*Methylococcus vinelandii*).
Tribe II. "Methylosineae"
Genus IV. "*Methylosinus*": "*M. trichosporium*," "*M. sporium*."
Genus V. "*Methylocystis*": "*M. parvus*," "*M. echinoides*."

1.5.2 Obligate and Restricted Facultative Methanol-Utilizing Bacteria

Family "Methylobacillaceae"
Genus I. *Methylobacillus*: M. glycogenes ("*Pseudomonas insueta*," "*Methylomonas methanolica*," "*Methylomonas methylovora*," "*Methanomonas methylovora*," "*Methanomonas methanolica*," "*Pseudomonas methanolica*").
Genus II. *Methylophaga*: M. marina, M. thalassica ("*Alteromonas thalasso-methanolica*," "*Methylomonas thalassica*").
Genus III. *Methylophilus*: M. methylotrophus ("*Pseudomonas methylotropha*," "*Methylomonas clara*").

1.5.3 Pink-Pigmented Facultative Methylotrophs

Genus I. *Methylobacterium*: M. organophilum, M. rhodium ("*Pseudomonas rhodos*"), M. radiotolerans ("*Pseudomonas radiora*"), M. mesophilicum ("*Pseudomonas mesophilica*"), M. extorquens ("*Protomonas extorquens*," "*Bacillus extorquens*," "*Vibrio extorquens*," "*Pseudomonas extorquens*," "*Flavobacterium extorquens*"), M. rhodezianum, M. zatmanii, M. fujisawaense.

The proposed scheme is not universal and only as stable as any system of classification can be; it should be improved by future experimental data. A more detailed analysis of the classification of methylotrophic bacteria as well as diagnoses of all taxa are given in *Systematics of Methylotrophic Bacteria* (Romanovskaya et al. 1991).

REFERENCES

Bergey's Manual of Determinative Bacteriology (1957) 7th ed. (Breed, R.S., ed.), Williams and Wilkins, Baltimore.
Bergey's Manual of Determinative Bacteriology (1974) 8th ed. (Buchanan, R.E., and Gibbons, N.E., eds.), Williams and Wilkins, Baltimore.
Bergey's Manual of Systematic Bacteriology (1984) vol. 1 (Holt, J.G., ed.), Williams and Wilkins, Baltimore.

Bousfield, I.J., and Green, P.N. (1985) *Int. J. Syst. Bacteriol.* 35 (2), 209–216.
Bulygina, E.S., Galchenko, V.F., Govorukhina, N.I., et al. (1989) *Molekulyarnaya Genetika, Mikrobiologiya i Virusologiya* (in Russian) 4, 18–24.
Byrom, D. (1981) in *Microbial Growth on C_1-Compounds* (Dalton, H., ed.), pp. 278–284, Heyden and Son, London.
Davies, S.L., and Whittenbury, R. (1970) *J. Gen. Microbiol.* 61 (2), 227–232.
Dijkhuizen, L., Arfman, N., Attwood, M.M., et al. (1988) *FEMS Microbiol. Lett.* 52 (3), 209–214.
Foster, J.W., and Davis, R.H. (1966) *J. Bacteriol.* 91 (5), 1924–1931.
Galchenko, V.F., Andreev, L.V., and Trotsenko, Y.A. (1986) *Taksonomiya i Identifikatsiya Obligatnykh Metanotrofnykh Bakteriy* (Taxonomy and identification of obligate methanotrophic bacteria) (in Russian). ONTY NTsBY AN SSSR, Pushchino.
Govorukhina, N.I., Doronina, N.V., Andreev, L.V., and Trotsenko, Y.A. (1989) *. Mikrobiologiya* (in Russian) 58 (2), 326–333.
Green, P.N., and Bousfield, I.J. (1983) *Int. J. Syst. Bacteriol.* 33 (4), 875–877.
Green, P.N., Bousfield, I.J., and Hood, D. (1988) *Int. J. Syst. Bacteriol.* 38 (1), 124–127.
Hazeu, W., Batenburg van der Vegte, W.H., and Bruyn, J.C. (1980) *Arch. Microbiol.* 124 (2), 211–220.
Hood, D., Dow, C.S., and Green, P.N. (1987) *J. Gen. Microbiol.* 133 (3), 709–720.
Janvier, M., Frenel, C., Grimont, F., et al. (1985) *Int. J. Syst. Bacteriol.* 35 (2), 131–139.
Jenkins, O., Byrom, D., and Jones, D. (1984) in *Microbial Growth on C_1-Compounds* (Crawford, R.L, and Hanson, R.S., eds.), pp. 255–261, ASM, Washington DC.
Jenkins, O., Byrom, D., and Jones, D. (1987) *Int. J. Syst. Bacteriol.* 37 (4), 446–448.
Krassilnikov, N.A. (1949) *Opredelitel bakterii i aktinomitsetov* (Discriminator of bacteria and actinomyces), AN SSSR, Moscow-Leningrad.
Lawrence, A.J., and Quayle, J.R. (1970) *J. Gen. Microbiol.* 63 (2), 371–374.
Lidstrom-O'Connor, M.E., Fulton, G.L., and Wopat, A.E. (1984) *J. Gen. Microbiol.* 129 (10), 3139–3148.
Lysenko, A.N., Galchenko, V.N., and Chernykh, N.A. (1988) *Mikrobiologiya* (in Russian) 57 (5), 816–822.
Makula, R.A. (1978) *J. Bacteriol.* 134 (3), 771–777.
Malashenko, Y.R. (1976) in *Microbial Production and Utilization of Gases* (Schlegel, H.G., Gottschalk, G., and Pfennig, N., eds.), pp. 293–300, E. Goltze KG, Göttingen.
Malashenko, Y.R., Romanovskaya, V.A., and Trotsenko, Y.A. (1978) *Metanokislyayushchie Mikroorganismi* (Methane-oxidizing microorganisms) (in Russian), Nauka, Moscow.
Meyer, J., Haubold, R., Heyer, J., et al. (1986) *J. Basic Microbiol.* 26 (3), 155–160.
Orla-Jensen, S. (1909) *Zentrabl. Bakteriol. Parasitenk. Infektionskr.* Abt. 2 (22), 305–346.
Patt, F.E., Cole, G.C., and Hanson, R.S. (1976) *Int. J. Sys. Bacteriol.* 26 (2), 226–229.
Romanovskaya, V.A., Malashenko, Y.R., and Bogachenko, V.N. (1978) *Mikrobiologiya* (in Russian) 47 (1), 120–130.
Romanovskaya, V.A. (1984) *Mikrobiologiya* 53 (5), 777–784.
Romanovskaya, V.A., Sadovnikov, Y.S., Malashenko, Y.R., et al. (1986) *Mikrobiologiya* (in Russian) 55 (2), 305–307.

Romanovskaya, V.A., and el Sayed, M. (1986) *Mikrobiologicheskii Zhurnal* (in Russian) 48 (2), 97–109.
Romanovskaya, V.A., Stolyar, S.M., and Malashenko, Y.R. (1991) in *Sistematika Metilotrofnich Bakteriy (Systematics of Methylotrophic Bacteria)* (in Russian), Naukova Dumka, Kiev (in press).
Sieburth, J. McN., Johnson, P.W., Eberhardt, M.A., et al. (1987) *Curr. Microbiol.* 14 (5), 285–293.
Söhngen, N.L. (1906) *Z. Bakteriol. Parasitenk* Abt. II 15, 513–517.
Stackebrandt, E., Murray, R.G.E., and Trüper, H.G. (1988) *Int. J. Syst. Bacteriol.* 38, 321–326.
Strand, S.E., and Lidstrom, M.E. (1984) *FEMS Microbiol. Lett.* 21 (2), 247–251.
Tsuji, K., Tsien, H.C., Hanson, R.S., et al. (1989) in *Microbial Growth on C_1-Compounds* Programme and Abstracts of the 6th Int. Symp., p. 117, Göttingen, FRG.
Uhlig, H., Karbaum, K., and Steudel, A. (1986) *Int. J. Syst. Bacteriol.* 36 (2), 317–322.
Urakami, T., and Komagata, K. (1984) *Int. J. Syst. Bacteriol.* 34 (1), 188–201.
Urakami, T., Tamaoka, J., and Komagata, K. (1985) *J. Gen. Appl. Microbiol.* 31 (3), 243–253.
Urakami, T., and Komagata, K. (1986a) *Int. J. Syst. Bacteriol.* 36 (4), 502–511.
Urakami, T., and Komagata, K. (1986b) *J. Gen. Appl. Microbiol.* 32 (4), 317–341.
Validation of the publication of new names and new combinations previously effectively published outside the IJSB (1984) *Int. J. Syst. Bacteriol.* 34 (2), 355–357.
Whittenbury, R., Phillips, K.C., and Wilkinson, J.F. (1970) *J. Gen. Microbiol.* 61 (2), 205–217.
Whittenbury, R., Dalton, H., Eccleston, M., et al. (1975) in *Microbial Growth on C_1-Compounds*, pp. 1–9, The Society of Fermentation Technology, Osaka.
Whittenbury, R., and Krieg, N.R. (1984) in *Bergey's Manual of Systematic Bacteriology*, vol. 1 (Krieg, N.R., ed.), pp. 256–261, Williams and Wilkinson, Baltimore.
Wolfrum, T., Gruner, G., and Stolp, H. (1986) *Int. J. Syst. Bacteriol.* 36 (1), 24–28.
Wolfrum, T., and Stolp, H. (1987) *Syst. Appl. Microbiol.* 9 (3), 273–276.
Yordy, J.R., and Weaver, T.L. (1977) *Int. J. Syst. Bacteriol.* 27 (3), 247–255.

CHAPTER 2

Systematics of Methylotrophic Yeasts

K. Komagata

Recent progress in applied microbiology has made it possible to produce many useful substances from a variety of raw materials. Since methanol is chemically pure, inexpensive, and easily mixed with ordinary liquid media, the large-scale production of single cell protein (SCP) on methanol has been intensively studied in the last two decades. Thus, a large number of methanol-utilizing yeasts have been isolated from a wide variety of natural sources. Ogata et al. (1969 and 1970) first reported the isolation of a methanol-utilizing yeast, and new methanol-utilizing yeast species have since been described. However, in spite of the fact that methanol-utilizing yeasts are the most promising microorganisms for the production of SCP from methanol, their systematics have not been studied in detail, and several yeasts with very similar characteristics have been assigned to different species or even to different genera.

In yeast systematics, morphological and physiological characteristics are usually employed as the major criteria, but it is often difficult to determine which species are related. Modern yeast systematics are based on DNA base composition, DNA-DNA homology, quinone systems, cell wall composition, other information, and relationships and differences among the yeast species have been revealed on the basis of this type of data. Previously, Phaff and Price (1979) pointed out the strengths and weaknesses of traditional criteria in yeast systematics as compared with nucleic acid studies, and Komagata (1979) stressed the usefulness of chemosystematics in the identification of the hydrocarbon-utilizing yeasts. Kurtzman and Phaff (1987) have reviewed recent progress in yeast taxonomy. Several species of methanol-utilizing yeasts show similar biological characteristics and are described below from the viewpoint of chemosystematics.

2.1 GENERAL CHARACTERISTICS OF METHANOL-UTILIZING YEASTS

Ogata et al. (1970) reported the methanol-utilizing ability of a new isolate, *Kloeckera* sp. No. 2201, and the nonutilization of methanol by 37 other strains maintained in their laboratory. Oki et al. (1972) screened 192 type and stock yeast strains for methanol utilization and found that none of them could utilize methanol. Hazeu et al. (1972) investigated methanol utilization of about 500 strains belonging to 422 yeast species that were maintained in the Centraalbureau voor Schimmelcultures in Delft, The Netherlands. They found that the type strains of only 15 species in the genera *Hansenula, Pichia, Torulopsis,* and *Candida* could utilize methanol. Strains of *Candida boidinii, Pichia pini,* and *Hansenula capsulata* were also found to utilize methanol consistently. Yokote et al. (1974) did not find any methanol-utilizing strain out of 156 stock cultures. These studies show that methanol is utilized by a rather limited number of yeast species.

The methanol-utilizing yeasts are all ascomycetous, rather than basidiomycetous yeasts, based on their biochemical characteristics, such as negative responses to urease and the diazonium blue B (DBB) reaction, their low GC content, and the presence of ubiquinones Q-7 and Q-8. Most methanol-utilizing yeasts ferment glucose (Table 2–1). Lee and Komagata (1980b and 1983) reported that the methanol-utilizing yeasts are found in only four genera, *Hansenula, Pichia, Candida,* and *Torulopsis*. Later, Yarrow and Meyer (1978) combined the genera *Candida* and *Torulopsis* into the genus *Candida*. Methanol-utilizing *Torulopsis* species were all transferred to the genus *Candida*. Kurtzman (1984) reclassified the genus *Hansenula* into the genus *Pichia* and the genus *Williopsis* based on DNA relatedness and phenotypic similarity. The methanol-utilizing *Hansenula* species were all transferred to the genus *Pichia*.

2.1.1 Identification and Nomenclature

Methanol-utilizing yeasts were originally classified into several species and genera, but some were reidentified as different taxa and others were found not to utilize methanol. Actually, *Kloeckera* sp. no. 2201 isolated by Ogata et al. (1970) and *Candida methanolica* (Oki et al. 1972) were later renamed as *Candida boidinii* (Lee and Komagata 1980b). *Torulopsis glabrata* was

2.1 General Characteristics of Methanol-Utilizing Yeasts

TABLE 2-1 Properties of Species of the Methanol-Utilizing Yeasts

Species	Fermentation of Glucose[1]	GC Content (mol%)	Quinone Homolog	PMR Spectrum Group
Group 1				
Candida boidinii	+	29.2[2], 33.0	Q-7	18-p
Group 2				
Pichia angusta	+	47.8, 48.3	Q-7	9-b
P. finlandica	−	45.4	Q-7	9-b
P. glucozyma	+, D	45.1	Q-7	9-b
P. henricii	D	49.6, 50.2	Q-7	9-b
P. kodamae	+			
P. methylovora[3]	+	35.8	Q-7	
P. minuta	+	46.8, 47.3	Q-7	9-b
P. minuta var. nonfermentans	−, D	45.3, 45.6	Q-7	9-b
P. naganishii	+	43.3[4], 46.1	Q-7	9-c
P. philodendra	−	44.0, 49.7	Q-7	9-b[4]
P. pini	+, −	42.9, 44.4	Q-7	9-b
P. trehalophila	+	34.6	Q-7	9-c
Hansenula ofunaensis	−	32.6		
Candida cariosilignicola	D	35.1	Q-7	9-b
C. entomophila	+	54.0, 56.3	Q-7	
C. maris	−	47.3	Q-7	9-b
C. methanolophaga[3]	+	41.2	Q-7	
C. methanosorbosa	+	34.3[4]	Q-7	9-c
C. methylica	+	35.6		
C. nemodendra	−	39.5, 43.2	Q-7	9-c
C. nitratophila	D	36.6	Q-7	9-b
C. ooitensis[3]	+	34.6	Q-7	
C. ovalis[3]	+	35.8	Q-7	
C. pignaliae	+	40.1[4], 43.7	Q-7	9-c
C. pinus	D	37.3	Q-7	9-b
C. sonorensis	+	36.0, 36.1	Q-7	6-f
C. succiphila	+	40.9	Q-7	9-a
Group 3				
Pichia methanolica	+	36.4, 36.9	Q-7, Q-8	9-d
Group 4				
Pichia capsulata	+	46.8, 47.2	Q-8	11-h
P. pastoris	+	40.2, 43.0	Q-8	6-b

[1] D, delayed fermentation. All data on fermentation and major part of the GC content are cited from Barnett et al. (1983, 1990).
[2] Data from Lee and Komagata 1980b.
[3] Data from Kumamoto et al. 1986.
[4] Data from Lee and Komagata 1983.

later found not to utilize methanol although its utilization of methanol was originally reported by Asthana et al. (1971) (Hazeu et al. 1972; Lee and Komagata 1980b). On the basis of several reviews and papers (Barnett et al. 1983; Kreger-van Rij 1984; Kurtzman 1984; Kumamoto et al. 1986), most of the known methanol-utilizing yeasts can be reclassified into only two genera, *Pichia* and *Candida*. In this chapter, yeast names are limited to those published validly and to those for which the type strains are available from culture collections. The species of the genus *Pichia* are as follows (the major synonym is given in parentheses):

P. angusta (*Hansenula polymorpha*)
P. capsulata (*Hansenula capsulata*)
P. finlandica (*Hansenula wickerhamii*)
P. glucozyma (*Hansenula glucozyma*)
P. henricii (*Hansenula henricii*)
P. kodamae
P. methanolica (*P. cellobiosa*)
P. methylovora
P. minuta (*Hansenula minuta; P. lindneri; Torulopsis methanolovescens*)
P. minuta var. *nonfermentans* (*Hansenula nonfermentans*)
P. naganishii
P. pastoris
P. philodendra (*Hansenula philodendra*)
P. pini
P. trehalophila
Hansenula ofunaensis

Torulopsis methanolovescens was described by Oki et al., and was transferred to the genus *Candida* as *Candida methanolovescens* (Yarrow and Meyer 1978). Later, this isolate was considered to be an anamorphic strain of *Pichia lindneri* (Henninger and Windisch 1975; Barnett et al. 1983). Kurtzman (1984) unified *Hansenula minuta* and *Pichia lindneri* based on DNA-DNA homology, and transferred them to *Pichia minuta*. Thus, *Candida methanolovescens* is treated as a synonym of *Pichia minuta* in this chapter. Kurtzman (1984) did not suggest the transfer of *Hansenula ofunaensis* to the genus *Pichia*, but this species should be transferred because it forms hat-shaped ascospores. *Pichia kodamae* was originally described as not utilizing methanol (van der Walt et al. 1982), but a delayed utilization of methanol by this strain was later described (Barnett et al. 1983). *Candida molischiana* is treated as the anamorph of *Hansenula capsulata* (Barnett et al. 1983; Kreger-van Rij 1984), and *Hansenula capsulata* is transferred to the genus *Pichia* as *Pichia capsulata* (Kurtzman 1984). Thus, *Candida molischiana* is regarded as the anamorph of *Pichia capsulata*. Lee and Komagata (1983) studied six strains of *Hansenula capsulata* and five strains of *Candida molischiana* and pointed out the heterogeneity of these strains in electrophoretic patterns of enzymes, assimilation of maltose, and DNA base composition, although all the strains tested contained Q-8, belonged to group 11 based on proton magnetic resonance (PMR) spectra of cell wall mannans, and utilized methanol. Furthermore, the utilization of methanol by strains of *Pichia capsulata (Hansenula capsulata)* was reported to be variable (Barnett et al. 1983).

The methanol-utilizing species of *Candida* are as follows (the major synonym is given in parentheses):

C. boidinii (Kloeckera sp. no. 2201)
C. cariosilignicola
C. entomophila
C. maris
C. methanolophaga
C. methanosorbosa (C. nagoyaensis; Torulopsis methanosorbosa)
C. methylica
C. nemodendra
C. nitratophila
C. ooitensis
C. ovalis
C. pignaliae
C. pinus
C. sonorensis
C. succiphila

Torulopsis methanosorbosa was described in 1974 (Yokote et al. 1974), and *T. nagoyaensis* in 1976 (Asai et al. 1976). Both species are methanol-utilizing yeasts. Later, the type strain of *T. methanosorbosa* was identified as *T. nagoyaensis,* and therefore the correct name should be *T. methanosorbosa* on the basis of priority (Barnett et al. 1979). Later, *T. methanosorbosa* was moved to the genus *Candida* as *Candida methanosorbosa* (Barnett et al. 1983).

A key for identification of methanol-utilizing yeasts based on physiological tests has been described by Barnett et al. (1990). *Candida nanaspora* described by Saez and Rodorigues de Miranda (1988) is included in the key but chemosystematic data of this species are not described. *Candida glucosophila* is included in the key but according to Barnett et al. (1990) the utilization of methanol by this species is not shown. The utilization of methanol by this species was not reported in the original description (Tokuoka et al. 1987). *Dipodascus albidus, D. ambrosiae, D. armillariae, D. australiensis, D. geniculatus, D. macrosporus, D. mangusii, D. ovetensis, D. spicifer, D. tetrasperma, Galactomyces geotrichum, G. reessii,* and *Geotrichum clavatum* are also included in the key of the methanol-utilizing yeasts but the utilization of methanol of these yeasts are not shown (Barnett et al.

1990). Thus, utilization of methanol by above-mentioned species except for *Candida nanaspora* is not clear.

2.1.2 DNA Base Composition

DNA base composition (GC content) is a common criterion for the description not only of bacteria but also of yeasts. The thermal denaturation temperature (Tm) of DNA has been widely employed for an indirect determination of the composition. In contrast, a modern direct method was developed using enzymatic digestion of DNA and reverse-phase high performance liquid chromatography (HPLC) (Tamaoka and Komagata 1984). Good agreement in the GC content of yeast species was found between values determined by Tm and those determined by HPLC (Hamamoto et al. 1986). The GC content of methanol-utilizing yeasts ranges from 29.2% (*Candida boidinii*) to 56.3% (*Candida entomophilia*) (Table 2–1); this indicates that methanol-utilizing yeasts belong to the ascosporogenous yeasts because generally the ascomycetous yeasts show lower GC contents than basidiomycetous yeasts. The data in Table 2–1 are mainly based on published data (Barnett et al. 1983, 1990), and the lowest and highest values of GC content reported for each species are shown.

2.1.3 Quinone Systems

The quinone system is commonly used for the characterization and description of bacteria and yeasts. Yeasts have a range of ubiquinone systems from Q-5 to Q-10. Most methanol-utilizing yeast species have Q-7, although *Pichia capsulata* and *P. pastoris* have Q-8 (Table 2–1). Interestingly, *P. cellobiosa* (Lee and Komagata 1980a), later reidentified as *P. methanolica* (Barnett et al. 1983), contains both Q-7 and Q-8, at a ratio of approximately 1:1 (Lee 1982). Recently, Billon-Grand (1989) reported that the ratio of Q-7 to Q-8 in *P. methanolica* varied slightly with cultural age.

2.1.4 Cell Wall Mannans

Gorin and Spencer (1970) recognized the systematic significance of PMR spectra of the cell wall mannans of yeasts, and on this basis they classified yeasts and related organisms into 24 groups. Lee and Komagata (1980b and 1983) found that most methanol-utilizing yeasts belong to group 9 in the classification suggested by Gorin and Spencer. Mannans produced by yeasts of group 9 are characterized by spectra having major signals at approximately τ 4.42 and τ 4.38 (Gorin and Spencer 1970), and this indicates a close relationship of methanol-utilizing yeasts with Q-7 to the genus *Pichia*. However, *Pichia capsulata* belongs to group 11, *Candida boidinii* belongs to group 18, and *Pichia pastoris* and *Candida sonorensis* belong to group 6 (Table 2–1).

2.1.5 Electrophoretic Comparison of Enzymes

Electrophoretic comparison of enzymes is regarded as a useful criterion in yeast systematics (Yamazaki and Komagata 1981). Lee and Komagata (1983) examined 13 enzymes in 53 strains of methanol-utilizing yeasts. *Candida boidinii* strains were divided into two closely related interspecific clusters based on electrophoretic patterns. Glucose-6-phosphate dehydrogenases, malate dehydrogenases, glutamate dehydrogenases, lactate dehydrogenases, and fumarases were characteristic for each species. Other enzymes showed inconsistent patterns. Generally speaking, methanol-utilizing yeasts seem to be related on the basis of the zymograms.

2.1.6 Immunological Characterization of Methanol Oxidases from Methanol-Utilizing Yeasts

Antigenic analyses have been extensively used for the classification and identification of bacteria. Tsuchiya et al. (1965 and 1974) studied the antigenic structure of a large number of yeast species and proposed a new classification system based on serological characteristics. London and Kline (1973) studied the aldolases of lactic acid bacteria immunologically in terms of the evolution of this group of bacteria. Lachance and Phaff (1979) reported the comparative study of the molecular size and structure of exo-β-glucanase from *Kluyveromyces* and other yeast genera and discussed their evolutionary and taxonomic implications. Lee (1982) determined the molecular weight of methanol oxidases from 24 strains of methanol-utilizing yeast and reported that their molecular weight ranged from 5.1×10^5 to 5.7×10^5 and that a majority of the strains had enzyme molecules of 5.1×10^5 to 5.2×10^5. The immunological relatedness of methanol-utilizing yeasts was determined by immunodiffusion experiments using antisera prepared from *Candida boidinii* MG 69 (= [Centraalbureau voor Schimmelcultures] CBS 2428, type strain), *Pichia angusta* MG 70 (= CBS 4732, *Hansenula polymorpha*), *P. methanolica* KL 1 (*P. cellobiosa*), and *P. capsulata* KL 3 (*Candida molischiana*). As a result, 28 methanol-utilizing yeast strains could be divided into four groups (Table 2–2), and this grouping agreed well with the grouping based on the chemosystematic data given in Table 2–1. Two *C. boidinii* strains showed identity against antiserum of *C. boidinii* MG 69, and other strains had one dominant antigen. *Pichia minuta,* nine other species, and one variety (11 strains) all showed identity against the antiserum to *P. angusta* MG 70; three species (three strains) showed partial identity; and the remaining 11 species (14 strains) showed one dominant antigen. *P. glucozyma,* 14 species, and one variety (16 strains) showed identity against *P. methanolica* KL 1, four species (four strains) showed partial identity, and five species (eight strains) showed one dominant antigen. *P. capsulata* showed identity against *P. capsulata* KL 3; *P. pastoris* showed partial identity, *Candida boidinii* and 11 species (13 strains) showed a dominant antigen, and the remaining 10 species and one variety (12 strains) showed

TABLE 2–2 **Immunodiffusion with Methanol Oxidases Against Antisera to Methanol Oxidases from Four Reference Strains**

		Reference Strain			
Species	Tested Strain	C. boidinii MG 69	P. angusta MG 70	P. methanolica KL 1	P. capsulata KL 3
C. boidinii	MG 69	=	*	*	*
	ATCC 26175	=	*	*	*
	AKU 4705	=	*	*	*
P. minuta	IFO 0975	*	=	+	x
P. angusta	MG 70	*	=	+	*
P. glucozyma	IFO 1472	*	=	=	x
P. finlandica	MG 73	*	=	=	x
P. philodendra	MG 75	*	=	=	x
P. minuta var. nonfermentans	IFO 1473	*	=	=	x
C. maris	CBS 5151	*	=	=	x
P. naganishii	IFO 1670	*	=	=	*
P. pini	MG 66	*	=	=	*
P. henricii	IFO 1477	*	=	=	x
C. pignaliae	CBS 6071	*	=	=	*
P. trehalophila	MG 76	*	+	=	*
C. methanosorbosa	CBS 6853	*	+	=	*
P. methanolica	KL 1	*	+	=	*
C. methanolovescens	ATCC 26176	*	*	=	x
P. minuta (P. lindneri)	DSM 70718	*	*	=	x
C. nemodendra	MG 72	*	*	=	x
C. sonorensis	UDC 71-148	*	*	=	x
C. cariosilignicola	KL 24	*	*	+	*
C. succiphila	KL 30	*	*	+	*
C. nitratophila	MG 68	*	*	*	*
C. pinus	IFO 1327	*	*	*	x
P. pastoris	MG 65	*	*	*	+
P. capsulata (C. molischiana)	KL 3	*	*	*	=
P. capsulata	MG 117	*	*	*	=

Symbols: =, Identity or apparent identity; +, partial identity; *, dominant antigen; x, non-identity.

nonidentity (Table 2-2). On the basis of immunodiffusion experiments using methanol oxidases, the methanol-utilizing yeasts seem to be closely related to one another. However, *Candida boidinii* and *Pichia pastoris* are less closely related than other species. *P. methanolica* was placed in group 3, based on its quinone system (Table 2-1), but this species showed similarity to other *Pichia* species. Furthermore, the grouping based on the immunodiffusion of methanol oxidases agreed well with the grouping based on the immunological distance of microcomplement fixation, using antisera from *Candida boidinii* MG 69, *Pichia angusta* MG 70, *P. methanolica* KL 1, *P. capsulata* KL 3, versus purified methanol oxidases from 19 species and one variety (24 strains).

2.1.7 Groups of Methanol-Utilizing Yeasts

As discussed above, methanol-utilizing yeasts are divided into four major groups on the basis of their DNA base composition, quinone system, PMR spectra of cell wall mannans, enzyme patterns, and immunological characteristics of the methanol oxidases (Table 2-1). *Pichia kodamae, Hansenula ofunaensis,* and *Candida methylica* are included in group 2 for convenience although chemotaxonomic data for them are insufficient.

2.1.7.1 Group 1. This group consists only of the strains of *Candida boidinii. C. boidinii* has a lower GC content than any other methanol-utilizing yeast, and two chemovars exist based on GC content of the DNA and zymograms (Lee and Komagata 1983). Strains of this species are the most commonly isolated methanol-utilizing yeast from natural sources. Lee and Komagata (1980b) identified 20 strains as *Candida boidinii* out of 31 methanol-utilizing yeasts, and Kumamoto et al. (1986) identified 135 *Candida boidinii* strains from 205 isolates.

2.1.7.2 Group 2. Most methanol-utilizing yeasts, except *Candida boidinii, Pichia methanolica, P. capsulata,* and *P. pastoris,* belong to this group. The GC content of the DNA of this group shows a rather broad distribution from 34.3 mol% (*Candida methanosorbosa*) to 56.3 mol% (*C. entomophila*). The dominant ubiquinone is Q-7, which is found in most species of the genera *Hansenula* and *Pichia* (Yamada et al. 1973). Gorin and Spencer (1970) suggested that yeasts belonging to group 9 of the PMR spectra of cell wall mannans have affinity to the genus *Pichia*. However, *Candida sonorensis* in PMR spectra group 6 is the only exception in this group. Methanol-utilizing yeasts in this group are therefore considered to belong to the genus *Pichia* and its anamorphs, and they are systematically closely related.

2.1.7.3 Group 3. Only *Pichia methanolica* belongs to this group. The PMR spectrum of its cell wall mannans and immunological characteristics of methanol oxidase are similar to those of group 2, but this species shows a characteristic ubiquinone content (Q-7 and Q-8 at a ratio of approximately 1:1) which is clearly different from any other methanol-utilizing yeasts.

2.1.7.4 Group 4. *Pichia capsulata* and *P. pastoris* belong to this group. Its dominant ubiquinone is Q-8, which differs from those of any other methanol-utilizing yeasts. These species have different PMR spectra of their cell wall mannans and different zymograms.

2.1.8 Ecology of the Methanol-Utilizing Yeasts

Hazeu et al. (1972) suggested that the numbers of methanol-utilizing yeasts may be high in the presence of a compound rich in methoxy groups (for example, lignin), because these yeasts can be isolated from the bark of trees or from insects living on trees. Furthermore, Lee and Komagata (1980b) reported that almost all of the known methanol-utilizing yeasts examined so far can assimilate pectin, from which methanol may be formed by hydrolysis of the methyl esters of pectin. Kouno and Ozaki (1975) mentioned the same consideration in the case of methanol-utilizing bacteria. From this point of view, methanol-utilizing microorganisms may play an important role in the decomposition of plant materials and the carbon cycle in nature.

2.2 METHANE-UTILIZING YEASTS

Five methane-utilizing yeasts were isolated by Wolf and Hanson (1979), and one strain was identified as *Sporobolomyces roseus*, one as *S. gracilis*, one as *Rhodotorula glutinis*, and two as *R. rubra*; these yeasts did not utilize methanol (Wolf 1981). However, the type strains of *Sporobolomyces gracilis*, *Rhodotorula glutinis*, and *Rhodotorula rubra*, and a strain of *Sporobolomyces roseus* did not utilize methane (Wolf and Hanson 1980).

2.3 APPLICATIONS OF METHANOL-UTILIZING YEASTS

The production of useful substances from methanol by the methanol-utilizing yeasts has been studied extensively (see Chapter 11). Methanol-utilizing yeasts are still promising tools for the production of single cell protein (SCP). Cartledge (1987) mentioned the advantages and disadvantages of using methanol as a raw material, and cited *Candida boidinii, Pichia angusta (Hansenula polymorpha)*, and other species as potentially useful yeasts for production of SCP. A thermotolerant strain of *P. angusta (H. polymorpha)* was isolated for the production of SCP at a high temperature (Urakami et al. 1983), and a strain of *P. angusta (H. polymorpha)* highly tolerant to sodium chloride was selected for production of SCP from methanol when sea water was used in medium preparation (Urakami and Takano 1984).

2.4 CONCLUSION

In 1948 Wickerham and Burton reported that methanol did not support the growth of yeasts. Twenty-one years later, however, Ogata et al. (1969) isolated and described a methanol-utilizing yeast and showed the potential usefulness of methanol as a raw material for microbial production. Since then, a large number of microbiologists have studied the methanol-utilizing yeasts, and a considerable number of new species have been described on the basis of traditional criteria in yeast systematics. Out of approximately 600 yeast species described to date (Barnett et al. 1990), only 30 species are known to utilize methanol, and are considered to inhabit particular niches in nature. Methanol-utilizing yeasts are therefore of special interest in studies of the ecology and phylogeny of yeasts.

Modern yeast systematics was developed based on cellular constituents and chemosystematics. The division, unification, and transfer of yeast taxa have recently been proposed. As a result, there is presently some confusion in yeast taxonomy. Because yeasts are important organisms for modern biotechnology, the correct names should be used not only in systematic studies but also in biotechnology in order to protect patent rights.

REFERENCES

Asai, Y., Makiguchi, N., Shimada, M., and Kurimura, Y. (1976) *J. Gen. Appl. Microbiol.* 22, 197–202.

Asthana, H., Humphrey, A.E., and Morita, V. (1971) *Biotechnol. Bioenging.* 13, 923–929.

Barnett, J.A., Payne, R.W., and Yarrow, D. (1979) *A Guide to Identifying and Classifying Yeasts,* Cambridge University Press, Cambridge.

Barnett, J.A., Payne, R.W., and Yarrow, D. (1983) *Yeasts: Characteristics and Identification,* Cambridge University Press, Cambridge.

Barnett, J.A., Payne, R.W., and Yarrow, D. (1990) *Yeasts: Characteristics and Identification,* 2nd ed., Cambridge University Press, Cambridge.

Billon-Grand, G. (1989) *J. Gen. Appl. Microbiol.* 35, 261–268.

Cartledge, T.G. (1987) in *Yeast Biotechnology* (Berry, D.R., Russel, I., and Stewart, G.G., ed.), pp. 311–342, Allen & Unwin, London.

Gorin, P.A.J., and Spencer, J.F.T. (1970) *Adv. Appl. Microbiol.* 13, 25–89.

Hamamoto, M., Sugiyama, J., and Komagata, K. (1986) *J. Gen. Appl. Microbiol.* 32, 215–223.

Hazeu, W., de Bruyn, J.C., and Bos, P. (1972) *Arch. Mikrobiol.* 87, 185–188.

Henninger, W., and Windisch, S. (1975) *Arch. Microbiol.* 105, 47–48.
Kato, K., Kurimura, Y., Makiguchi, N., and Asai, Y. (1974) *J. Gen. Appl. Microbiol.* 20, 123–127.
Komagata, K. (1979) in *Single-cell Protein, Safety for Animals and Human Feeding* (Garattini, S., Paglialunga, S., and Scrimshaw, N.S., ed.), pp. 39–43, Pergamon Press, Oxford.
Kouno, K., and Ozaki, A. (1975) in *Proceedings of the International Symposium on Microbial Growth on C_1-compounds* (The Organizing Committee ed.), pp. 11–21, Society of Fermentation Technology, Osaka.
Kreger-van Rij, N.J.W. (ed.) (1984) *The Yeasts, a Taxonomic Study, 3rd revised and enlarged ed.*, Elsevier Science Publishers B.V., Amsterdam.
Kumamoto, T., Yamamoto, M., Seriu, Y., et al. (1986) *Trans. Mycol. Soc. Japan* 27, 387–397.
Kurtzman, C.P. (1984) *Antonie van Leeuwenhoek.* 50, 209–217.
Kurtzman, C.P., and Phaff, H.J. (1987) in *Biology of Yeasts* vol. 1, 2nd ed. (Rose, A.H., and Harrison, J.S., ed.), pp. 63–94, Academic Press, London.
Lachance, M.-A., and Phaff, H.J. (1979) *Int. J. Syst. Bacteriol.* 29, 70–78.
Lee, J.-D. (1982) Ph.D. thesis, Faculty of Agriculture, The University of Tokyo, Tokyo.
Lee, J.-D., and Komagata, K. (1980a) *Int. J. Syst. Bacteriol.* 30, 514–519.
Lee, J.-D., and Komagata, K. (1980b) *J. Gen. Appl. Microbiol.* 26, 133–158.
Lee, J.-D., and Komagata, K. (1983) *J. Gen. Appl. Microbiol.* 29, 395–416.
London, J., and Kline, K. (1973) *Bacteriol. Rev.* 37, 453–478.
Ogata, K., Nishikawa, H., and Ohsugi, M. (1969) *Agric. Biol. Chem.* 33, 1519–1520.
Ogata, K., Nishikawa, H., Ohsugi, M., and Tochikura, T. (1970) *J. Ferment. Technol.* 48, 389–396.
Oki, T., Kouno, K., Kitai, A., and Ozaki, A. (1972) *J. Gen. Appl. Microbiol.* 18, 295–305.
Phaff, H.J., and Price, C.W. (1979) in *Single-cell Protein—Safety and Human Feeding* (Garattini, S., Paglialunga, S., and Scrimshaw, N.S., ed.), pp. 1–12, Pergamon, Oxford.
Saez, H., and Rodorigues de Miranda, H. (1988) *Bull. Soc. Mycol. France* 104, 213–215.
Tamaoka, J., and Komagata, K. (1984) *FEMS Microbiology Letters* 25, 125–128.
Tokuoka, K., Ishitani, T., Goto, S., and Komagata, K. (1987) *J. Gen. Appl. Microbiol.* 33, 1–10.
Tsuchiya, T., Fukazawa, Y., and Kawakita, S. (1965) *Mycopath. Mycol. Appl.* 26, 1–15.
Tsuchiya, T., Fukazawa, Y., Taguchi, M., Nakase, T., and Shinoda, T. (1974) *Mycopath. Mycol. Appl.* 53, 77–91.
Urakami, T., Terao, I., and Nagai, I. (1983) *J. Ferment. Technol.* 61, 221–232.
Urakami, T., and Takano, S. (1984) *J. Ferment. Technol.* 62, 111–115.
van der Walt, J.P., Yarrow, D., Opperman, A., and Halland, L. (1982) *J. Gen. Appl. Microbiol.* 28, 155–160.
Wickerham, L.J., and Burton, K.A. (1948) *J. Bacteriol.* 56, 363–371.
Wolf, H.J. (1981) in *Proceedings of the International Symposium on Microbial Growth on C_1-compounds* (Dalton, H., ed.), pp. 202–210, Heyden & Son Ltd., London.
Wolf, H.J., and Hanson, R.S. (1979) *J. Gen. Microbiol.* 114, 187–194.

Wolf, H.J., and Hanson, R.S. (1980) *FEMS Microbiology Letters* 7, 177–179.
Yamada, Y., Okada, T., Ueshima, O., and Kondo, K. (1973) *J. Gen. Appl. Microbiol.* 19, 189–208.
Yamazaki, M., and Komagata, K. (1981) *Int. J. Syst. Bacteriol.* 31, 361–381.
Yarrow, D., and Meyer, S. (1978) *Int. J. Syst. Bacteriol.* 28, 611–615.
Yokote, Y., Sugimoto, M., and Abe, S. (1974) *J. Ferment. Technol.* 52, 201–209.

CHAPTER 3

Ultrastructure of Methylotrophic Microorganisms

Thomas E. Jensen
William A. Corpe

Methylotrophic microorganisms are characterized by their ability to derive both energy and carbon from C_1 compounds more reduced than CO_2. It is perhaps remarkable that these properties are shared by such a broad array of species including yeasts and filamentous fungi on one hand and Gram-positive and Gram-negative bacteria on the other. A thorough study of the methylotrophs as a group using electron microscopy has not been done; however, even the fragmentary information available on the ultrastructure of various methylotrophs emphasizes that there is little variation from the "basic plan" of organization for prokaryotic and eukaryotic cells.

The utilization of methane and other C_1 compounds as a sole carbon and energy source requires two major biochemical processes: (1) those concerned with the oxidation of C_1 substrate to CO_2 and the coupling of the process to ATP synthesis; and (2) pathways by which C_1 compounds are assimilated into cell material. These processes have been the main focus of the study of methylotrophs over the past 20 years. When Whittenbury et al. (1970a) recognized by electron microscopy the extensive internal membrane arrays characteristic of the methane-oxidizing bacteria, their main interest was the role of these arrays in the biochemical processes. Other ultrastructural features of the obligate and facultative methylotrophs as cited by Whittenbury and Dalton (1980), just now are beginning to emerge.

Most of this chapter will cover the ultrastructure of membranes and discuss some aspects of technique that may be of particular use in the study of cell structure.

3.1 PROKARYOTIC CELLS

3.1.1 Surface Layers

3.1.1.1 Extracellular Polymers and Holdfast Structures. Extracellular polysaccharides are produced by many of the methylotrophic species from C_1 compounds. Very little information is available concerning the ultrastructure of the voluminous exopolysaccharide, except for reports by Jones et al. (1969) and Cagle (1974). Recent work deals mainly with the production from C_1 compounds of the polymers that may be most useful (Hou et al. 1978; Huq et al. 1978; Jarman and Pace, 1984; Linton et al. 1986; Southgate and Goodwin 1989).

Hyphomicrobium species are a numerically important group of prosthecate-bearing methylotrophs that reproduce by budding (Harder and Attwood 1978). They colonize water-solid interfaces by means of a monopolar excretion of holdfast material by swarm cells. The composition of the fibrous holdfast is probably polysaccharide. In laboratory culture, single cells may become joined through the holdfast to form rosettes (Conti and Hirsch 1965).

Capsules produced by vegetative cells of methane-oxidizing *Methylosinus trichosporium* (Figure 3–1) anchored by holdfast material develop a spore at the opposite pole, which buds off and is enveloped by more capsule (Whittenbury et al. 1970b). Both spore and cell capsule display a fibrous character as shown with the electron microscope using both negatively stained and metal-shadowed preparations. While the holdfast material tested positively for polysaccharide (Whittenbury et al. 1970b), capsular material did not. Sutherland (1988) has pointed out the importance of polysaccharides in the morphogenesis and development of some bacteria, but notes that the chemical nature and specific role of *Methylosinus trichosporium* polymer has not been determined.

3.1.1.2 Cell Walls. Most cell walls or envelopes of Gram-negative methylotrophs have a cross-sectional appearance similar to that of other Gram-negative bacteria; however, *Methylobacterium* species studied by the authors (Corpe et al. 1986) have multilayered outer membranes. Six isolates were studied which represent at least four species of *Methylobacterium* (Green et al. 1988). Cell wall profiles of these organisms showed three structural

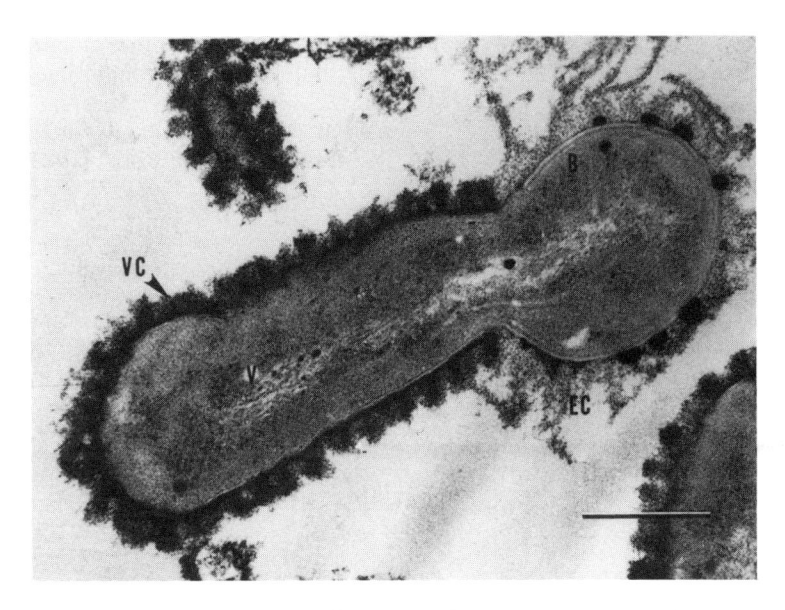

FIGURE 3-1 Early stage of *Methylosinus trichosporium* exospore formation. The vegetative cell (V) shows a budlike enlargement at one end (B) which is surrounded by the exospore capsule (EC). The vegetative portion of the cell has capsular material (VC) that is distinct from the exospore capsule. Bar, 0.5 μm. Reproduced with permission from Titus et al. (1982).

layers, external to the cytoplasmic membranes. Differences between the isolates examined were related to the thickness and density of the outer layers (Figures 3-2, 3-3, and 3-4). A periplasmic space was observed in most sectioned cells; there is a considerable variation in its volume and density. The precise location of the rigid layer was not established for any isolate.

Studies of the chemical composition of the outer membrane of a strain of methanol-grown *Methylobacterium organophilum* was described by Hancock and Williams (1986). The outer membrane contained all of the pink carotenoid pigment of the cell and an unusually small amount of phospholipid and lipopolysaccharide. The outer membrane was resistant to 2% sodium dodecylsulfate at 50°C. The authors suggested that the resistance to detergent may be due in part to the presence of unidentified, polar, phosphate-free lipids, possibly related to hopane polyols (Rohmer 1987). Electron microscopy of the cell envelopes of thin sections of *Methylobacterium* species showed that many cells had a tendency to buckle, distorting them and giving them an irregular appearance (Corpe et al. 1986). This phenomenon might also be related to the partial destruction of phospholipids by an endogenous phospholipase, which was described by Hancock and Williams (1986).

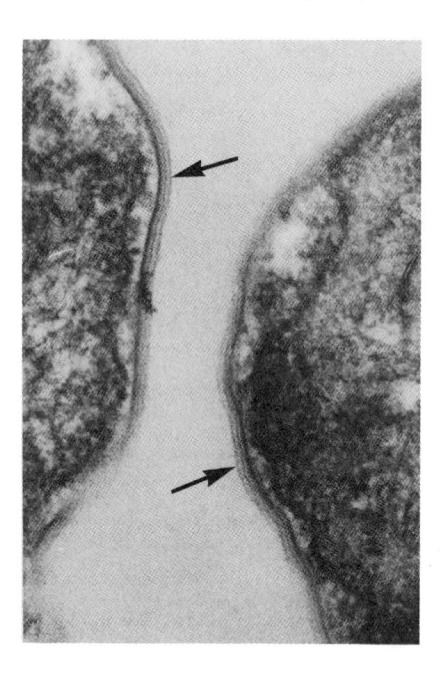

42 Ultrastructure of Methylotrophic Microorganisms

FIGURE 3-2 Thin section of part of two cells of the methylotrophic bacterium MIM-2. Note distinct layers in the cell walls at arrows. Reproduced with permission from Jensen and Corpe (1988).

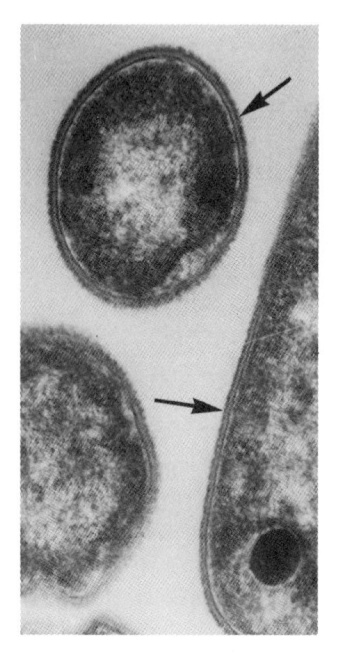

FIGURE 3-3 Thin section of SAL. Note granular material at arrows on the outer layer of the cell walls. Reproduced with permission from Jensen and Corpe (1988).

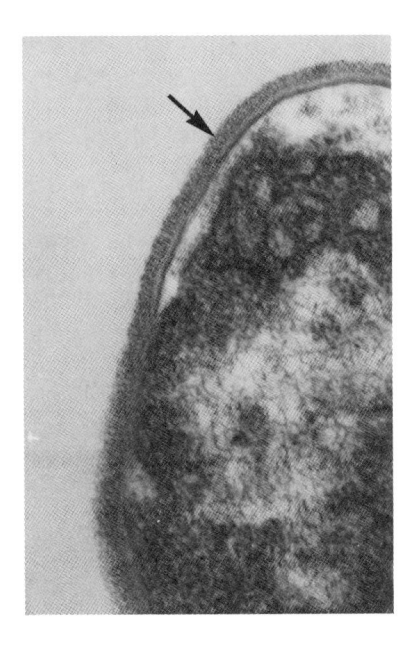

FIGURE 3-4 Electron micrograph of a thin section of a cell of *Methylobacterium* strain AA10. Note the thick layer of fibrillar material (arrow) on the outer layer of the cell wall. Reproduced with permission from Jensen and Corpe (1988).

3.1.1.3 Cytoplasmic Membrane Arrangements. A number of different membrane arrangements have been described in the C_1-utilizing bacteria. The arrangement of the membrane may appear very concise in an organism grown on one given C_1 compound, but when growth conditions are varied, this precision is generally lost. The internal membranes of methylotrophs have been widely studied, and many investigators use them as a standard. A variety of terms have been applied to name these membranes such as tubules (Whittenbury 1969), saccules (Smith et al. 1970), vesicles (Davies and Whittenbury 1970), internal membrane, intracycloplasmic membrane, and membrane-bound vesicles (Corpe et al. 1986).

One of the more recent methods to be applied to the study of membranes in cells is morphometric analysis (Weibel and Bolender 1973). This approach is especially useful in that it allows the determination of the total surface area of membranes in a cell in μm^2. The surface area is estimated from S_v, the surface density of structure in tissues, where $S_v = 2I/L$. I equals the number of line intercepts of the sampling grid crossing the surface of interest, and L equals the length of the line. This number is divided by $10^4 \times$ the magnification of print (Sicko-Goad et al. 1977). Using randomly selected cells and a randomly placed grid pattern allows an accurate determination of the surface area of membranous structures in cells. The method has been used quite widely in studies on cyanobacteria and protista to determine heavy metal effects (Sicko-Goad et al. 1977; Rachlin et al. 1982) and by Jensen and Corpe (1988) on a number of methylotrophic bacteria grown under different conditions.

In the following discussion of membranes in C_1-utilizing bacteria, we have not tried to cover all of the literature; instead, we have attempted to select those papers which we feel best illustrate the details we wish to emphasize.

3.1.1.4 Absence or Presence of Mesosomes. These inclusions have been reported in several C_1-utilizing bacteria. Dahl et al. (1972), in a study of a new obligate methylotroph, reported the organism contained bodies of the mesosome type. No micrographs of the inclusion were presented. The isolate would only grow with methanol or methylamine as the carbon source. Reed and Dugan (1987) found mesosomes in a study of the facultative methylotroph *Mycobacterium* strain ID-Y. However, they only presented micrographs of the organism grown on glucose. Few cells had the typical bacterial mesosomes, but other cells had plasma membrane invaginations that formed membranous whorls in the cell. The authors suggested that mesosomes might perform the same function as the intracytoplasmic membrane system reported in other methane-oxidizing bacteria. The problem certainly needs to be more thoroughly investigated using methanol-grown cells. In a study of the fine structure of methane and other hydrocarbon-utilizing bacteria, Davies and Whittenbury (1970) reported mesosomes. However they were only present when isolates were cultured in C_2 to C_4 gaseous alkanes.

From the above, it can be seen that little solid information exists on mesosomes in the methylotrophic bacteria. These inclusions have been studied extensively in other bacteria, especially for the genus *Bacillus*. Several authors have presented evidence that, for this genus, the elaborate folding of the plasma membrane seen in electron microscopy is artifactual (Silva et al. 1976; Nanninga 1971; Ebersold et al. 1981; Aldrich et al. 1987; Ryter 1988). The new methods of freeze-substitution, freeze-fracturing, and freeze-slamming (Phillips and Boyne 1984) must be used before one can be certain of the true nature of the inclusion.

3.1.1.5 Methane Utilizers. Of all the C_1-utilizing microbes, those which oxidize methane have been the most widely studied. Many photographs of the ultrastructure of methylotrophic bacteria were published in a book by Suzina and Fikhte (1986). (The text, however, is in Russian.) One of the first investigations on the fine structure of these bacteria was by Proctor et al. (1969), who looked at *Methylococcus capsulatus* and seven other isolates. All had membranes in which the inner faces were in close contact (paired) so little or no intramembranous space was present. In some strains, the membranes were arranged in layers of three or four around the periphery of the cell. In others, the paired membranes ran across the cell at right angles to the plasma membrane, and in others they were irregularly arranged in the cytoplasm. Smith and Ribbons (1970) investigated the fine structure of

Methanomonas methanooxidans (Figure 3-5). This organism had many membranes arranged in stacks of about six layers parallel to the plasma membrane. No intramembranous space was present. In a similar study, Smith et al. (1970) looked at the fine structure of *Methylococcus capsulatus* (Figure 3-6). They found numerous stacked arrays of membranes running across the cell at right angles to the plasma membrane. An intramembranous space of about 160 Å was present and the flattened membranes retained a space of about 80 Å between the individual units. This probably represents the true morphology of the membranes that are stacked in the methane users.

Davies and Whittenbury (1970) carried out an extensive study on the fine structure of the methane-oxidizing bacteria. Based on their data, they suggested that only two types of membrane organization occur in this group. Type I consisted of pairs of membranes which either extended throughout the organism or were arranged at the periphery where they ran parallel to the cytoplasmic membrane. They found rod-coccoid organisms of the genera *Methylomonas, Methylococcus,* and *Methylobacter* had type I membranes. All other membrane oxidizers, the vibrioid organisms of the genera *Methylosinus* and *Methylocystis,* had a different internal membrane arrangement termed type II. The type II structure consisted of vesicular discs of membrane organized into bundles which were distributed throughout the cytoplasm. This report by Davies and Whittenbury has frequently been cited in studies involving methane oxidizers, and many other authors tried to

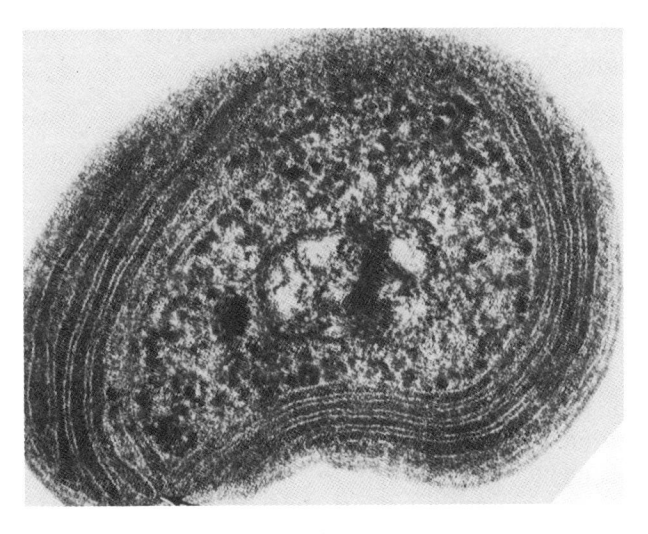

FIGURE 3-5 Profile of a cell of *Methylomonas methanooxidans* showing concentric intracytoplasmic membranes terminating as blind-ending saccules (arrow). The dense components of the membrane cannot be distinguished from the surrounding cellular constituents. Reproduced with permission from Smith and Ribbons (1970).

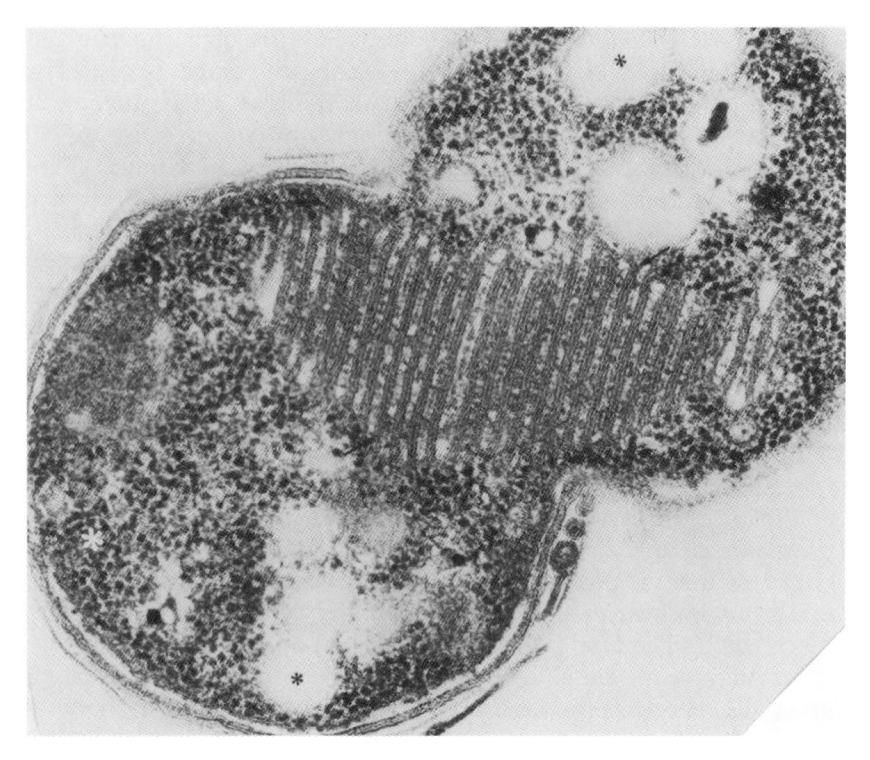

FIGURE 3-6 Profile of *Methylococcus capsulatus* in the process of cell division showing stacks of parallel membranes (between arrows). Lipid deposits are indicated by black asterisks and ribosomes by white asterisks. Reproduced with permission of Longman Ltd. from Smith et al. (1970).

determine if the membrane arrangement in the cells of organisms they were studying fit either type. Anthony (1982) reported that type I bacteria have bundles of disc-shaped vesicles which appear to be formed by invaginations of the cytoplasmic membrane, while type II cells have a system of paired peripheral membranes. This has led to much confusion in the literature. As we shall see in the studies cited below, the arrangement of the internal membranes of methane oxidizers can vary greatly, depending on growth conditions. Thus the designation of type I or type II does not hold. Perhaps it is now time to drop this once-very-useful descriptive scheme. Other workers have presented data indicating that all organisms with type I membrane arrangements metabolized methane via the serine pathway while organisms with type II membrane arrangements metabolized methane via the ribulose-monophosphate (RuMP) pathway. We shall see that this also has failed to hold for all organisms.

DeBoer and Hazeu (1972) present a very fine analysis of the membrane arrangement in *Methylomonas* species. The membranes were arranged in stacks and generally were at right angles to the plasma membrane, in a very similar fashion to the one observed in *Methylococcus capsulatus* by Smith and Ribbons (1970). By using serial sectioning, DeBoer and Hazeu (1972) were able to show that the membranes fill nearly the entire cell. In contrast, *Methylococcus mobilis* had flattened membrane vesicles when grown on methane (Hazeu et al. 1980). Patt et al. (1976), in a study of *Methylobacterium organophilum* grown on methane, found three to four layers of membrane parallel to the plasma membrane. When the same strain was grown on methanol or glucose, no membranes were present. In an earlier study, they reported the same morphology in an unnamed methane-oxidizing strain (Patt et al. 1974). Weaver and Dugan (1975), in a freeze-etch study of *Methylosinus trichosporium,* found the same arrangement of membranes as described above. They clearly showed that the membranes enclose cavities within the cytoplasm. DeBont (1976) carried out a study on three strains of *Methylosinus* able to grow on methane and fix nitrogen. Two strains had membranes running parallel to the plasma membrane, and one strain had vesicles. These isolates proved difficult to maintain and only one isolate survived for a year. This study suggests that perhaps the membranes may be important for nitrogen fixation. However, when nitrate was added to the medium, no change occurred in the membranes.

The effect of copper sulfate on *Methanomonas margaritae* was studied by Ohtomo et al. (1977). When the organism was grown on methane with copper sulfate, it had concentric membranes parallel to the plasma membrane in about 40% of the cells and vesicles in about 60% of the cells. When grown in the absence of copper sulfate, only large, membrane-limited vesicles were observed in the cells. Similar studies by Takeda and Tanaka (1980) showed that the large vesicles originated from the concentric membranes, some were elongate and others spherical and extended nearly across the cell. Patel et al. (1978) found in *Methylobacterium* R6 membranes in concentric layers, four layers, next to the plasma membrane. The organism used the serine pathway for methane metabolism and had a complete tricarboxylic acid (TCA) cycle. Zhao and Hanson (1984) report that, in *Methylomonas* strain 761M, a complete TCA cycle was found and α-ketoglutarate dehydrogenase was also present. Lynch et al. (1980) also reported the presence of concentric membrane layers in methane-grown *Methylobacterium ethanolicum* and *M. hypolimneticum*. When these organisms were grown heterotrophically, no membranes were present. Cavanaugh et al. (1987) reported the isolation of a methylotrophic bacterium which grew symbiotically with deep-sea mussels. Enzymes of the RuMP pathway were detected. Two morphological types of bacteria were present in the gill tissue: cells 1.6 μm in diameter with numerous membranes projecting into the cytoplasm at right angles to the plasma membrane, and cells 0.4 μm in diameter with no membranes.

Patt and Hanson (1978) carried out a study on the internal membranes in *Methylobacterium organophilum* as a function of growth conditions. They found that the number of layers of membranes parallel to the plasma membrane was three to four pairs when grown on methane with very little O_2 but one and two pairs were present when it was grown on methane with high O_2 level. No membranes were found when the organism was grown on methanol or glucose. Hyder et al. (1979) carried out a study on the internal membranes of *Methylococcus capsulatus* in relation to substrate and growth phase. When methane-grown cells in exponential phase were observed, the cells had parallel stacks of membranes at right angles to the plasma membrane. Cells in early stationary phase showed the parallel stacks less frequently. In late stationary phase, intracellular membrane sacs are seen and the orderly parallel sacs were seldom observed. The majority of the cells had no internal membranes.

Chetina et al. (1981) used morphometric methods to quantify and thus explain contradictory results in regard to internal membranes in *Methylomonas methanica*. They found that with low partial pressure of O_2 or methane or during exponential growth, the membranes increased. If the partial pressure of O_2 or methane was increased or if the concentration of Mg^{2+} or of PO_4^{3-} was changed, the membrane decreased. High respiratory activity also increased the amount of membrane. Ammonium and nitrate sources of nitrogen and Cu^{2+} had no effect.

A similar study was carried out by Scott et al. (1981) on *Methylosinus trichosporium* OB3b. When the organism was grown in a chemostat, few membranes were present. However when grown in a shake flask at a 5:1 ratio of methane to O_2, approximately five layers of concentric membranes were observed. When the ratio was shifted to 1:5 methane to O_2, few membranes were observed (Figures 3–7 through 3–10). In continuous culture at 22°C, 90–95% of the cells showed the concentric membrane arrangement. Well-developed, peripheral, concentric membranes were only observed in low-O_2 shake culture. Here also the number of membranes was reduced when more O_2 was available (Figures 3–11 through 3–15). This study, as well as the previous one, shows the clear relationship between the membranes and the culture conditions.

Suzina et al. (1983) also investigated the effect of culture conditions on the internal membranes of *Methylocystis echinoides*. They reported that the membranes were layered around the periphery of the cell parallel to the plasma membrane. The most membranes were present in the stationary growth phase. When grown on low partial O_2 pressure, their number was greatly reduced. These authors suggest that the presence of membranes increases oxidative phosphorylation.

A few investigators have reported organisms which have few membranes when grown on methane. Reed and Dugan (1987) grew *Mycobacterium* ID-Y on methane and observed only mesosome-like inclusions. Also, no difference in internal structure was observed when it was grown on glucose or in a nutrient broth.

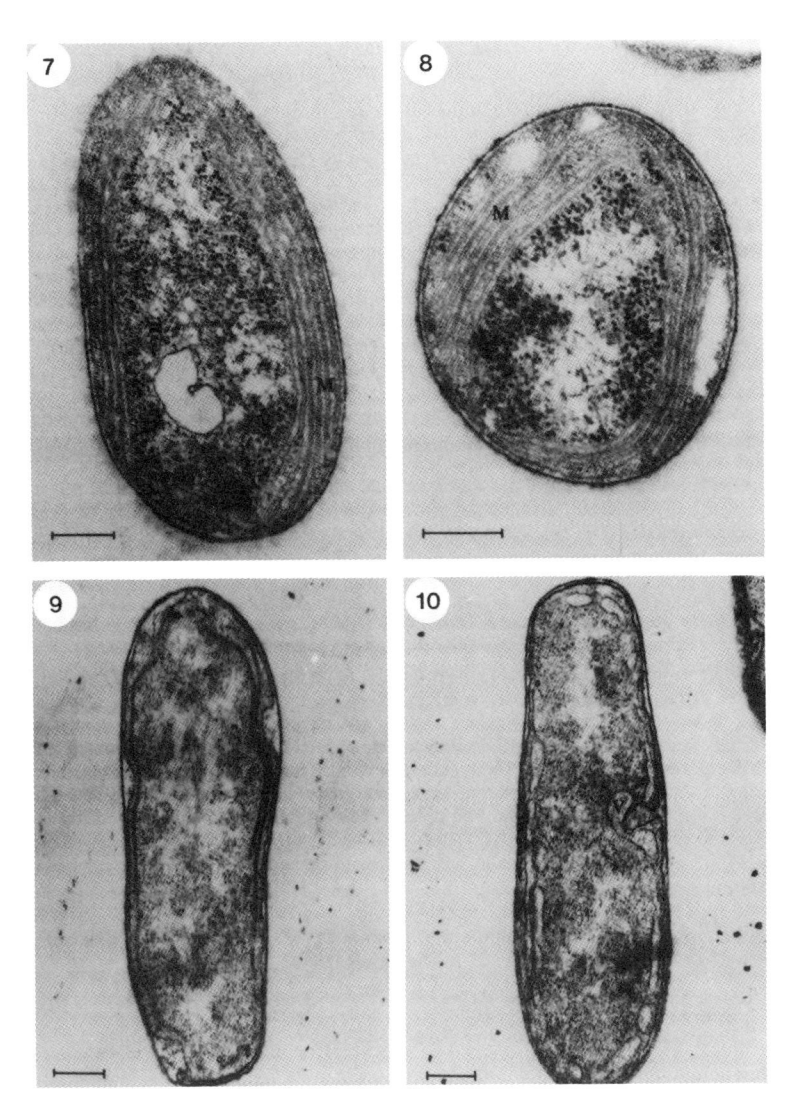

FIGURE 3-7 through 3-10 Electron micrographs of ultra-thin sections of *Methylosinus trichosporium* OB3b grown in shake flasks under different conditions. (3-7 and 3-8). Organisms grown under methane/air (1:5, v/v) showing fewer membranes and some vesicles (V). (3-9 and 3-10). Bars, 0.2 μm. Reproduced with permission from Scott et al. (1981).

We can see from the above partial survey of the literature on intracytoplasmic membranes in the methane oxidizers that a complete explanation of their function is lacking. Two basic pathways for C_1 uptake have been observed: the RuMP pathway and the serine pathway. However, this does

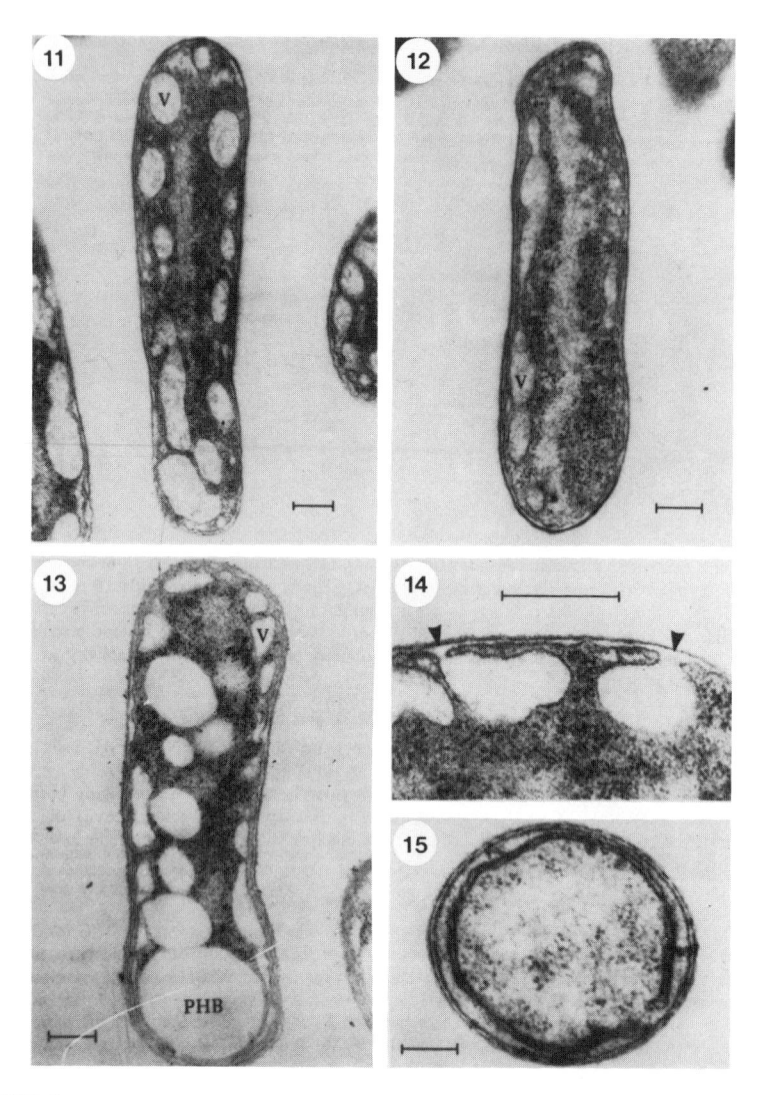

FIGURE 3-11 through 3-15 Electron micrographs of ultra-thin sections of *Methylosinus trichosporium* OB3b grown in continuous culture under different conditions (3-11 and 3-12). Organisms grown at 30°C under a variety of gas mixtures contain vesicles (V), often membrane-bound and lack a peripheral membrane system (3-13). Organisms grown at 30°C under nitrogen-limiting and methane-excess conditions had large granules of poly-β-hydroxybutyrate (PHB) in addition to vesicles (V) (3-14). Vesicles appeared to be formed by the invagination of the cytoplasmic membrane (arrows) (3-15). A small percentage (5–10%) of the organisms grown at 22°C showed a poorly developed membrane system. Bars, 0.2 μm. Reproduced with permission from Scott et al. (1981).

not hold for all methane oxidizers (Shishkina et al. 1975; Zhao and Hanson 1984; Ohtomo et al. 1977). The correlation of membrane type with pathway also does not hold, as pointed out above. The function of the membranes has also not been shown. Chetina and Trotsenko (1981) found methane monooxygenase only in the membrane fraction of broken cells of *Methylomonas methanica.* Scott et al. (1981) used *Methylosinus trichosporium* OB3b in a study designed to determine the location of methane monooxygenase (MMO). In oxygen-limited shake flask cultures, extensive peripheral membranes were observed and substantial particulate MMO activity was present. Organisms from shake flask cultures grown with a high proportion of oxygen in the gas phase contained only soluble MMO activity. As the examination of cells at the electron microscope level showed that membranes were limited or absent, these authors concluded that the exact role of intracytoplasmic membranes in *M. trichosporium* OB3b has not yet been determined. They are clearly not obligatory for growth on methane. The membrane may increase the phosphorylating ability of cells, and Chetina and Trotsenko (1986) have demonstrated that membrane preparations from several methylotrophic bacteria synthesize ATP. During most growth conditions, this organism does not have paired intracytoplasmic membranes.

Chetina et al. (1986) used cytochemical methods to determine the location of several enzymes in *Methylomonas methanica* 12 grown on methane. They found methanol, formaldehyde, succinate dehydrogenase, and NAD-independent formate dehydrogenase in the periplasm, plasma membrane, and intracytoplasmic membranes and vesicles. NADH dehydrogenase was found only on intracytoplasmic membranes while ATPase and NAD-dependent formate dehydrogenase were found only on intracytoplasmic membranes and vesicles. NADH dehydrogenase was found only on intracytoplasmic membranes while ATPase and NAD-dependent formate dehydrogenase were found on intracytoplasmic membranes and vesicles. Suzina et al. (1985) have shown that the membranes in several methane-utilizing bacteria are covered by hexagonally arrayed globular structures with a lattice spacing of approximately 9 nm. The authors present a model of how the seven subunits forming the hexagonal arrays with a central subunit may arrange themselves when associated with the membrane or free in the cytoplasm.

The use of the protein A antibody procedure is also useful in determining the exact location of the enzymes. Debette and Prensier (1989) recently used this procedure to demonstrate that esterases are located in the outer membrane of *Xanthomonas maltophilia.* It seems unlikely that a cell would expend the energy required to construct these membrane units if they did not serve a useful function.

3.1.1.6 Methanol Utilizers. Best and Higgins (1981) grew *Methylosinus trichosporium* on methanol under various conditions and then determined the membrane configuration in the cells. When grown in a shake flask with 1.0 or 0.1% (v/v) methanol, no membranes were observed. When grown in a chemostat on 0.25% methanol, numerous vesicles were observed initially. In the mid- and late-exponential phase of growth, the number of vesicles increased. These vesicles were not plasma membrane extensions. In the stationary phase, the membranes became appressed and project across the cell at nearly right angles to the plasma membrane. In this growth condition, the membrane arrangement is very similar to that reported by Weaver and Dugan (1975) for *Methylosinus trichosporium* grown on methane. However, the number of membranes was not nearly as great. When the organism was grown in a steady state culture, very few peripheral membranes were observed but numerous vesicles filled a large part of the cell, and these were continuous with the plasma membrane. When the temperature was raised to 30°C, numerous vesicles were observed and some of these were appressed and at right angles to the plasma membrane. When cells were grown in batch cultures at 22°C, 22% of the cells possessed membranes at right angles to the plasma membrane. Cells grown in continuous culture at 22°C and dilution rates of 0.025 or 0.013 per hr, the same membrane arrangement was observed as in the 30°C continuous culture.

This study clearly shows that particular membrane arrangements are determined by particular environmental conditions. Best and Higgins (1981) found that when methylotrophic bacteria are grown on methanol and no membranes are observed, this should be regarded with caution, as only one set of environmental conditions may not determine this cellular feature. This may be the case for *Methylococcus capsulatus;* when it was grown on methanol, it possessed virtually no membranes (Hyder et al. 1979). Rokem et al. (1978) also reported no internal membranes in *Pseudomonas methylotropha, Pseudomonas* 1, *Pseudomonas* 135, and *Methylomonas methanolica* when they were grown on methanol. Stieglitz and Mateles (1973) also report no membranes in methanol-grown *Pseudomonas* C. Linton and Vokes (1978) also studied membranes in *Methylococcus* NC 113 11083 when grown on methanol. They found that under methanol limitation, the organism grew well and had numerous internal paired membranes at right angles to the plasma membrane. These membranes were retained over a 50-day growth period in chemostat culture. The ability to oxidize methane was retained in these cells. Sato and Shimizu (1979) carried out a study on the interesting methylotroph *Protaminobacter ruber.* When it was grown on 1,2-propanediol as the sole carbon and energy source, it formed bacteriochlorophyll. The cells showed spherical membrane vesicles which look nearly identical to the chromatophores in other photosynthetic bacteria. *Methanomonas margaritae,* when grown on methanol, possesses a number of vesicles of different size arranged in a chain along the cell periphery. Suzina et al. (1983), in a study of *Methylocystis echinoides,* also observed numerous membranes around the periphery of the cell parallel to the plasma membrane, when it was grown on methanol. The membranes were greatly reduced in number when the oxygen partial pressure was lowered.

Prior and Dalton (1985) grew *Methylococcus capsulatus* on methanol at several different copper (Cu) concentrations. When Cu was increased, methane monooxygenase activity increased also. It was always associated with the particulate fraction. When Cu was increased, the number of paired membranes at right angles to the plasma membrane also increased. If no Cu was present, no membranes were observed.

Hancock and Williams (1986) in a study of *Methylobacterium organophilum* grown on methanol found 20% of the methanol dehydrogenase associated with the inner plasma membrane, cell fraction. The other 80% was found in the soluble fraction. Hancock and Williams (1986) separated the membranes of *Methylobacterium organophilum* into an outer membrane fraction and an inner fraction, which included the plasma membrane and internal membranes. They found 20% of the methanol dehydrogenase activity associated with the latter fraction. In *Paracoccus denitrificans,* however, Alefounder and Ferguson (1981) found that the enzyme was located in periplasmic space. Chetina and Trotsenko (1981) in a study of enzyme location in *Methylomonas methanica* also found that the enzymes responsible for the oxidation of methanol and formaldehyde were distributed between the periplasm, cytoplasm, and the fractions of the cytoplasmic and intracytoplasmic membranes.

Corpe et al. (1986), in a study of five pink-pigmented facultative methylotrophic isolates (PPFMs), found membrane-bound vesicles in all strains (Figure 3–16). They were present in varying numbers when grown in meth-

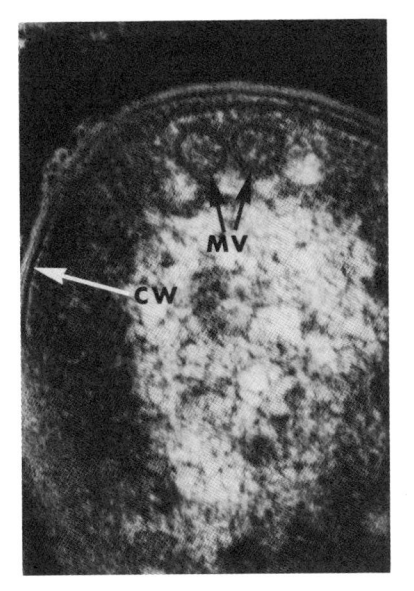

FIGURE 3-16 Thin section of strain SAL (GB11) showing the membrane vesicles (MV), nuclear region (N), and cell wall. Reproduced with permission from Corpe et al. (1986).

anol, formate, or glycerol-ammonium salts solution and glycerol-peptone broth. The vesicles were spherical with a diameter of about 0.03 μm. In a follow-up study, Jensen and Corpe (1988) determined the surface area of the membrane-bound vesicles in seven strains of PPFMs grown on methanol. Two strains, M_1R_1 and SAL, did not produce measurable membrane-bound vesicles under the growth conditions used. The surface area of the membranous vesicles of the other strains varied from 1.99 μm^2 per cell to 0.13 μm^2 per cell. By way of comparison, in strain AA10, the surface area of the membrane vesicles varied depending on the carbon source as follows: methanol, 2.08 μm^2 per cell; sodium formate, 1.6 μm^2; and glycerol, none.

Smith and Ribbons (1970) found extensive membranes in *Methylomonas methanoxidans* when grown on methanol. The membranes were arranged in stacks and were parallel to the plasma membrane. No intramembranous space was present.

Gal'chenko et al. (1985) studied 11 strains of seven obligate methanotrophs that grew in low concentrations of methanol. The addition of CO_2 allowed growth of some strains at higher methanol concentrations. Gal'chenko et al. presented a micrograph of cells of *Methylomonas methanica* which showed a number of intracytoplasmic membranes at right angles to the plasma membrane, very similar to the membrane arrangement of the organism was on methane.

We can see that much needs to be done in regard to membranes in methanol users. It is not enough to grow an organism under one set of conditions in an attempt to clarify the membranous arrangement. The effect of a number of growth conditions must be explored. Oxygen tension seems to play an important role in membrane organization. In addition, it is difficult to determine the amount and configuration of membranes in cells by examining a few micrographs. Flattened membranes appear to have less surface area than those which are greatly expanded. The use of morphometric techniques allows one to determine surface area with great accuracy. This approach has great utility when studying organisms under different growth conditions, permitting correlations between amount of membrane and enzyme content or other biochemical property. We can also see that isolating cell fractions in an attempt to determine the location of key enzymes has not resulted in clarifying the function of membranes. It appears that the enzymes in many cases are "loosely attached" to membranes. Thus in the isolation procedures, they may be lost or moved. We suggest the antibody procedure using protein A-gold with thin sections may be a better approach to determining where the enzymes are located.

3.1.1.7 Methylamine Utilizers. Conti and Hirsch (1965) carried out an extensive study at the electron microscope level of 14 strains of *Rhodomicrobium* and *Hyphomicrobium* species grown with methylamine as the carbon source. In most of the *Hyphomicrobium* strains, the internal membranes were from plasma membrane invaginations. Others possessed spherical or ellipsoidal vesicles. In a few isolates, paired membranes ran parallel to the plasma membrane in about six layers. These membranes had inner layers in contact with each other so no intramembranous space was present. Some of the *Rhodomicrobium* species had membranes arranged parallel to the plasma membrane but generally they were not as abundant. In addition, these organisms contained vesicles. In two strains, no membranes were observed. When *Blastobacter aminooxidans* was grown on methylamine, no membranes were observed (Doronina et al. 1983).

Very little work has been done at the electron microscope level on the methylamine-oxidizing bacteria. Much more needs to be done to explore the effect of growth conditions on the membrane arrangement in methylotrophs.

3.1.1.8 Carbon Monoxide Utilizers. Very few studies have pursued the ultrastructure of this bacterial group. Dashekvicz and Uffen (1979) found when *Rhodopseudomonas gelatinosa* was grown on CO that the cells had numerous plasma membrane invaginations. Wakin and Uffen (1983) have shown the CO oxidation system in this organism was associated with the plasma membrane fraction. Kim and Kim (1986) report also that carbon monoxide dehydrogenases were loosely attached to the cytoplasmic membrane fraction in *Pseudomonas carboxydovorans* and *Acinetobacter* sp. 1.

3.1.2 Appendages

3.1.2.1 Pili (Fimbrae). Six groups or types of pili have been recognized by Ottow (1975), taking into account their properties and presumed functions. The only methylotrophs that have been shown to bear pili are the "star-forming" cells of *Pseudomonas rhodos (Methylobacterium rhodium)* which become polarly attached to each other. As the result of contractions of their tubular pili (Ottow, 1975), they are pulled together in star-forming clusters (Mayer and Schmitt 1971; Moll and Ahrens 1970).

3.1.2.2 Spinae. Moll and Ahrens (1970) first reported large "fimbria" (spinae) in an *Agrobacterium* species isolated from brackish water. The fimbrie were 1–3 μm long and 40–60 nm in width. The only report of spinae in C_1-utilizing bacteria was by Suzina and Fikhte (1977) and Suzina et al. (1983) in *Methylocystis echinoides* (Figures 3–17 through 3–20). They called the structures "tubules," but they have the same basic morphology as spinae, so we will use this term in describing them.

Suzina et al. (1983) reported the effect of culture conditions on the spinae surface structure in *Methylocystis echinoides*. Based on these studies, the

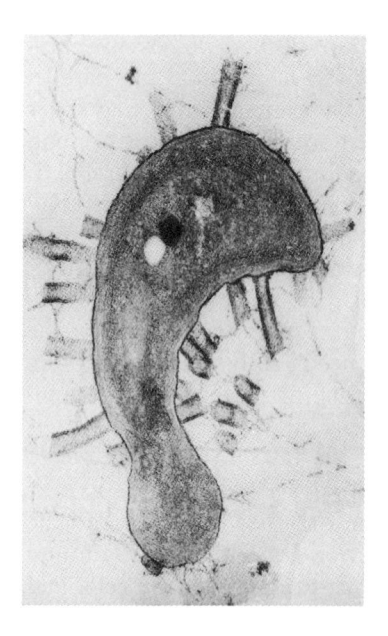

FIGURE 3-17 Thin section of a methane-oxidizing bacterium showing tubules (spinae). Reproduced with permission from Suzina and Fikhte (1977).

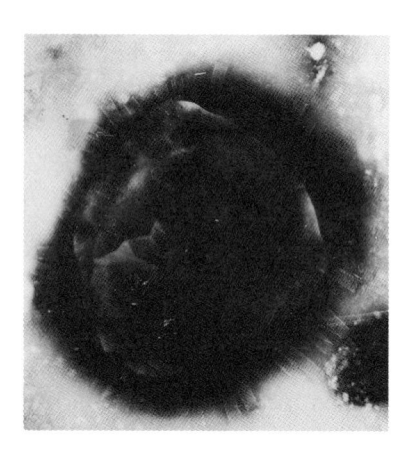

FIGURE 3-18 Negative-stain preparation of a methane-oxidizing bacterium showing spinae (tubules) on the surface. Reproduced with permission from Suzina and Fikhte (1977).

authors suggested that spinae may participate in oxidation processes. This was indirectly corroborated by the localization of the product of oxidation of 3,3′-diaminobenzidine in the presence of H_2O_2, on the outer surface of the tubules (Suzina et al. 1982). These structures have not been widely reported in the bacteria, nor are their cellular functions known. The presence of spinae on a methylotrophic bacterium, a marine pseudomonad (Easterbrook and Sperker 1982; Easterbrook and Alexander 1983) and also the cyanobacteria (Sarokin and Carpenter 1981; Easterbrook and Subba Rao 1984) suggest that they are probably widespread.

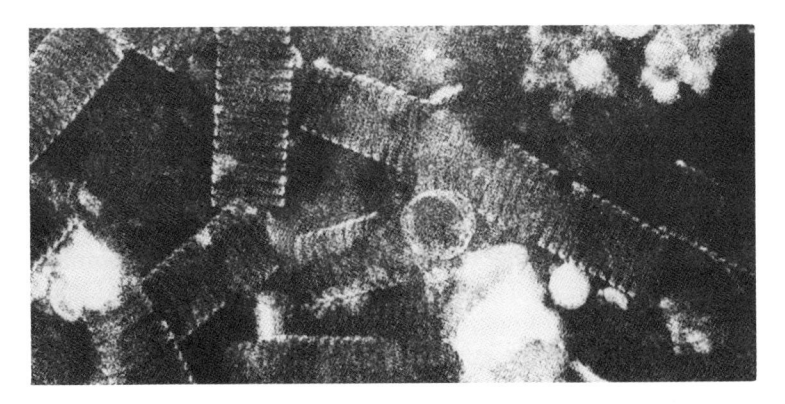

FIGURE 3-19 Negative stain preparation of spinae showing the subunits forming a spiral. Reproduced with permission from Suzina and Fikhte (1977).

3.1.2.3 Flagella. Most of the methylotrophic bacteria are motile, presumably by the flagella which are revealed by light or electron microscopic techniques. The methane-oxidizing bacteria isolated by Whittenbury et al. (1970a) were placed in five groups on the basis of flagellar type. Isolates motile by means of single or tufts of polar flagella were species of the genus *Methylosinus*. Two of the other groups, species of *Methylocystis* and *Methylococcus,* were reported to be nonmotile. However, four of nine *Methylococcus* species have been found to be flagellated (Romanovskaya et al. 1978). Hazeu et al. (1980) isolated *Methylococcus mobilis* sp. nov., which was described as a coccoid, methane-assimilating organism that grew in sarcina-like clusters and had a single or occasionally two flagella. No ultrastructural details were provided.

FIGURE 3-20 Replica of a methane-oxidizing bacterium showing the spinae. Reproduced with permission from Suzina and Fikhte (1977).

The facultative methanotrophs as represented by *Methylobacterium organophilum* are Gram-negative rods with a type II membrane arrangement that assimilate C_1 compounds by the serine pathway. The genus *Methylobacterium* currently contains eight species, all of which are motile by means of polar, subpolar, or lateral flagella (Green et al. 1988). Several strains of *Methylobacterium rhodium,* previously known as *Pseudomonas rhodos* (Heumann 1968), were shown to possess flagella with two different types of surface structure (Lowy and Hanson 1965). One type was similar to the flagella of other bacteria. The second had a unique structural appearance due in part to a sheath of helically wound bands. Studies by Schmitt et al. (1974) confirmed that the filament of the plain (unsheathed) and complex (sheathed) flagella are entirely different in structure and composition; each one has a different mode of action. It would be useful to know whether other facultative methylotrophic species have similar types of flagella. The question of how a rotary flagellum functions when sheathed and most particularly if the sheath is of cell surface origin as was discovered in *Bdellovibrio bacteriovorans* (Seidler and Starr 1968), should be investigated.

3.1.2.4 Prostheca. Prostheca are cellular appendages extending from a prokaryotic cell. They have a diameter less than that of a mature cell and are bounded by the cell wall (Staley 1968). Figure 3–21 shows a section of a *Hyphomicrobium vulgare* cell with a short hyphal prostheca. The function of the prostheca in *Hyphomicrobium* is obscure (Harder and Attwood 1978) but it has been suggested that prothecae represent specialized organelles providing increased surface area for the uptake and transport of nutrients from dilute environments.

3.1.3 Cytoplasmic Inclusions

3.1.3.1 Membranous Sheets. These inclusions have only been observed in a few methylotrophic bacterial isolates (Corpe et al. 1986; Jensen and Corpe 1988). These pink-pigmented bacteria were isolated from plant surfaces, and all cell isolates possessed the inclusion. Several of the isolates produced membranous sheets in sufficiently large amounts under certain growth conditions to determine their surface area. When strain MIM-2 was grown with methanol it contained 1.43 μm^2 of membranous sheets per cell; and 0.13 μm^2 when methanol was only diffused into the medium (Jensen and Corpe 1988). Strain AA10 contained 0.35 μm^2 when grown on glycerol-peptone agar. In all other growth conditions, the strains did not produce measurable amounts of this inclusion. When the sheets are cut in cross-section they often look like a unit membrane arranged in a circular form (Figures 3–22 and 3–23). The distance of one osmiumophilic boundary to the next varied from 7–24 nm, with most being at 15 nm. Their boundaries

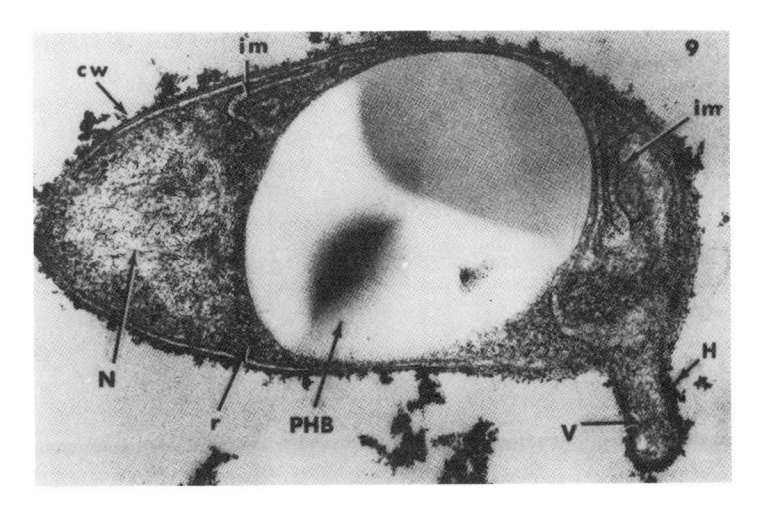

FIGURE 3-21 *Hyphomicrobium* strain B-522; longitudinal section illustrating the general features of the hyphomicrobial cell. The cell wall and cytoplasmic membrane of the rod portion of the cell and the hyphal prostheca (H) are continuous. The major structural elements of the cell are: cell wall (CW), cytoplasmic membrane (CM), internal membranes (im), nucleoplasm (N), ribosomes (r), and a poly-β-hydroxybutyrate granule (PHB). Note the membrane-bounded vesicle within the prostheca. Reproduced with permission from Conti and Hirsch (1965).

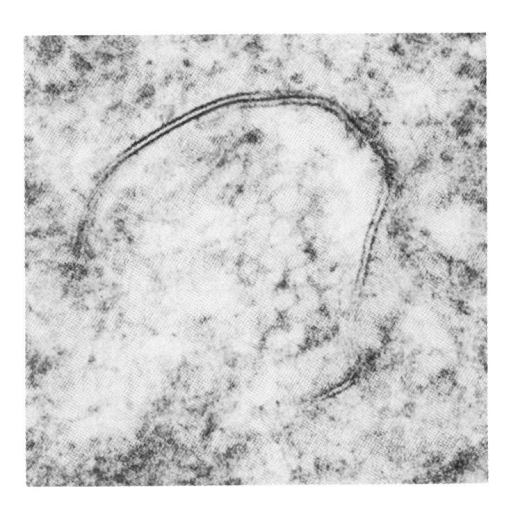

FIGURE 3-22 Thin section of part of a cell of the methylotrophic bacterium MIM-2 showing a circular membraneous sheet. Reproduced with permission from Corpe et al. (1986).

are not always clear and they usually display definite ends. Corpe et al. (1986) suggested that they may be composed of simple sheets that are commonly arranged in pairs, giving the appearance of a double-layered membrane.

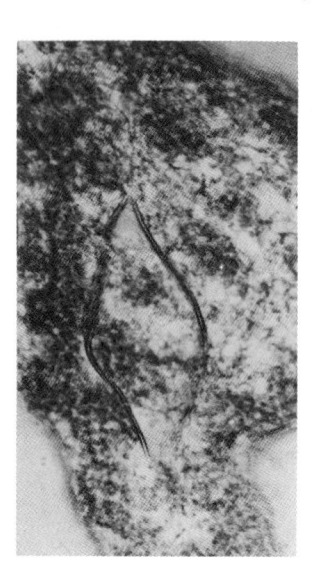

FIGURE 3-23 Thin section of part of a cell of the methylotrophic bacterium MIM-2 showing a membraneous sheet.

3.1.3.2 Polar Organelles—Polar Bodies.

DeBoer and Hazeu (1972) described a polar body in an isolate of *Methylomonas* (Figure 3-24). The body appeared in two forms. In one form, the inclusion is 20 nm thick with the inner layer filled with electron-dense spikes perpendicular to the bordering layers. In the second form, it is 45 nm thick with two layers parallel to the plasma membrane. Between the two layers were electron-dense spikes 30 nm in thickness. This is the only C_1-utilizing bacterium in which this structure has been described. It is possible that it occurs with greater frequency, but may not be distinguished from membranes which are generally found in the cells.

The term "polar organelle" seems to be the most commonly used name for this inclusion in other bacteria. However, "polar body" (Koval and Jarrell 1986) and "plate organelle" (Adams and Ghiorse 1986) were also used. It is generally found associated with the flagellar apparatus and is considered part of it. The inclusion was first reported in *Spirillum serpens* by Murray and Birch-Andersen (1963) and in *Nitrosocystis oceanus* by Murray and Watson (1963).

3.1.3.3 Crystalloids and Rods.

The first report of a crystalloid inclusion in a methylotroph was by Hyder et al. (1979). They frequently found a crystalloid in *Methylococcus capsulatus* in the late stationary phase of cells grown on methane. The crystalloid has a regular spacing of about 10 nm. They suggest it is proteinaceous. In the micrograph presented, the body covers an area 0.4 μm wide and at least 1.7 μm in length. We suggest this crystalloid is composed of spherical units about 27 nm in diameter arranged

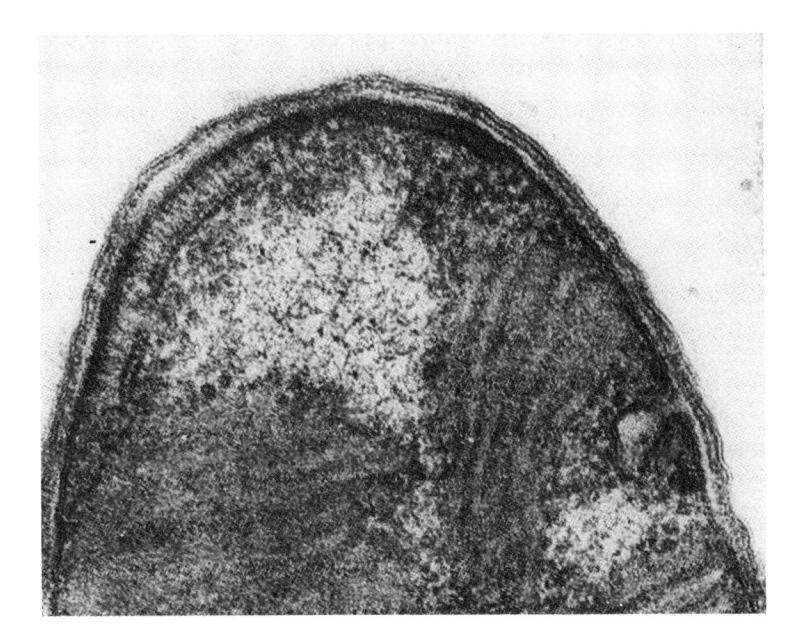

FIGURE 3-24 The polar organelles of a *Methylomonas* species. In the left organelle, the two layers parallel the cytoplasmic membrane and the spikes in the interspace are clearly visible. The structure of the right organelle is similar to those found in other Gram-negative bacteria. Reproduced with permission from DeBoer and Hazeu (1972) and the Dept. of Microbiology and Enzymology, Kluyver Laboratory of Biotechnology, Delft University of Technology.

in rows to form the regular pattern. This inclusion, on a structural basis, is very similar to the vesicular crystalloid described in the cyanobacterium *Microcystis aeruginosa* (Jensen and Baxter 1981; Reynolds et al. 1981) and *Calothrix pulvinata* (Fliesser and Jensen 1981).

Crystalloid inclusions were also observed by Corpe et al. (1986) in cells of some methylotrophic bacteria isolated from plant surfaces (Figure 3-25). They were commonly cuboidal in shape, with 0.1-μm faces. Osmiophilic striations in the bodies generally give them a cross-hatched appearance and a 125-nm periodicity. The frequency of their occurrence in cells varies with the culture conditions. None of the sectioned cells of strain M_1R_1 grown in methanol contained crystalloids but 15% of the sections of cells grown in glycerol contained them. In contrast, none of the sections of AA10 cells grown in glycerol contained them, but 4% of the sections of methanol-grown cells did. The chemical characterization and function of these inclusions must await further study. Shively (1974) reviewed the occurrence of these bodies in bacteria.

Hazeu et al. (1980) has described tubular (rod) structures in *Methylococcus mobilis* when grown in the presence of peptone. They have a similar

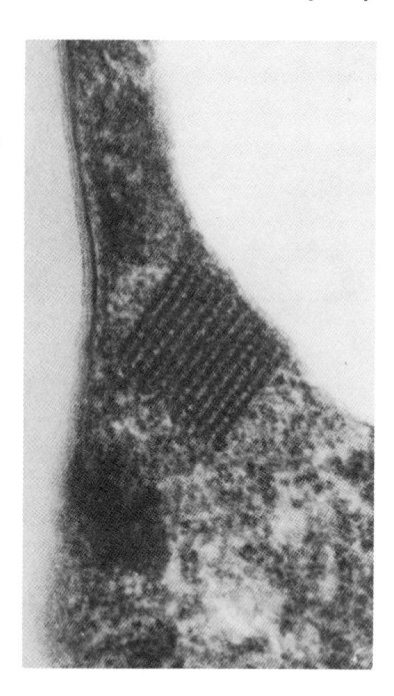

FIGURE 3-25 Thin section of a cell of the methylotrophic bacterium MIM-2 showing a crystalloid.

morphological appearance to the hexagonal tubules described in *Methanosarcina barkeri* by Archer and King (1983). The organism is a methanogen capable of growth on H_2/CO_2, acetate, methanol, and methylamine. These hexagonally packed tubules were about 50 nm in diameter and up to 0.5 μm in length and are always located next to the plasma membrane. The individual tubules appear to fuse together in the structure. Only about 2% of the cells seemed to have the inclusion. They were not observed in cells which possessed gas vacuoles.

3.1.3.4 Gas Vacuoles. Gas vacuoles composed of gas vesicles (Bowen and Jensen 1964) have been reported in only two strains of methylotrophic bacteria. Zhilina (1971) described gas vacuoles in a *Methanosarcina* isolate grown on agar-containing methanol medium. In this isolate, they were 70-80 nm in diameter and 200-400 nm in length with conical ends (Figures 3-26 and 3-27). Collapsed gas vesicles were also observed. These dimensions are nearly the same as the 75-nm diameter reported for the gas vacuoles of the cyanobacterium *Aphanenomenon flos-aquae* by Bowen and Jensen (1964). The vacuoles were bounded by a membrane monolayer 2 nm in thickness as previously reported. Archer and King (1983) also observed gas vacuoles in *Methanosarcina barkeri* grown on an acetate-containing medium. Zhilina and Zavarzin (1979), based on a cytological comparison between *M. barkeri* strain MS and *Methanosarcina* biotype 2, considered the

3.1 Prokaryotic Cells   63

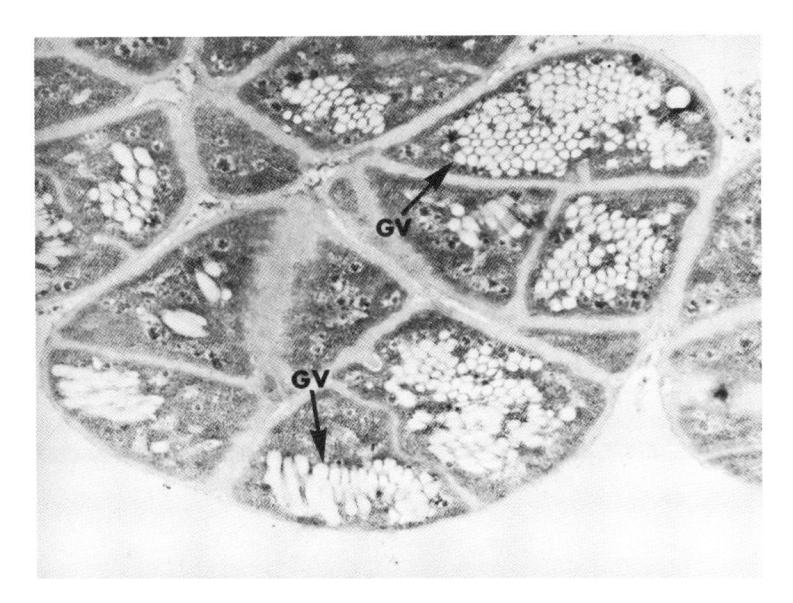

FIGURE 3-26 Ultra-thin section of a single packet of cells in *Methanosarcina*. Numerous gas vesicles (GV) can be seen in the cells. Reproduced with permission from Zhilina (1971).

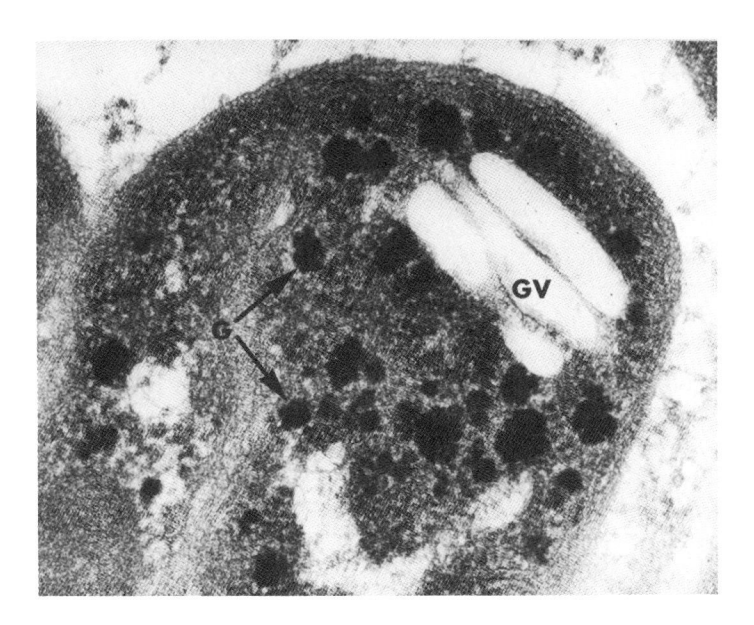

FIGURE 3-27. Portion of a cell of *Methanosarcina* showing gas vesicles (GV) and glycogen granules (G). Reproduced with permission from Zhilina (1971).

gas vacuole isolate as a separate species *Methanosarcina vacuolata* sp. nova Zhilina. Staley (1968) described a new prosthecate freshwater bacterium two strains of which could grow using formate as their carbon source. Both strains possessed gas vacuoles. They suggested the names *Prosthecomicrobium pneumaticum* and *Ancalomicrobium* for these isolates which included organisms grown on other carbon sources.

Konopka et al. (1977) prepared antisera against the protein from *Microcystis aquaticus* S1. It cross-reacted with the gas vesicle protein from *Prosthecomicrobium pneumaticum, Nostoc muscorum,* and *Anabaena flosaquae.* This indicates that it is a highly conserved protein in prokaryotes. Walsby and his associates have carried out many excellent investigations of the structure, function, and organization of gas vacuoles (Walsby 1975, 1978, and 1985). Shively et al. (1988) also reviewed the literature on gas vacuoles. Aspects of the gas vesicles of *Ancylobacter aquaticus* are reviewed by Raj (1989).

3.1.3.5 Storage Polymers.

Poly-β-hydroxyalkanoates (PHA). Globules or granules in bacteria stained with Sudan black B were shown by Williamson and Wilkinson (1958) to be associated with poly-β-hydroxybutyrate (PHB), a form of poly-β-hydroxyalkanoate (PHA). The function of PHA for endogenous storage of carbon and energy was suggested by the work of Macrae and Wilkinson (1958) and confirmed by Doudoroff and Stanier (1959). The importance of PHA as an energy-reserve polymer in bacteria has been reviewed by Dawes and Senior (1973). PHA-containing globules have been recognized in species of over 30 bacterial genera (Preiss 1989). That number is probably a bit conservative since there are some startling omissions, mainly from among methylotrophs. However, 11 strains of methylotrophic bacteria representing three genera were shown to yield polymers containing both β-hydroxyvalerate and PHB (Findley and White, personal communication).

Pink-pigmented, facultative methylotrophs (PPFMs) isolated by Corpe and Basile (1982) and obtained from culture collections (Green and Bousfield 1982) readily formed PHA globules. The globules were small in cells initiating cell division and large in cells completing cell division (Corpe et al. 1986). Several strains of PPFMs (*Methylobacterium* species) grown in methanol were compared for their content of PHA (Jensen and Corpe 1988). The percentage of cell volume occupied by PHA ranged from 4–15% when cells were grown in 1% methanol-ammonium salts solution (Fig. 3–28). When methanol was diffused into the inoculated medium during growth of the culture, only 1% of the cell volume or less was occupied by PHA.

Pfister and Lundgren (1964) demonstrated in thin sections prepared for electron microscopy that PHA granules in *Bacillus cereus* are surrounded by a single, non-unit membrane. Membrane-bounded PHA granules were

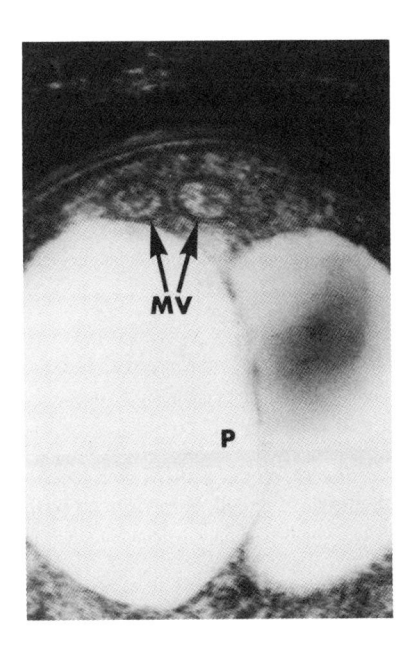

FIGURE 3-28 Thin section of the methylotrophic strain SAL (GB11) showing PHB (P) and membrane vesicles.

also observed in thin sections of *Rhodospirillum rubrum* cells (Boatman 1964), and in those of the methylotroph *Hyphomicrobium* strain B-522 (Conti and Hirsch 1965). This feature of PHA granules can be seen in Figure 3-21.

The amount of PHA produced by bacteria can be determined by using dry weight or by the percentage of the cell volume occupied by the granules. The latter method can provide a reasonably simple means of comparing or evaluating organisms and cultural conditions optimum for identifying which organism can produce the greatest yield. Both methane and methanol are inexpensive carbon substrates (Swift 1983) which can be converted to a number of potentially marketable products. Besides the PHA described above, high-grade food or feed proteins and vitamin B_{12} may soon become available.

PHA has a tensile strength and thermal properties similar to those of the common industrial polymers (Asenjo and Suk 1986) but has the advantage of being biodegradable. They could be used in packaging food products and in biochemical applications.

Polyphosphate Bodies. These inclusions have frequently been observed in C_1-utilizing organisms, as well as in other Monera and Protista. Polyphosphate bodies (PPBs) are identical to the metachromatic bodies, or volutin bodies, reported in light microscope studies (Harold 1966; Kulaev and Vagabov 1983). Polyphosphate bodies are electron dense after most fixing procedures (Figures 3-3 and 3-29). However some fixing and stain methods

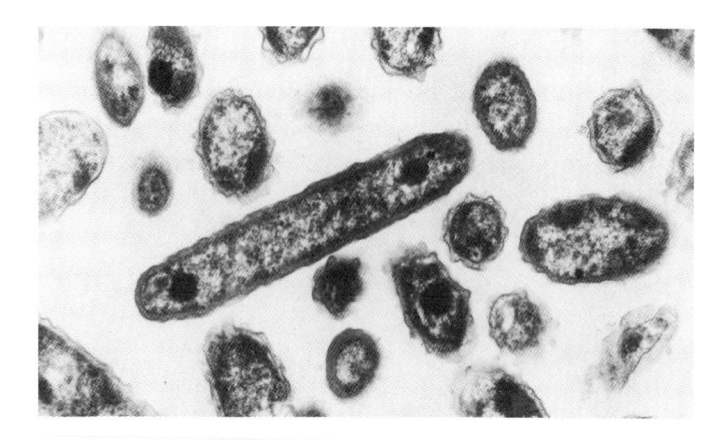

FIGURE 3-29 Ultra-thin section of *Methylobacillus (Methylophilus) methanolovorus*. Polyphosphate bodies can be seen near both ends of the cell. Reproduced with permission from Loginova and Trotsenko (1981).

render the PPBs more electron transparent (Jensen et al. 1977). These spherical inclusions range in size from about 20 nm to several micrometers in diameter. In some bacterial cells, one large body will be located in the central area of the cell with DNA. In other cases, one large body is found at each end of the cell (Figure 3-29). In the fungi, PPBs are generally located in vacuoles (Urech et al., 1978; Doonan et al. 1979). However Vorisek et al. (1982) using *Saccharomyces cerevisiae* showed that 2 hr after being exposed to "overplus" conditions (see below), the cell wall also contained much polyphosphate; furthermore, in late exponential phase, polyphosphate associated with the plasma membrane, endoplasmic reticulum, mitochondria, and the nucleus, as well as in the vacuoles. If a culture is starved of phosphorus or sulfur for a period of time and then exposed to an excess of phosphorus, the phosphorus is taken up rapidly and in amounts far greater than during any regular growth phase. This phenomenon has been termed the "polyphosphate overplus phenomenon" (Liss and Langen 1962). Unfortunately, no studies of this nature have been carried out on C_1-utilizing microbes.

The only study in which volume relationships were determined for a C_1-using organism was carried out by Jensen and Corpe (1988) on various methylotrophic bacteria. In this study only log phase cells were utilized. PPBs were present in numbers large enough to use morphometric analysis for two growth conditions and for two strains. Strain AA10 grown in glycerol had 2% of the cell volume occupied by PPB, strain SAL grown in the methanol had 3%, and strain SAL grown in glucose had 1% of the cell volume occupied by PPB.

The use of the electron microscope, either a scanning electron microscope or a transmission electron microscope, which has scanning transmission (STEM) capability in conjunction with an energy dispersive x-ray spectrometer (EDX) allows for in situ analysis. However, fixation has been shown to produce the loss of some elements and the enhancement of others (Baxter and Jensen 1980a). These authors found that freeze-drying from a liquid nitrogen slush or air-drying gave consistent results. The only other preparation method available at this time is the use of cryosectioned material, which has been freeze-dried or air-dried. If sections are very thin, their mass will not be great enough to generate sufficient x-rays for analysis. Only two studies using these methods have been carried out on C_1 users and in both, methylotrophs were studied. Corpe et al. (1986) analyzed the bodies in cells of pink-pigmented organisms isolated from leaf surfaces. They found Mg, P, and K as the main elements. We can see from studies such as these and others (Scherer and Bochem 1983a and 1983b; Baxter and Jensen 1980a; Jones and Chambers 1975; Sicko-Goad and Jensen 1976; Nissen et al. 1987; Baxter and Jensen 1986) that the use of the EDX system is a very good means of determining the elemental composition of these bodies. Undoubtedly they will vary in composition under different conditions. The use of windowless or thin window detectors on EDX systems that allow determination of all elements should provide additional information on the composition of the bodies in situ.

The exploitation of PPBs for industrial purposes has recently received some attention. Fuhs and Chen (1975) describe the parameters needed to use microbes in sewage plants to remove phosphorus; Wiechers (1983) is editor of a volume resulting from a seminar on the biological removal of phosphorus in the sewage treatment process. A number of studies have shown that PPBs sequester large amounts of various heavy metals, which are major pollutants in certain environments (Baxter and Jensen 1980b; Rachlin et al. 1984).

Phosphorus, polyphosphate, and PPBs have been the subject of several very fine reviews. Kulaev (1975) reviews the biochemistry of inorganic polyphosphates, Harold (1966) reviews polyphosphates, Beever and Burns (1980) review phosphorus uptake, storage, and utilization in fungi; and Kulaev and Vagabov (1983) and Preiss (1989) have recently reviewed all aspects of polyphosphates in microbes.

Glycogen Storage. Hazeu et al. (1980) describe the formation of large deposits of glycogen when *Methylococcus mobilis* is grown on methane and methanol. Jones et al. (1977) in a study of *Methanococcus vannielii* cells grown anaerobically on formate observed rosettes of glycogen. Cells from aerobic liquid culture did not have these inclusions. In a study of various isolates of pink-pigmented methylotrophic bacteria isolated from plant surfaces, only cells grown on glucose as a carbon source formed glycogen (Corpe et al. 1986). The SAL strain had 2% of its cell volume occupied by glucose under this growth condition (Jensen and Corpe 1988). Preiss (1989) provides a partial list of the occurrence of glycogen in bacteria. About 50 species of bacteria have been shown to store glycogen. He also provides a very fine synopsis of the structure, synthesis, and utilization of this storage polymer. Shively et al. (1988) also review the literature on glycogen granules.

Glycogen granules can be recognized on the basis of their morphology. The granules are short rods about 29–100 nm in diameter with uneven, rough-appearing surfaces. The elongated granules are often arranged in rosettes (Figure 3–27). The glycogen consists of a polymer of D-glucose monomers linked by an α-1,4-glucosidic bond and branched through 1,6-glucosidic bonds. The branching pattern produces the elongated shape.

3.1.4 Resting Cells

Many of the methane-oxidizing bacteria isolated and studied by Whittenbury et al. (1970b) have been shown to develop two types of resting stages: the exospores observed in the genus *Methylosinus* and cysts. The cysts occur in three forms also: (1) the "Azotobacter type," seen in *Methylobacter* species; (2) the "lipid cysts," seen in *Methylocystis* species; and (3) "immature *Azotobacter*-type" cysts in *Methylomonas* and *Methylococcus* species. The third type is not desiccation-resistant.

The best-documented and most thoroughly studied of these resting stage cells are the exospores of the genus *Methylosinus*. *M. trichosporium* forms rosettes joined at one pole of the vegetative cells, and the exospores are formed by external budding at the other unoccupied pole of the rod- or pear-shaped vegetative cell (Figure 3–1). When the exospores were sectioned and examined with the electron microscope, they showed no well-defined cortex or exosporium (Figure 3–30), in contrast to the endospores of *Bacillus* and *Clostridium* species (Gould and Hurst 1969). Instead, they have a thick, electron-dense exospore wall and a well-defined fibrous capsule (Reed et al. 1980). An exospore antigen occurs in the exospore region of the vegetative

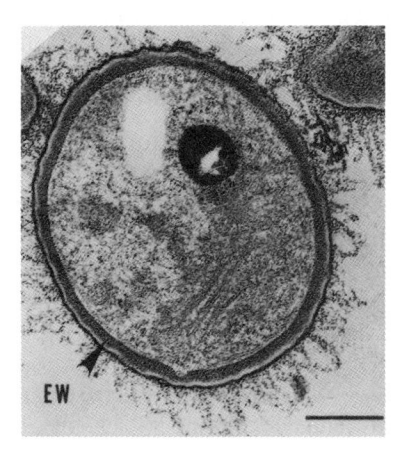

FIGURE 3–30 Mature *Methylosinus trichosporium* exospore showing a nearly spherical shape and a thick exospore wall (EW). Bar, 0.5 μm. Reproduced with permission from Titus et al. (1982).

cells of *Methylosinus* but it is not found in nonsporulating vegetative cells (Reed et al. 1980). Cup-shaped exospores observed in some preparations are thought to be an early stage in exospore development, since they can be made to change to their spherical form by merely extending the incubation period of the stationary phase of the sporulating culture (Titus et al. 1982). They are fully viable and germinate in fresh medium in the presence of methane. At germination, the exospore wall begins to disappear and the intracytoplasmic membranes extend out from the exospore into the germ tube in a parallel array (Reed et al. 1980). *Methylosinus sporium* does not possess a capsule surrounding the exospore but otherwise its spores are similar (Whittenbury et al. 1970b). Exospores have also been observed in *Rhodomicrobium vannielii* (Whittenbury and Dow 1978).

All of the *Methylobacter* species were reported to produce cysts indistinguishable in microscopic appearance from those produced by *Azotobacter* (Beamon et al. 1968): rod-shaped cells round up, increase in size, and eventually become refractile; and become desiccation-resistant. Immature cysts were observed in *Methylomonas* and *Methylococcus* groups. They rounded up and increased in size like *Azotobacter* cysts and had a wall structure intermediate in appearance between a vegetative cell and a mature cyst (Whittenbury et al. 1970b).

Lipid cysts were formed by only one methanotrophic group, represented by *Methylocystis parvus*. These strains produced large lipid inclusions and increased in size. The wall of the lipid cyst became more complex and the cell lost its membrane system.

It seems clear that the details of cyst formation and their properties should be thoroughly explored, especially if they remain important cornerstones of the taxonomy of obligate methanotrophs.

3.2 EUKARYOTIC MICROBES

A number of microbes primarily in the yeast group of fungi have been shown to utilize C_1 compounds. Studies on their ultrastructure have shown all isolates to be typical eukaryotic cells with a nucleus, Golgi, endoplasmic reticulum system, and, in the yeast group, a single mitochondrion. We have already discussed the location of PPBs in the vacuole and in some cases other locations in fungal cells. In the study of C_1 utilization by the fungi, most authors have concentrated on the microbodies that are also called peroxisomes. The microbodies in this group are generally from about 0.2–1.5 μm in diameter, are limited by a single membrane, and often possess a crystalloid core of varying dimensions (Avers 1971). They may either be spherical or nearly cuboidal (Figures 3–31 through 3–34). In *Pichia pastoris,* they are cuboidal and the crystalloid fills the entire microbody (Hazeu et al. 1975). The crystalloid cores have a 12.5-nm repeat. In *Hansenula polymorpha,* the microbodies are cuboidal (Veenhuis et al. 1976 and 1989).

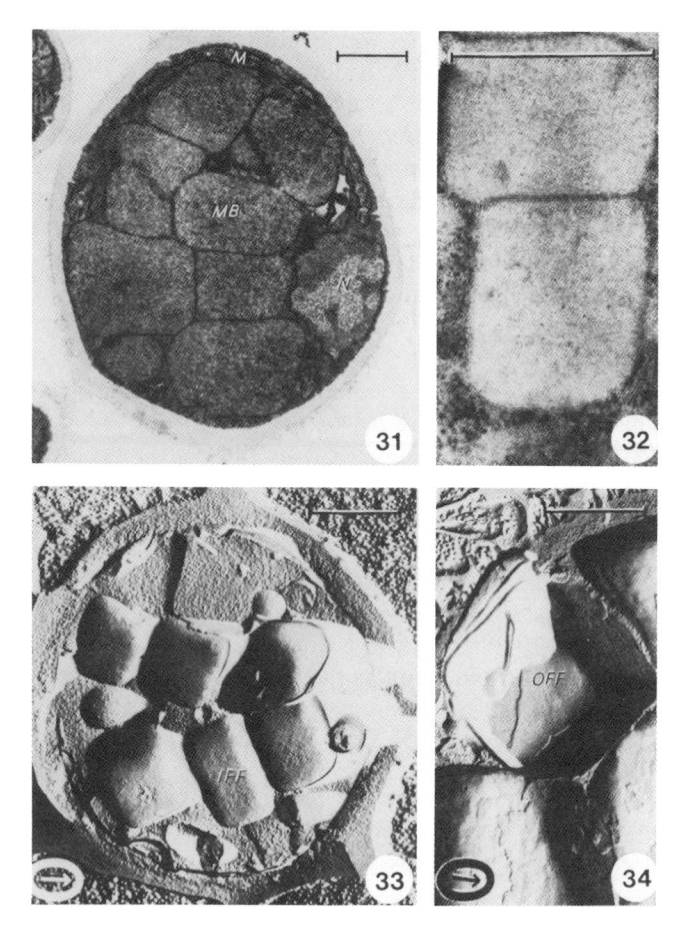

FIGURE 3-31 through 3-34 Thin sections of *Hansenula polymorpha*. Bars, 0.5 μm. (3-31) Section through a cell from a chemostat culture showing many microbodies (*MB*). The nucleus (*N*) and some mitochondria (*M*) are also visible. (3-32) Section through a cell from a batch culture showing two microbodies surrounded by a unit membrane.

Freeze-etched preparations of cells of *H. polymorpha* from a chemostat culture. The direction of shadowing is indicated by the arrows. (3-33) Several microbodies having the typical rectangular shape with rounded edges are present. The smooth inner fracture face (IFF) of a surrounding membrane is also visible. (3-34) The smooth outer fracture face (OFF) can be seen. Reproduced with permission from Van Dijken et al. (1975).

A number of studies have been carried out to determine the composition of the microbodies. Microbodies have been shown to contain catalase, alcohol oxidase, amine oxidase, D-amino acid oxidase, L-α-hydroxyacid oxidase, and dihydroxyacetone synthase (Fukui et al. 1975; Roggenkamp et al. 1975; Veenhuis et al. 1976 and 1981; Tanaka et al. 1977; Tani and Yamada 1980; Sharyshev et al. 1987).

Veenhuis et al. (1989) have presented good evidence for the functional heterogeneity of microbodies in *Candida boidinii* and *Hansenula polymorpha*. When these organisms were grown on ammonium sulfate and then placed in a medium containing methylamine, the new amine oxidase went into small microbodies at the perimeter of the cell. These small microbodies may be developing microbodies. In the case of cells grown on glucose and then transferred to methanol/methylamine-containing medium, the new alcohol oxidase, amine oxidase, and catalase were incorporated into existing microbodies. These investigators used antibodies against the enzymes and then the immunogold procedure on thin sections to demonstrate the location of the enzymes. This is a very good and precise procedure for such studies.

REFERENCES

Adams, L.F., and Ghiorse, W.C. (1986) *Arch. Microbiol.* 145, 126–135.
Aldrich, H.C., Beimborn, D.B., and Schonheit, P. (1987) *Can. J. Microbiol.* 33, 844–849.
Alefounder, P.R., and Ferguson, S.J. (1981) *Biochem. Biophys. Res. Comm.* 98, 778–784.
Anthony, C. (1982) *The Biochemistry of Methylotrophs,* Academic Press, New York.
Archer, D.B., and King, N.R. (1983) *FEMS Microbiol. Letters* 16, 217–223.
Asenjo, J.A., and Suk, J.S. (1986) *J. Ferment. Technol.* 64, 271–278.
Avers, C.J. (1971) *Sub-Cell. Biochem.* 1, 25–37.
Baxter, M., and Jensen, T.E. (1980a) *Arch. Microbiol.* 126, 213–215.
Baxter, M., and Jensen, T.E. (1980b) *Protoplasma* 104, 81–89.
Baxter, M., and Jensen, T.E. (1986) *Cytobios* 45, 147–159.
Beamon, B.L., Jackson, L.S., and Shankel, D.M. (1968) *J. Bacteriol.* 96, 266–269.
Beever, R.E., and Burns, D.J.W. (1980) *Adv. Bot. Res.* 8, 127–219.
Best, D.J., and Higgins, I.J. (1981) *J. Gen. Microbiol.* 125, 73–84.
Boatman, E.S. (1964) *J. Cell Biol.* 20, 297–311.
Bowen, C.C., and Jensen, T.E. (1964) *Science* 147, 1460–1462.
Cagle, G.D. (1974) *Can. J. Microbiol.* 21, 395–408.
Cavanaugh, C.M., Levering, P.R., Maki, J.S., Mitchell, R., and Lidstrom, M.E. (1987) *Nature* (London) 325, 346–348.
Chetina, E.V., Suzina, N.E., Fikhte, B.A., and Trotsenko, Yu.A. (1981) *Mikrobiologiya* 51, 247–254.
Chetina, E.V., Suzina, N.E., Fikhte, B.A., and Trotsenko, Yu.A. (1986) *Mikrobiologiya* 57, 125–130.
Chetina, E.V., and Trotsenko, Yu.A. (1981) *Mikrobiologiya* 50, 446–452.
Chetina, E.V., and Trotsenko, Yu.A. (1986) *Mikrobiologiya* 55, 539–542.
Conti, S.F., and Hirsch, P. (1965) *J. Bacteriol.* 89, 503–512.
Corpe, W.A., and Basile, D.V. (1982) *Dev. Indust. Microbiol.* 23, 438–493.
Corpe, W.A., Jensen, T.E., and Baxter, M. (1986) *Arch. Microbiol.* 145, 107–112.
Dahl, J.S., Mehta, R.J., and Hoare, D.S. (1972) *J. Bacteriol.* 109, 916–921.

Dashekvicz, M.P., and Uffen, R.L. (1979) *Int. J. Syst. Bacteriol.* 29, 145–148.
Davies, S.L., and Whittenbury, R. (1970) *J. Gen. Microbiol.* 61, 227–232.
Dawes, E.A., and Senior, P.J. (1973) *Adv. Microbiol. Physiol.* 10, 136–257.
Debette, J., and Prensier, G. (1989) *Appl. Environ. Microbiol.* 55, 233–239.
DeBoer, W.E., and Hazeu, W. (1972) *Antonie van Leeuwenhoek* 38, 33–47.
DeBont, J.A.M. (1976) *Antonie van Leeuwenhoek* 42, 245–253.
Doonan, B.B., Crang, R.E., Jensen, T.E., and Baxter, M. (1979) *J. Ultra. Res.* 69, 232–238.
Doronina, N.V., Govorukhina, N.I., and Trotsenko, Yu.A. (1983) *Mikrobiologiya* 52, 709–715.
Doudoroff, M., and Stanier, R.Y. (1959) *Nature* (London) 123, 1440–1442.
Easterbrook, K.B., and Sperker, S. (1982) *Can. J. Microbiol.* 28, 130–136.
Easterbrook, K.B., and Alexander, S.A. (1983) *Can. J. Microbiol.* 29, 476–487.
Easterbrook, K.B., and Subba Rao, D.V. (1984) *Can. J. Microbiol.* 30, 716–718.
Ebersold, H.R., Cordier, J.L., and Luthy, P. (1981) *Arch. Microbiol.* 130, 19–22.
Fliesser, S., and Jensen, T.E. (1982) *Cytobios* 33, 203–222.
Fukui, S., Kawamoto, S., Yasukara, S., and Tanaka, A. (1975) *Eur. J. Biochem.* 59, 561–566.
Fuhs, G.W., and Chen, M. (1975) *Microbiol. Ecol.* 2, 119–138.
Gal'chenko, V.F., Namsaroev, B.B., Mshenskii, Yu.N., Nesterov, A.I., and Ivanov, M.V. (1985) *Mikrobiologiya* 53, 724–730.
Gould, G.W., and Hurst, A. (1969) *The Bacterial Spore,* Academic Press, New York.
Green, P.N., and Bousfield, I.J. (1982) *J. Gen. Microbiol.* 128, 623–638.
Green, P.N., Bousfield, I.J., and Hood, D. (1988) *J. Syst. Bacteriol.* 38, 124–127.
Hancock, I.C., and Williams, K.M. (1986) *J. Gen. Microbiol.* 132, 599–610.
Harder, W., and Attwood, M.M. (1978) *Adv. Microbiol. Physiol.* 17, 303–356.
Harold, F.M. (1966) *Bacteriol. Rev.* 30, 772–794.
Hazeu, W., Batenburg-van der Vegte, W.H., and Nieuwdorp, P.J. (1975) *Experientia* 31, 926–927.
Hazeu, W., Batenburg-van der Vegte, W.H., and de Bruyn, J.C. (1980) *Arch. Microbiol.* 124, 211–220.
Heumann, W. (1968) *Biol. Zentralbl.* 81, 341–354.
Hou, C.T., Laskin, A.I., and Patel, R.N. (1978) *Appl. Environ. Microbiol.* 37, 800–804.
Huq, M.N., Ralph, B.J., and Rickard, P.A.D. (1978) *Aust. J. Biol. Sci.* 31, 311–316.
Hyder, S.L., Meyers, A., and Cayer, M.L. (1979) *Tissue Cell* 11, 597–610.
Jarman, J.R., and Pace, G.W. (1984) *Arch. Microbiol.* 137, 231–235.
Jensen, T.E., Sicko-Goad, L., and Ayala, R.P. (1977) *Cytologia* 42, 357–359.
Jensen, T.E., and Baxter, M. (1981) *Cytobios* 32, 129–137.
Jensen, T.E., and Corpe, W.A. (1988) *Cytobios* 53, 159–171.
Jones, H.C., Roth, I.L., and Saunders, W.M. (1969) *J. Bacteriol.* 99, 316–325.
Jones, H.E., and Chambers, L.A. (1975) *J. Gen. Microbiol.* 89, 67–72.
Jones, J.B., Bowers, B., and Stadtman, T.C. (1977) *J. Bacteriol.* 130, 1357–1363.
Kim, S.W., and Kim, Y.M. (1986) *Korean J. Microbiol.* 24, 270–275.
Konopka, A.E., Lara, J.C., and Staley, J.T. (1977) *Arch. Microbiol.* 112, 133–140.
Koval, S.F., and Jarrell, K.F. (1986) *J. Bacteriol.* 169, 1298–1306.
Kulaev, I.S. (1975) *Rev. Physiol. Biochem. Pharmacol.* 73, 131–158.
Kulaev, I.S., and Vagabov, V.M. (1983) *Adv. Microbial. Physiol.* 24, 83–171.
Linton, J.D., and Vokes, J. (1978) *FEMS Microbiol. Letters* 4, 125–128.

References

Linton, J.D., Watts, P.D., Austin, R.M., Haugh, D.E., and Niekus, H.G.D. (1986) *J. Gen. Microbiol.* 132, 779-788.
Liss, E., and Langen, P. (1962) *Arch. Microbiol.* 41, 383-392.
Loginova, N.V., and Trotsenko, Y.A. (1981) *Mikrobiologiya* 50 (1), 21-28.
Lowy, J., and Hanson, J. (1965) *J. Mol. Biol.* 11, 293-313.
Lynch, M.J., Wopat, A.E., and O'Connor, M.L. (1980) *Appl. Environ. Microbiol.* 40, 400-407.
Macrae, R.M., and Wilkinson, J.F. (1958) *J. Gen. Microbiol.* 19, 210-222.
Mayer, F., and Schmitt, R. (1971) *Arch. Mikrobiol.* 79, 311-327.
Moll, G., and Ahrens, R. (1970) *Arch. Mikrobiol.* 70, 361-368.
Murray, R.G.E., and Birch-Andersen, A. (1963) *Can. J. Microbiol.* 9, 393-401.
Murray, R.G.E., and Watson, S.W. (1963) *Nature* (London) 197, 211-212.
Nanninga, N. (1971) *J. Cell Biol.* 48, 219-224.
Nissen, H., Heldal, M., and Norland, S. (1987) *Can. J. Microbiol.* 33, 583-588.
Ohtomo, T., Iizuka, H., and Takeda, K. (1977) *Eur. J. Appl. Microbiol.* 4, 267-272.
Ottow, J.C.G. (1975) *Ann. Rev. Microbiol.* 29, 79-108.
Patel, R.N., Hou, C.T., and Felix, A. (1978) *J. Bacteriol.* 136, 352-358.
Patt, T.E., Cole, G.C., Bland, J., and Hanson, R.S. (1974) *J. Bacteriol.* 120, 955-964.
Patt, T.E., Cole, G.C., and Hanson, R.S. (1976) *J. Syst. Bacteriol.* 26, 226-229.
Patt, T.E., and Hanson, R.S. (1978) *J. Bacteriol.* 134, 636-644.
Pfister, R.M., and Lundgren, D.G. (1964) *J. Bacteriol.* 88, 1119-1129.
Phillips, T.E., and Boyne, A.F. (1984) *J. Electron Micro. Tech.* 1, 9-29.
Preiss, J. (1989) in *Bacteria in Nature,* vol. 3, (Poindexter, J.S., and Leadbetter, E.R., eds.) Plenum Publishing Corp., New York.
Prior, S.D., and Dalton, H. (1985) *J. Gen. Microbiol.* 131, 155-164.
Proctor, H.M., Norris, J.R., and Ribbons, D.W. (1969) *J. Appl. Bacteriol.* 32, 118-121.
Rachlin, J.W., Jensen, T.E., Baxter, M., and Jani, V. (1982) *Arch. Environ. Contam. Toxicol.* 11, 323-333.
Rachlin, J.W., Jensen, T.E., and Warkentine, B. (1984) *Arch. Environ. Contam. Toxicol.* 13, 143-151.
Raj, H.D. (1989) *Crit. Rev. Microbiology* 17, 89-106.
Reed, W.M., Titus, J.A., Dugan, P.R., and Pfister, R.M. (1980) *J. Bacteriol.* 141, 908-913.
Reed, W.M., and Dugan, P.R. (1987) *J. Gen. Microbiol.* 133, 1389-1395.
Reynolds, C.S., Jaworski, G.H.M., Cmiech, H.A., and Leedale, G.F. (1981) *Phil. Trans. Royal Soc.* London 293, 419-477.
Roggenkamp, R., Sahm, H., Hinkelmann, W., and Wagner, F. (1975) *Eur. J. Biochem.* 59, 231-236.
Rohmer, M. (1987) in *Surface Structures of Microorganisms and their Interaction with the Mammalian Host* (Schrinner, E., Richmond, M.H., Seibert, G., and Schwarz, U., eds.), pp. 225-242, VCH Publishers, New York.
Rokem, J.S., Reichler, J., and Goldberg, I. (1978) *Antonie van Leeuwenhoek* 44, 123-127.
Romanovskaya, V.A., Malashenko, Yu.R., and Bogachenko, V.N. (1978) *Microbiology* 47, 96-103.
Ryter, A. (1988) *Ann. Inst. Pasteur Microbiol.* 139, 33-44.
Sarokin, D.J., and Carpenter, E.J. (1981) *Botanica Marina* 24, 389-392.

Sato, K., and Shimizu, S. (1979) *Agric. Biol. Chem.* 43, 1669–1676.
Scherer, P.A., and Bochem, H.P. (1983a) *Current Microbiol.* 9, 187–194.
Scherer, P.A., and Bochem, H.P. (1983b) *Can. J. Microbiol.* 29, 1190–1199.
Schmitt, R., Raska, I., and Mayer, F. (1974) *J. Bacteriol.* 117, 844–857.
Scott, D., Brannan, J., and Higgins, I.J. (1981) *J. Gen. Microbiol.* 125, 63–72.
Seidler, R.J., and Starr, M.P. (1968) *J. Bacteriol.* 95, 1952–1955.
Sharyshev, A.A., Sysoev, O.V., Bystrykh, L.V., Krauzova, V.I., and Trotsenko, Yu.A. (1987) *Biokhimiya* 53, 483–490.
Shishkina, V.N., Yurchenko, V.V., Romanovskaya, V.A., Malaskenko, Yu.R., and Trotsenko, Yu.A. (1975) *Mikrobiologiya* 45, 417–419.
Shively, J.M. (1974) *Ann. Rev. Microbiol.* 28, 167–187.
Shively, J.M., Bryant, D.A., Fuller, R.C., et al. (1988) *Int. Rev. Cytol.* 113, 35–99.
Sicko-Goad, L., and Jensen, T.E. (1976) *Am. J. Bot.* 63, 183–188.
Sicko-Goad, L., Stoermer, E.F., and Ladewski, B.G. (1977) *Protoplasma* 93, 147–163.
Silva, M.T., Sousa, J.C.F., Polonia, J.J., Macedo, M.A.E., and Parente, A.M. (1976) *Biochemica et Biophysica Acta* 443, 92–105.
Smith, U., and Ribbons, D.W. (1970) *Arch. Mikrobiol.* 74, 116–122.
Smith, U., Ribbons, D.W., and Smith, D.S. (1970) *Tissue Cell* 2, 513–520.
Southgate, G., and Goodwin, P.M. (1989) *J. Gen. Microbiol.* 135, 2859–2867.
Staley, J.T. (1968) *J. Bacteriol.* 95, 1921–1942.
Stieglitz, B., and Mateles, R.I. (1973) *J. Bacteriol.* 114, 390–398.
Sutherland, I.W. (1988) *Int. Rev. Cytol.* 113, 187–231.
Suzina, N.E., Dmitriev, V.V., and Fikhte, B.A. (1982) *Microbiology* 51, 650–653.
Suzina, N.E., Chemina, E.V., Trotsenko, Yu.A., and Fikhte, B.A. (1983) *Microbiology* 54, 214–221.
Suzina, N.E., Chetina, E.V., Trotsenko, Y.A., and Fikhte, B.A. (1985) *FEMS Microbiol. Lett.* 30, 111–114.
Suzina, N.E., and Fikhte, B.A. (1977) *Doklady Akad. Nauk. SSSR* 234(2), 470–471.
Suzina, N.E., and Fikhte, B. (1986) *Ultrastructural Organization of Methanotrophic Bacteria*, Pushchino, Division of the Sciences, Scientific Center for Biological Studies, Academy of Science, Moscow.
Swift, H.E. (1983) *Am. Scientist* 71, 616–670.
Takeda, K., and Tanaka, K. (1980) *Antonie van Leeuwenhoek* 46, 15–25.
Tanaka, A., Yasuhara, S., Osumi, M., and Fukui, S. (1977) *Eur. J. Biochem.* 80, 193–197.
Tani, Y., and Yamada, H. (1980) *Biotechnol. Bioengineering* 22 (suppl. 1), 163–175.
Titus, J.A., Reed, W.M., Pfister, R.M., and Dugan, P.R. (1982) *J. Bacteriol.* 149, 354–360.
Urech, K., Durr, M., Boller, T.H., and Wiemken, A. (1978) *Arch. Microbiol.* 116, 275–278.
Van Dijken, J.P., Veenhuis, M., Kreger-Van Rij, N.J.W., and Harder, W. (1975) *Arch. Microbiol.* 102, 41–44.
Veenhuis, M., Van Dijken, J.P., and Harder, W. (1976) *Arch. Microbiol.* 111, 123–135.
Veenhuis, M., Zwart, K.B., and Harder, W. (1981) *Arch. Microbiol.* 129, 35–41.
Veenhuis, M., Sulter, G., Van Der Klei, I., and Harder, W. (1989) *Arch. Microbiol.* 151, 105–110.
Vorisek, J., Knotkova, A., and Kotyk, A. (1982) *Mikrobiol.* 137, 421–432.

Wakin, B.T., and Uffen, R.L. (1983) *J. Bacteriol.* 153, 571–573.
Walsby, A.E. (1975) *Ann. Rev. Pl. Physiol.* 26, 429–439.
Walsby, A.E. (1978) *Symp. Soc. J. Microbiol.* 28, 327–358.
Walsby, A.E. (1985) *Algological Studies* 38/39 (Arch. Hydrobiol./suppl. 71), 303–304.
Weaver, T.L., and Dugan, P.R. (1975) *J. Bacteriol.* 121, 704–710.
Weibel, E.R., and Bolender, R.P. (1973) in *Principles and Techniques of Electron Microscopy* (Hayat, M.A., ed.), pp. 239–296, Van Nostrand-Reinhold, New York.
Whittenbury, R. (1969) *Process Biochem.* 4, 51–56.
Whittenbury, R., Phillips, K.C., and Wilkinson, J.F. (1970a) *J. Gen. Microbiol.* 61, 205–218.
Whittenbury, R., Davies, S.L., and Davey, J.F. (1970b) *J. Gen. Microbiol.* 61, 219–226.
Whittenbury, R., and Dow, C.S. (1978) in *Companion to Microbiology* (Bull, A.T., and Meadow, P.M., eds.), pp. 221–263, Longman Group Ltd., New York and London.
Whittenbury, R., and Dalton, H. (1980) in *The Prokaryotes*, vol. 1, (Starr, M.P., Stolp, K., Truper, H., Balows, A., and Schlegel, A.G., eds.), Springer-Verlag, Berlin, FRG.
Wiechers, H.N.S. (1983) *Water Science Technol.*, vol. 15, no. 3/4, pp. 105–115, Pergamon Press, New York.
Williamson, D.H., and Wilkinson, J.F. (1958) *J. Gen. Microbiol.* 19, 198–209.
Zhao, S.J., and Hanson, R.S. (1984) in *Microbial Growth on C_1 Compounds* (Crawford, R.L., and Hanson, R.S., eds.), pp. 262–268, American Society for Microbiology, Washington, DC.
Zhilina, T.N. (1971) *Mikrobiologiya* 40, 674–680.
Zhilina, T.N., and Zavarzin, G.A. (1979) *Mikrobiologiya* 48, 223–228.

PART II

Growth and Metabolism

CHAPTER 4

Assimilation of Carbon by Methylotrophs

C. Anthony

As the context of this chapter is that of biotechnology, the substrates to be considered here are exclusively methane and methanol, and the microbes are the (mainly) aerobic bacteria and yeasts. The particular assimilation pathway operating in a given organism does not depend on the growth substrate; bacteria using the serine pathway on methane will also use it during growth on methanol. The potential of most carbon substrates in biotechnology often depends merely on their availability or price; the biochemistry involved in their utilization may not be of special interest or relevance. This is not, however, the case with the substrates methane and methanol. The wide range of their potential uses reflects the diversity and novelty of the biochemistry of their utilization as well as their relatively low cost and availability.

The diversity of assimilation pathways found within the methylotrophs is as marked as the diversity of biological types described. A knowledge of these pathways is necessary in order to compare and evaluate particular methylotrophs for their use in biotechnology, whether it be for production of single cell protein (SCP), for production of useful enzymes, or for overproduction of valuable metabolites and storage compounds.

There are four different pathways by which aerobic methylotrophs assimilate carbon substrate into cell material and each has at least two potential variants. One route, the *serine pathway,* involves carboxylic acids and amino acids as intermediates whereas the other three all involve carbohydrate intermediates (the *ribulose bisphosphate pathway,* the *ribulose monophosphate pathway,* and the *dihydroxyacetone pathway*). The latter pathway is found only in yeasts. In this chapter, the pathways first will be described, together with their enzymes, and later the distribution of the pathways within the methylotrophs will be discussed.

This chapter aims to provide the basic framework for discussion in later chapters of aspects more related to biotechnology, such as the regulation, genetics, and exploitation of metabolic pathways. The elucidation of these pathways was more or less complete by the end of the 1970s, and complete descriptions of the methods used for their elucidation and complete references to work on all aspects of methylotrophy prior to 1981 have been previously published (Anthony 1982). More recent work on yeast metabolism has been reviewed by Harder and Veenhuis (1988). For this reason, and to increase clarity, references to original work are not given.

Because references to the original work leading to elucidation of the pathways is not provided here, it is important to record that, except for the Calvin cycle, our understanding of all the pathways summarized here has rested to a great extent on the work of J.R. Quayle and his collaborators.

4.1 THE RIBULOSE BISPHOSPHATE (RuBP) PATHWAY

The ribulose bisphosphate (RuBP) pathway, also known as the Calvin cycle, is the least important of the pathways in methylotrophs. It is included because it is the main pathway for assimilation in the closely related chemolithotrophic (autotrophic) bacteria, and because many of its enzymes are similar to those found in the more typical methylotrophs.

4.1.1 Overview of the RuBP Pathway

In the RuBP pathway (Figures 4–1 and 4–2), cell carbon is assimilated at the level of carbon dioxide, produced by oxidation of the reduced one-carbon (C_1) substrate. The overall reaction for the production of glyceraldehyde phosphate (GAP) is as follows:

$3CO_2 + 6NAD(P)H + 6H^+ + 9ATP \rightarrow$
$\qquad GAP + 6NAD(P)^+ + 9ADP + 8Pi$

For comparison with energetics of other methylotrophic pathways this may be summarized as producing phosphoglycerate (PGA):

$3CO_2 + 5NAD(P)H + 5H^+ + 8ATP \rightarrow$
$\qquad PGA + 5NAD(P)^+ + 8ADP + 7Pi$

4.1 The Ribulose Bisphosphate (RuBP) Pathway

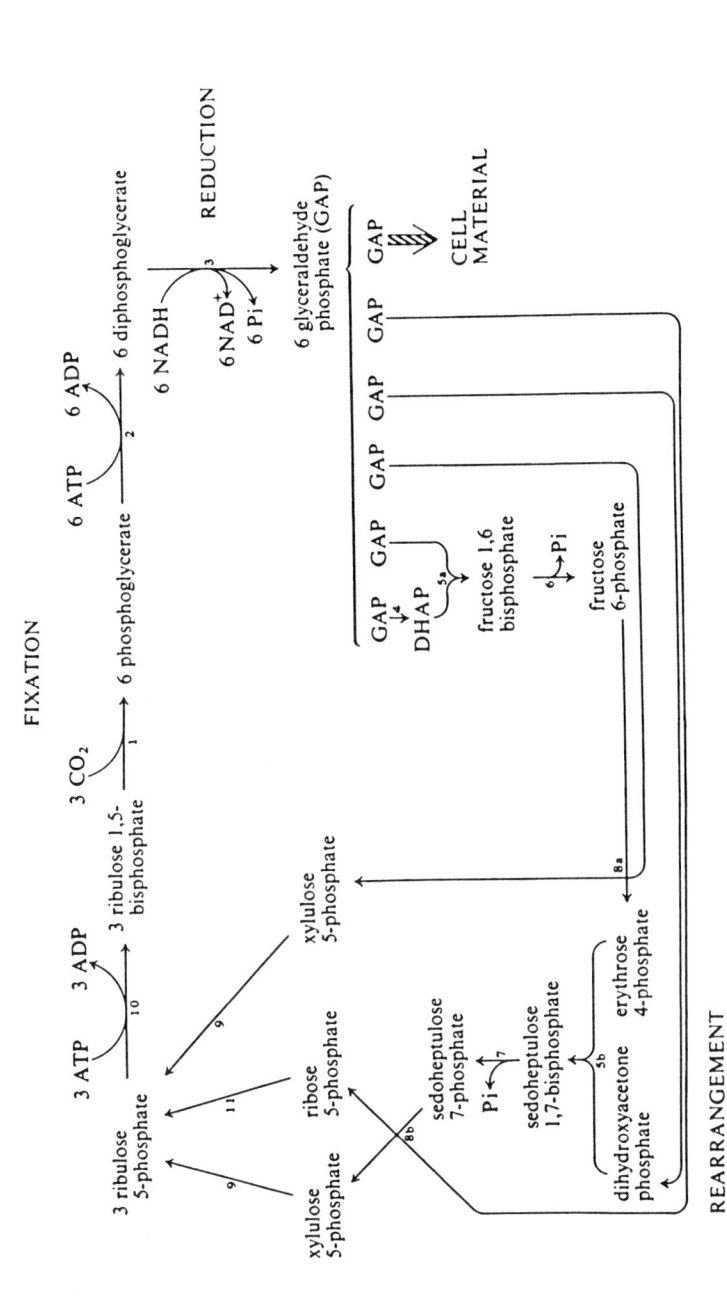

FIGURE 4-1 RuBP pathway of CO_2 fixation (the sedoheptulose bisphosphatase variant). The enzyme numbers shown in the figure correspond to those given in the text: (1) ribulose bisphosphate carboxylase; (2) phosphoglycerate kinase; (3) glyceraldehyde phosphate dehydrogenase; (4) triose phosphate isomerase; (5a and 5b) aldolase; (6) fructose bisphosphatase; (7) sedoheptulose bisphosphatase; (8) transketolase; (9) pentose phosphate epimerase; (10) phosphoribulokinase; (11) pentose phosphate isomerase. Reproduced with permission from Anthony (1982).

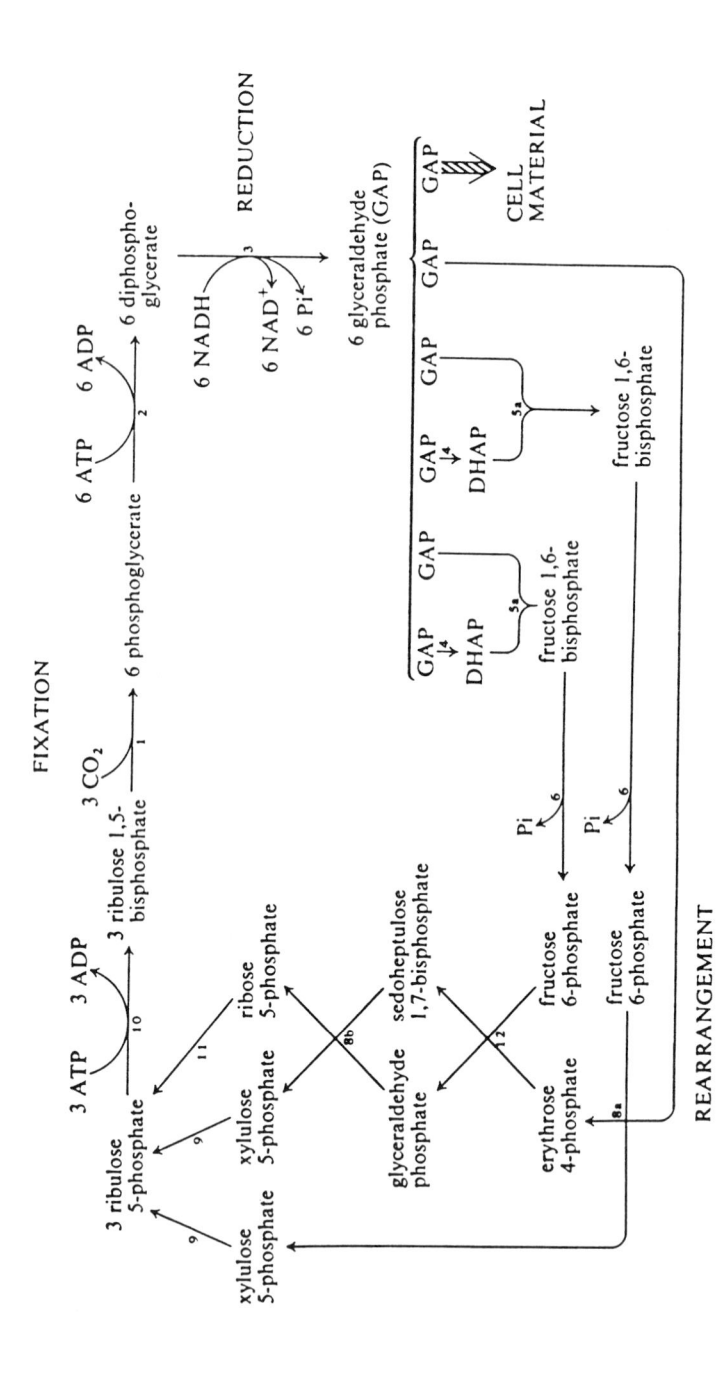

FIGURE 4-2 The RuBP pathway of CO_2 assimilation (the transaldolase variant). The enzyme numbers shown in the figure correspond to those in the text: (1) ribulose bisphosphate carboxylase; (2) phosphoglycerate kinase; (3) glyceraldehyde phosphate dehydrogenase; (4) triose phosphate isomerase; (5a and 5b) aldolase; (6) fructose bisphosphatase; (8) transketolase; (9) pentose phosphate epimerase; (10) phosphoribulokinase; (11) pentose phosphate isomerase; (12) transaldolase. Reproduced with permission from Anthony (1982).

4.1 The Ribulose Bisphosphate (RuBP) Pathway

The *fixation* part of the pathway involves the carboxylation of 3 molecules of ribulose bisphosphate to 6 molecules of 3-phosphoglycerate, which are then phosphorylated to 1,3-diphosphoglycerate. These are then *reduced* to 6 molecules of glyceraldehyde 3-phosphate, one of which is the endproduct of the pathway. The third part of the pathway consists of *rearrangement* reactions, which regenerate 3 molecules of ribulose bisphosphate. At least two variants of the cycle occur which differ in their rearrangement reactions. The sedoheptulose bisphosphatase variant involves this enzyme (see Figure 4–1), which is not involved in the transaldolase variant (see Figure 4–2).

4.1.2 Reactions of the RuBP Pathway

This section describes the individual reactions of the pathway. Most information on these reactions derives mainly from work on autotrophic bacteria and plants. The numbers in parentheses correspond to the numbers given to the enzymes in Figures 4–1 and 4–2.

Ribulose Bisphosphate Carboxylase (Carboxydismutase) (1). This enzyme is usually induced during autotrophic and methylotrophic growth, being present at much lower levels during heterotrophic growth on multicarbon compounds. It catalyzes the carboxylation of ribulose 1,5-bisphosphate in a Mg^{++}-dependent reaction. It yields 2 molecules of 3-phosphoglycerate by way of a number of enzyme-bound intermediates:

$$
\begin{array}{c}
CH_2O-P \\
| \\
C=O \\
| \\
H-C-OH \\
| \\
H-C-OH \\
| \\
CH_2O-P
\end{array}
\longrightarrow
\left\{
\begin{array}{c}
CH_2O-P \\
| \\
C-OH \\
| \\
C-OH \\
| \\
H-C-OH \\
| \\
CH_2O-P
\end{array}
\xrightarrow{*CO_2}
\begin{array}{c}
CH_2O-P \\
| \\
HOOC^*-C-OH \\
| \\
C=O \\
| \\
H-C-OH \\
| \\
CH_2O-P
\end{array}
\xrightarrow{H_2O}
\begin{array}{c}
CH_2O-P \\
| \\
HOOC^*-C-OH \\
| \\
HO-C-OH \\
| \\
H-C-OH \\
| \\
CH_2O-P
\end{array}
\right\}
\begin{array}{c}
CH_2O-P \\
| \\
H-C-OH \\
| \\
^*COOH \\
COOH \\
| \\
H-C-OH \\
| \\
CH_2O-P
\end{array}
$$

ribulose 1,5-bisphosphate   3-phosphoglycerate (2 molecules)

A second reaction catalyzed by carboxylase is the oxygenation of the substrate to yield one molecule of phosphoglycollate as well as a molecule of 3-phosphoglycerate. This ribulose bisphosphate oxygenase reaction is the basis for the photorespiration pathway found in some plants.

Phosphoglycerate Kinase (2).

$$\begin{array}{c} \text{COOH} \\ | \\ \text{H}-\text{C}-\text{OH} \\ | \\ \text{CH}_2\text{O}-\text{P} \end{array} + \text{ATP} \longleftrightarrow \begin{array}{c} \text{COO}-\text{P} \\ | \\ \text{H}-\text{C}-\text{OH} \\ | \\ \text{CH}_2\text{O}-\text{P} \end{array} + \text{ADP}$$

3-phosphoglycerate   1,3-diphosphoglycerate

Glyceraldehyde Phosphate Dehydrogenase (3). In this pathway this enzyme functions as a reductase, the reducing agent being either NADH or NADPH:

$$\begin{array}{c} \text{COO}-\text{P} \\ | \\ \text{H}-\text{C}-\text{OH} \\ | \\ \text{CH}_2\text{O}-\text{P} \end{array} + \text{H}^+ + \text{NAD(P)H} \longleftrightarrow \begin{array}{c} \text{CHO} \\ | \\ \text{H}-\text{C}-\text{OH} \\ | \\ \text{CH}_2\text{O}-\text{P} \end{array} + \text{NAD(P)}^+ + \text{Pi} + \text{H}_2\text{O}$$

1,3-diphosphoglycerate   glyceraldehyde 3-phosphate

Triose Phosphate Isomerase (4).

$$\begin{array}{c} \text{CHO} \\ | \\ \text{H}-\text{C}-\text{OH} \\ | \\ \text{CH}_2\text{O}-\text{P} \end{array} \longleftrightarrow \begin{array}{c} \text{CH}_2\text{OH} \\ | \\ \text{C}=\text{O} \\ | \\ \text{CH}_2\text{O}-\text{P} \end{array}$$

glyceraldehyde   dihydroxyacetone
3-phosphate   phosphate

Aldolase (5a and 5b). Aldolase catalyzes the condensation of an aldose with a ketose sugar (dihydroxyacetone phosphate) giving a larger ketose sugar. The first such reaction (5a) in the RuBP pathway uses glyceraldehyde 3-phosphate and yields fructose 1,6-bisphosphate:

$$\begin{array}{c} \text{CH}_2\text{O}-\text{P} \\ | \\ \text{C}=\text{O} \\ | \\ \text{CH}_2\text{OH} \end{array} + \begin{array}{c} \text{CHO} \\ | \\ \text{H}-\text{C}-\text{OH} \\ | \\ \text{CH}_2\text{O}-\text{P} \end{array} \longleftrightarrow \begin{array}{c} \text{CH}_2\text{O}-\text{P} \\ | \\ \text{C}=\text{O} \\ | \\ \text{HO}-\text{C}-\text{OH} \\ | \\ \text{H}-\text{C}-\text{OH} \\ | \\ \text{H}-\text{C}-\text{OH} \\ | \\ \text{CH}_2\text{O}-\text{P} \end{array}$$

dihydroxyacetone   glyceraldehyde   fructose-
phosphate   3-phosphate   1,6-bisphosphate

The following second example of this reaction (5b) only occurs in the sedoheptulose bisphosphate variation of the pathway (Figure 4–1); it uses erythrose 4-phosphate as substrate and produces sedoheptulose 1,7-bisphosphate:

$$
\begin{array}{c}
CH_2O-P \\
| \\
C=O \\
| \\
CH_2OH
\end{array}
+
\begin{array}{c}
CHO \\
| \\
H-C-OH \\
| \\
H-C-OH \\
| \\
CH_2O-P
\end{array}
\longleftrightarrow
\begin{array}{c}
CH_2O-P \\
| \\
C=O \\
| \\
HO-C-OH \\
| \\
H-C-OH \\
| \\
H-C-OH \\
| \\
H-C-OH \\
| \\
CH_2O-P
\end{array}
$$

dihydroxyacetone phosphate — erythrose 4-phosphate — sedoheptulose 1,7-bisphosphate

Fructose Bisphosphatase (6).

$$
\begin{array}{c}
CH_2O-P \\
| \\
C=O \\
| \\
HO-C-H \\
| \\
H-C-OH \\
| \\
H-C-OH \\
| \\
CH_2O-P
\end{array}
+ H_2O \longrightarrow
\begin{array}{c}
CH_2OH \\
| \\
C=O \\
| \\
HO-C-H \\
| \\
H-C-OH \\
| \\
H-C-OH \\
| \\
CH_2O-P
\end{array}
+ P_i
$$

fructose 1,6-bisphosphate — fructose 6-phosphate

Sedoheptulose Bisphosphatase (7). This phosphatase reaction is only necessary in bacteria having the sedoheptulose bisphosphate variant (Figure 4–1); it is not known if the reaction is catalyzed by fructose bisphosphatase (above) or if a separate enzyme is required.

Transketolase (8a and 8b). This enzyme catalyzes the transfer of a glycolaldehyde moiety from a donor ketose to an acceptor aldose (glyceraldehyde 3-phosphate) to give xylulose 5-phosphate. In the first reaction (8a) in the pathway the donor is fructose 6-phosphate and second product is erythrose 4-phosphate:

$$
\begin{array}{c}
\text{*CH}_2\text{OH} \\
| \\
\text{*C}=\text{O} \\
| \\
\text{HO}-\text{C}-\text{H} \\
| \\
\text{H}-\text{C}-\text{OH} \\
| \\
\text{H}-\text{C}-\text{OH} \\
| \\
\text{CH}_2\text{O}-\text{P}
\end{array}
\;+\;
\begin{array}{c}
\text{CHO} \\
| \\
\text{H}-\text{C}-\text{OH} \\
| \\
\text{CH}_2\text{O}-\text{P}
\end{array}
\longleftrightarrow
\begin{array}{c}
\text{*CH}_2\text{OH} \\
| \\
\text{*C}=\text{O} \\
| \\
\text{HO}-\text{C}-\text{H} \\
| \\
\text{H}-\text{C}-\text{OH} \\
| \\
\text{CH}_2\text{O}-\text{P}
\end{array}
\;+\;
\begin{array}{c}
\text{CHO} \\
| \\
\text{H}-\text{C}-\text{OH} \\
| \\
\text{H}-\text{C}-\text{OH} \\
| \\
\text{CH}_2\text{O}-\text{P}
\end{array}
$$

fructose 6-phosphate   glyceraldehyde 3-phosphate   xylulose 5-phosphate   erythrose 4-phosphate

In the second reaction (8b) in the pathway the donor is sedoheptulose 7-phosphate and the second product is ribose 5-phosphate:

$$
\begin{array}{c}
\text{*CH}_2\text{OH} \\
| \\
\text{*C}=\text{O} \\
| \\
\text{HO}-\text{C}-\text{H} \\
| \\
\text{H}-\text{C}-\text{OH} \\
| \\
\text{H}-\text{C}-\text{OH} \\
| \\
\text{H}-\text{C}-\text{OH} \\
| \\
\text{CH}_2\text{O}-\text{P}
\end{array}
\;+\;
\begin{array}{c}
\text{CHO} \\
| \\
\text{H}-\text{C}-\text{OH} \\
| \\
\text{CH}_2\text{O}-\text{P}
\end{array}
\longleftrightarrow
\begin{array}{c}
\text{*CH}_2\text{OH} \\
| \\
\text{*C}=\text{O} \\
| \\
\text{HO}-\text{C}-\text{H} \\
| \\
\text{H}-\text{C}-\text{OH} \\
| \\
\text{CH}_2\text{O}-\text{P}
\end{array}
\;+\;
\begin{array}{c}
\text{CHO} \\
| \\
\text{H}-\text{C}-\text{OH} \\
| \\
\text{H}-\text{C}-\text{OH} \\
| \\
\text{H}-\text{C}-\text{OH} \\
| \\
\text{CH}_2\text{O}-\text{P}
\end{array}
$$

sedoheptulose 7-phosphate   glyceraldehyde 3-phosphate   xylulose 5-phosphate   ribose 5-phosphate

Pentose Phosphate Epimerase (9).

$$
\begin{array}{c}
\text{CH}_2\text{OH} \\
| \\
\text{C}=\text{O} \\
| \\
\text{HO}-\text{C}-\text{H} \\
| \\
\text{H}-\text{C}-\text{OH} \\
| \\
\text{CH}_2\text{O}-\text{P}
\end{array}
\longleftrightarrow
\begin{array}{c}
\text{CH}_2\text{OH} \\
| \\
\text{C}=\text{O} \\
| \\
\text{H}-\text{C}-\text{OH} \\
| \\
\text{H}-\text{C}-\text{OH} \\
| \\
\text{CH}_2\text{O}-\text{P}
\end{array}
$$

xylulose 5-phosphate   ribulose 5-phosphate

Phosphoribulokinase (10). This enzyme regenerates the CO_2 acceptor, ribulose bisphosphate, from ribulose phosphate in an irreversible reaction requiring ATP:

4.2 The Ribulose Monophosphate (RuMP) Pathway

$$\begin{array}{c} CH_2OH \\ | \\ C=O \\ | \\ H-C-OH \\ | \\ H-C-OH \\ | \\ CH_2O-P \end{array} \quad + \quad ATP \quad \longrightarrow \quad \begin{array}{c} CH_2O-P \\ | \\ C=O \\ | \\ H-C-OH \\ | \\ H-C-OH \\ | \\ CH_2O-P \end{array} \quad + \quad ADP$$

ribulose 5-phosphate   ribulose 1,5-bisphosphate

Pentose Phosphate Isomerase (11).

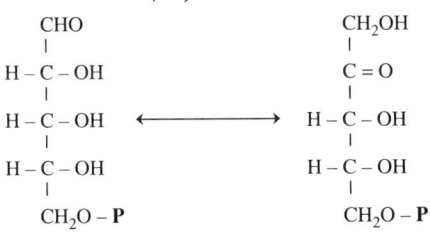

$$\begin{array}{c} CHO \\ | \\ H-C-OH \\ | \\ H-C-OH \\ | \\ H-C-OH \\ | \\ CH_2O-P \end{array} \quad \longleftrightarrow \quad \begin{array}{c} CH_2OH \\ | \\ C=O \\ | \\ H-C-OH \\ | \\ H-C-OH \\ | \\ CH_2O-P \end{array}$$

ribose 5-phosphate   ribulose 5-phosphate

Transaldolase (12).   This enzyme is only involved in the transaldolase variant of the pathway (see Figure 4–2):

$$\begin{array}{c} *CH_2OH \\ | \\ *C=O \\ | \\ HO-C-H \\ | \\ H-C-OH \\ | \\ H-C-OH \\ | \\ CH_2O-P \end{array} \; + \; \begin{array}{c} CHO \\ | \\ H-C-OH \\ | \\ H-C-OH \\ | \\ CH_2O-P \end{array} \quad \longleftrightarrow \quad \begin{array}{c} CHO \\ | \\ H-C-OH \\ | \\ CH_2O-P \end{array} \; + \; \begin{array}{c} *CH_2OH \\ | \\ *C=O \\ | \\ HO-C-H \\ | \\ H-C-OH \\ | \\ H-C-OH \\ | \\ H-C-OH \\ | \\ CH_2O-P \end{array}$$

fructose 6-phosphate   erythrose 4-phosphate   glyceraldehyde 3-phosphate   sedoheptulose 7-phosphate

4.2 THE RIBULOSE MONOPHOSPHATE (RuMP) PATHWAY

In general principle the RuMP pathway is similar to the RuBP pathway (above), except that the reduced carbon substrate is oxidized only to formaldehyde, prior to assimilation. Indeed it has been argued that the RuMP pathway for formaldehyde assimilation might have been an evolutionary precursor of the RuBP pathway for assimilation of carbon dioxide. There are two variants of this pathway: one occurs primarily in obligate methylotrophs, and the other in facultative methylotrophs.

4.2.1 Synthesis of Three-Carbon Compounds from Formaldehyde

In the RuMP pathway (Figures 4–3 and 4–4), all cell carbon is assimilated at the level of formaldehyde (HCHO), produced by the oxidation of reduced C_1 compounds. In this pathway, one molecule of pyruvate or dihydroxyacetone phosphate are synthesized from three molecules of formaldehyde.

In the first or *fixation* part of the pathway, formaldehyde (3 molecules) is condensed by an aldol condensation reaction with ribulose 5-phosphate (3 molecules) to give hexulose phosphate, which is then isomerized to fructose 6-phosphate (FMP) (3 molecules).

In the *cleavage* part of the cycle, one molecule of FMP is converted either to 2-keto 3-deoxy 6-phosphogluconate (KDPG) or to the bisphosphate (FBP); these molecules are then cleaved by aldolases to glyceraldehyde 3-phosphate plus the "product" of the pathway, which is either pyruvate (from KDPG) or dihydroxyacetone phosphate (from FBP).

In the *rearrangement* part of the pathway, the glyceraldehyde phosphate and FMP undergo a series of reactions which lead to regeneration of ribulose 5-phosphate. There are two possible rearrangement variants: the sedoheptulose bisphosphatase variant and the transaldolase variant. Because there are two possible rearrangement sequences and two possible cleavage enzymes, there is a total of four potential variants of the RuMP pathway. Only two of these appear to be important and these are presented in Figures 4–3 and 4–4.

The most common variant is the KDPG aldolase/transaldolase one (see Figure 4–3), whose overall summary equation for synthesis of 3-phosphoglycerate (PGA) is as follows:

$$3HCHO + NAD^+ + 2ATP \rightarrow PGA + NADH + H^+ + 2ADP + Pi$$

The less common variant is the FBP aldolase/SBP phosphatase variant (see Figure 4–4), whose overall assimilation equation is as follows:

$$3HCHO + NAD^+ + ATP \rightarrow PGA + NADH + H^+ + ADP$$

It is essential to produce both triose phosphate and pyruvate for biosynthesis, so bacteria using the KDPG aldolase variant must also contain enzymes for the synthesis of triose phosphate from pyruvate. Likewise, bacteria with the FBP aldolase variant must have all the "glycolytic" enzymes for production of pyruvate from triose phosphate.

4.2.2 Biosynthesis of Carbohydrates

It might be thought that when an early intermediate in a pathway is a carbohydrate there would be no special pathways required for carbohydrate biosynthesis. This is not the case however with the RuMP pathway because

4.2 The Ribulose Monophosphate (RuMP) Pathway

FIXATION

```
                    3 HCHO
    3 ribulose  ─────────→  3 hexulose  ─────→  fructose
    5-phosphate      13      6-phosphate   14    6-phosphate
                                                      │
                                                    16│
                                                      ↓
                                                  glucose
                                                  6-phosphate
                                                      │    ╱NAD(P)+
                                                    17│
                                                      ↓    ╲NAD(P)H
                                                  6-phosphogluconate
                                                      │
                                                    18│ ╲H₂O
                                                      ↓
                                                   KDPG
                                                      │
                                                    19│        pyruvate
                                                               ╱(2 ATP)
                                                            20
                                                               ╲(2 ADP+ Pi)
                                                            PEP
                                                               ╱H₂O
                                                         21,22↓
                                                  glyceraldehyde
                                                  ─ phosphate    PHOSPHOGLYCERATE
```

REARRANGEMENT   CLEAVAGE

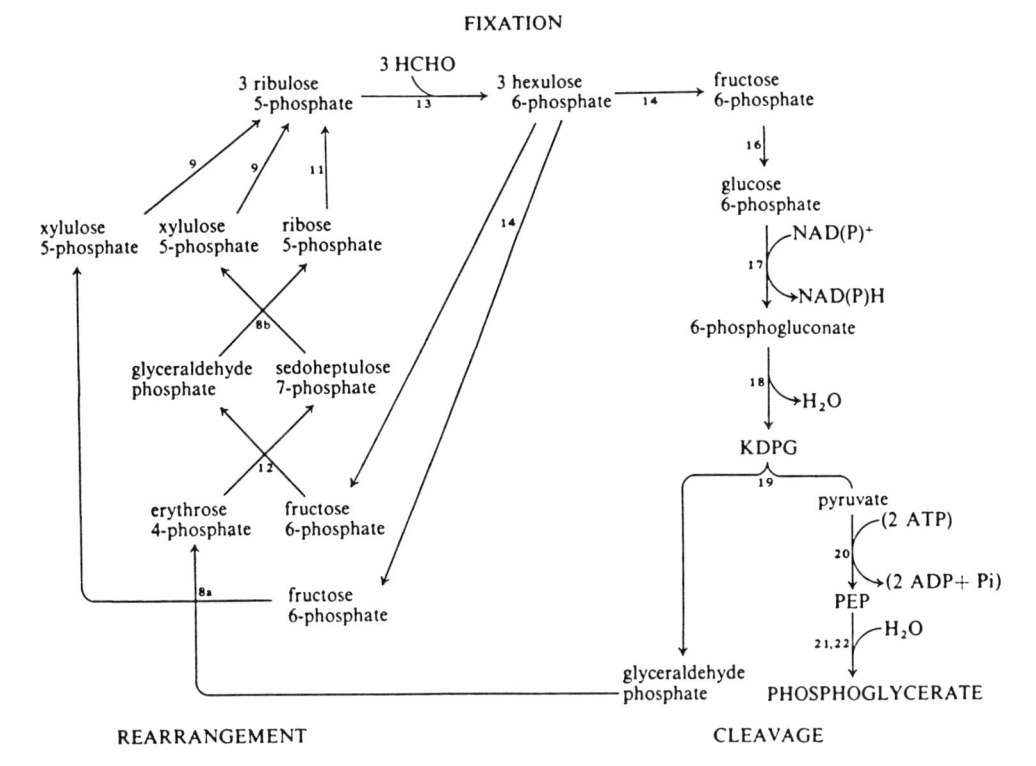

FIGURE 4-3 The RuMP pathway for formaldehyde assimilation (the KDPG aldolase/transaldolase variant). This variant occurs predominantly in obligate methylotrophs. The enzyme numbers shown in the figure correspond to those listed in the text: (8a) transketolase; (9) pentose phosphate epimerase; (11) pentose phosphate isomerase; (12) transaldolase; (13) hexulose phosphate synthase; (14) hexulose phosphate isomerase; (16) glucose phosphate isomerase; (17) glucose phosphate dehydrogenase; (18) phosphogluconate dehydrogenase; (19) 2-keto 3-deoxy 6-phosphogluconate (KDPG) aldolase; (20) phosphoenol pyruvate synthetase or equivalent enzyme(s); (21) enolase; (22) phosphoglyceromutase. Reproduced with permission from Anthony (1982).

the end-product of the pathway is a three-carbon (C_3) compound and hexose sugars must be synthesized from these, as illustrated in Figure 4–5.

Pentose sugars for nucleic acid biosynthesis and other uses are produced from glyceraldehyde phosphate by a modification of the main cycle with the cleavage reaction omitted (Figure 4–6); hexoses are then produced by addition of another formaldehyde molecule. Erythrose 4-phosphate, which is required for biosynthesis of aromatic amino acids, is produced from glyceraldehyde phosphate plus fructose 6-phosphate or sedoheptulose 7-phosphate in reactions catalyzed by transketolase (8) and transaldolase (12) (Figure 4–6).

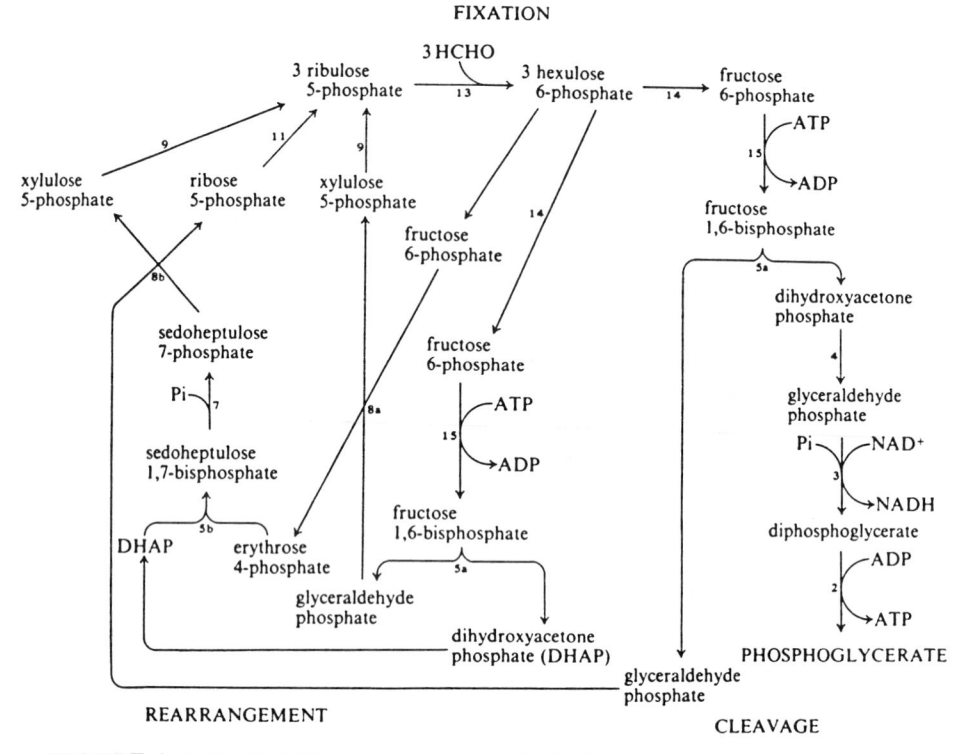

FIGURE 4-4 The RuMP pathway for formaldehyde assimilation (the fructose bisphosphate aldolase/sedoheptulose bisphosphatase variant). This variant occurs predominantly in facultative methylotrophs. The enzyme numbers shown in the figure correspond to those listed in the text: (2) phosphoglycerate kinase; (3) glyceraldehyde phosphate dehydrogenase; (4) triose phosphate isomerase; (5) aldolase; (7) sedoheptulose bisphosphatase; (8) transketolase; (9) pentose phosphate epimerase; (11) pentose phosphate isomerase; (13) hexulose phosphate synthase; (14) hexulose phosphate isomerase; (15) phosphofructokinase. Reproduced with permission from Anthony (1982).

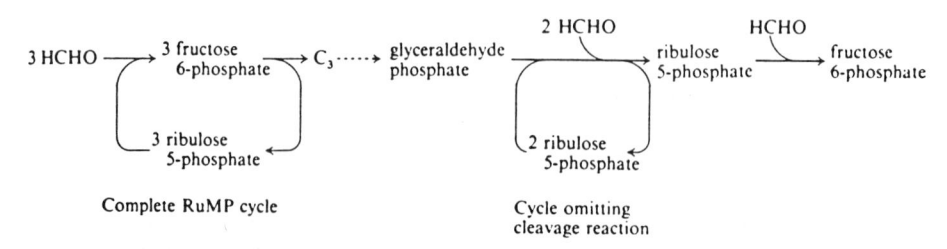

FIGURE 4-5 The biosynthesis from formaldehyde of triose, pentose, and hexose. This route can operate in bacteria having either variant of the RuMP pathway. Reproduced with permission from Anthony (1982).

4.2 The Ribulose Monophosphate (RuMP) Pathway

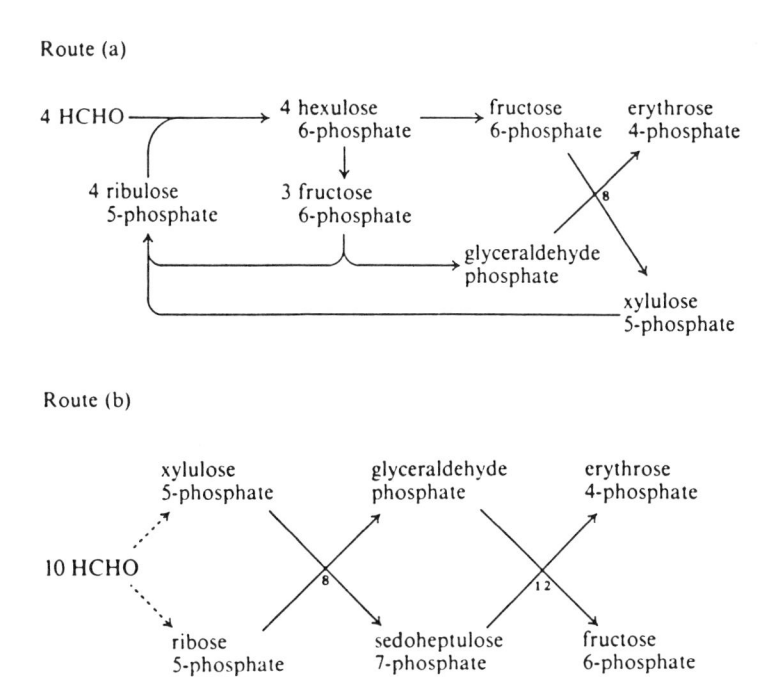

Route (a)

```
4 HCHO ──────► 4 hexulose ──────► fructose      erythrose
               6-phosphate        6-phosphate   4-phosphate
       │              │                      ╲  ╱
       │              ▼                       ╲╱  8
  4 ribulose    3 fructose                    ╱╲
  5-phosphate   6-phosphate                  ╱  ╲
       ▲              │       glyceraldehyde
       │              └──────► phosphate
       │                                        ╲
       └────────────────────────────────────► xylulose
                                              5-phosphate
```

Route (b)

```
           xylulose       glyceraldehyde    erythrose
           5-phosphate    phosphate         4-phosphate
              ↗                ╲   ╱            ↗
              ╱                 ╲ ╱             ╱
10 HCHO ─────                  8 ╳            ╳ 12
              ╲                 ╱ ╲             ╲
              ↘                ╱   ╲            ↘
           ribose          sedoheptulose     fructose
           5-phosphate     7-phosphate       6-phosphate
```

FIGURE 4-6 The biosynthesis of erythrose 4-phosphate and fructose 6-phosphate from formaldehyde. Route (a) is used in bacteria with either variant of the RuMP pathway. Route (b) operates in bacteria using only the transaldolase variant. Reproduced with permission from Anthony (1982).

4.2.3 Biosynthesis of Lipids and Amino Acids

The necessary precursors of these cellular constituents are predominantly acetyl-coenzyme A, phosphoenolpyruvate (PEP), and the keto acids pyruvate, oxaloacetate, and 2-oxoglutarate. These must all be synthesized from the products of the RuMP pathway. Figure 4-7 summarizes the enzymes necessary for synthesis of most of these metabolites. The biosynthesis of 2-oxoglutarate from acetyl-CoA and oxaloacetate is by way of citrate synthase and other enzymes of the tricarboxylic acid (TCA) cycle.

4.2.4 Oxidation of Formaldehyde by a Dissimilatory Cycle

Bacteria having the KDPG aldolase variant of the cycle are able to convert fructose 6-phosphate to 6-phosphogluconate; these bacteria usually also have 6-phosphogluconate dehydrogenase and so are able to oxidize formaldehyde completely to carbon dioxide by a cyclic dissimilatory pathway (Figure 4-8). The two dehydrogenases are able to reduce both NAD^+ and $NADP^+$, or there may be two separate enzymes for each of these two coenzymes; this

92 Assimilation of Carbon by Methylotrophs

FBP aldolase variant

3 HCHO
↓
glyceraldehyde phosphate ⇌ $\underset{PEP}{\begin{array}{c} COOH \\ | \\ C-O-P \\ \| \\ CH_2 \end{array}}$

KDPG aldolase variant

3 HCHO
↓
$\underset{PYR}{\begin{array}{c} COOH \\ | \\ C=O \\ | \\ CH_3 \end{array}}$ → acetyl-CoA, CO_2 + NADH

(20) AMP + Pi / AMP + PPi ↔ ATP / ATP + Pi
(28) ... (27) ADP / ATP

(24) CO_2 / (26) CO_2, ADP → ATP, Pi
(25) CO_2, ATP → ADP + Pi

$\begin{array}{c} COOH \\ | \\ C=O \\ | \\ CH_2COOH \end{array}$
OAA

FIGURE 4–7 The interconversion of phosphoenolpyruvate and oxaloacetate. The enzyme numbers shown correspond to those listed in the text: (20) PEP synthetase; (24) PEP carboxylase; (25) pyruvate carboxylase; (26) PEP carboxykinase; (27) pyruvate kinase; (28) pyruvate phosphate dikinase. Reproduced with permission from Anthony (1982).

ribulose 5-phosphate —(13)—[HCHO]→ hexulose 6-phosphate —(14)→ fructose 6-phosphate

NAD(P)H ← CO_2 / NAD(P)+ (23)

6-phosphogluconate ←—(17)— glucose 6-phosphate (16)

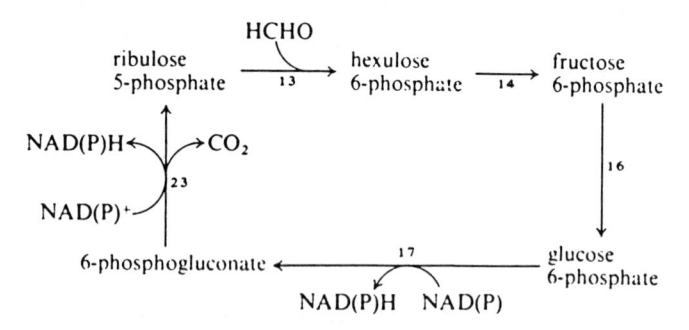

NAD(P)H   NAD(P)

FIGURE 4–8 The oxidation of formaldehyde to CO_2 by a dissimilatory RuMP cycle. The enzymes of this cycle are those of the assimilatory RuMP pathway plus the dissimilatory enzyme 6-phosphogluconate dehydrogenase (reaction 23). Other enzymes: (13) hexulose phosphate synthase; (14) hexulose phosphate isomerase; (16) glucose phosphate isomerase; (17) glucose 6-phosphate dehydrogenase. Reproduced with permission from Anthony (1982).

step is an important step in the regulation of the RuMP pathway and the oxidative cycle in obligate methylotrophs growing on methanol.

The reaction catalyzed by 6-phosphogluconate dehydrogenase (reaction 23 in Figure 4–8) is not a straightforward oxidation reaction but is an irreversible oxidative decarboxylation:

$$\begin{array}{c}\text{COOH}\\|\\\text{H}-\text{C}-\text{OH}\\|\\\text{HO}-\text{C}-\text{H}\\|\\\text{H}-\text{C}-\text{OH}\\|\\\text{H}-\text{C}-\text{OH}\\|\\\text{CH}_2\text{O}-\text{P}\end{array} + \text{NAD(P)}^+ \longrightarrow \begin{array}{c}\text{CH}_2\text{OH}\\|\\\text{C}=\text{O}\\|\\\text{H}-\text{C}-\text{OH}\\|\\\text{H}-\text{C}-\text{OH}\\|\\\text{CH}_2\text{O}-\text{P}\end{array} + \text{H}^+ + \text{NAD(P)H} + \text{CO}_2$$

6-phosphogluconate → ribulose 5-phosphate

4.2.5 Reactions of the RuMP Pathway

This section describes the individual reactions of all the variants of the RuMP pathway. Extensive, detailed information on each enzyme is given in Anthony (1982). The numbers in parentheses correspond to the numbers given to each reaction in the figures.

4.2.5.1 Fixation Phase Enzymes. These enzymes are the same in all variants of the RuMP pathway. The reactions that they catalyze are shown below.

3-Hexulose Phosphate Synthase (13). This key enzyme of the RuMP pathway catalyzes the aldol condensation of formaldehyde with ribulose 5-phosphate to produce a novel hexulose phosphate (D-arabino-3-hexulose 6-phosphate):

$$\text{HCHO} + \begin{array}{c}\text{CH}_2\text{OH}\\|\\\text{C}=\text{O}\\|\\\text{H}-\text{C}-\text{OH}\\|\\\text{H}-\text{C}-\text{OH}\\|\\\text{CH}_2\text{O}-\text{P}\end{array} \longrightarrow \begin{array}{c}\text{CH}_2\text{OH}\\|\\\text{HO}-\text{C}-\text{H}\\|\\\text{C}=\text{O}\\|\\\text{H}-\text{C}-\text{OH}\\|\\\text{H}-\text{C}-\text{OH}\\|\\\text{CH}_2\text{O}-\text{P}\end{array}$$

ribulose 5-phosphate → hexulose 6-phosphate

In *Methylococcus capsulatus* (an obligate methanotroph), in which this enzyme was first described by Quayle and his colleagues, it is very large and membrane-bound; in other bacteria (methanol utilizers) such as *Methylophilus methylotrophus*, it is relatively small and is a soluble enzyme.

3-Hexulose Phosphate Isomerase (14). This enzyme isomerizes the product of the initial fixation reaction to give fructose 6-phosphate:

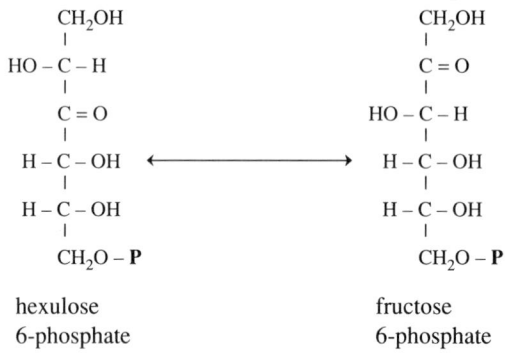

$$\begin{array}{c}CH_2OH \\ | \\ HO-C-H \\ | \\ C=O \\ | \\ H-C-OH \\ | \\ H-C-OH \\ | \\ CH_2O-P\end{array} \quad \longleftrightarrow \quad \begin{array}{c}CH_2OH \\ | \\ C=O \\ | \\ HO-C-H \\ | \\ H-C-OH \\ | \\ H-C-OH \\ | \\ CH_2O-P\end{array}$$

hexulose 6-phosphate   fructose 6-phosphate

This soluble enzyme was the first known example of an enzyme catalyzing the isomerization of a 3-ketulose to a 2-ketulose.

4.2.5.2 Cleavage Phase Enzymes in the KDPG Aldolase Variant.

The KDPG aldolase variant pathway shown in Figure 4–3 uses the four following enzymes:

Glucose Phosphate Isomerase (16).

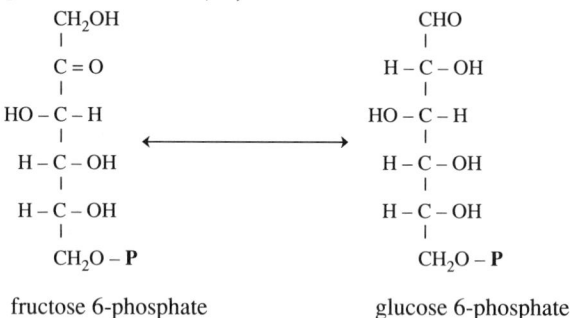

$$\begin{array}{c}CH_2OH \\ | \\ C=O \\ | \\ HO-C-H \\ | \\ H-C-OH \\ | \\ H-C-OH \\ | \\ CH_2O-P\end{array} \quad \longleftrightarrow \quad \begin{array}{c}CHO \\ | \\ H-C-OH \\ | \\ HO-C-H \\ | \\ H-C-OH \\ | \\ H-C-OH \\ | \\ CH_2O-P\end{array}$$

fructose 6-phosphate   glucose 6-phosphate

Glucose 6-Phosphate Dehydrogenase (17). 6-Phosphogluconolactone is an intermediate in this reaction, and the enzyme is usually equally active with NAD^+ and $NADP^+$:

4.2 The Ribulose Monophosphate (RuMP) Pathway

```
    CHO                                    COOH
     |                                      |
  H – C – OH                             H – C – OH
     |                                      |
 HO – C – H    + H₂O + NAD(P)⁺  ⟶    HO – C – H    + NAD(P) + H⁺
     |                                      |
  H – C – OH                             H – C – OH
     |                                      |
  H – C – OH                             H – C – OH
     |                                      |
   CH₂O – P                              CH₂O – P

glucose 6-phosphate                   6-phosphogluconate
```

6-Phosphogluconate Dehydrase (18). This enzyme produces KDPG from 6-phosphogluconate:

```
    COOH                        COOH
     |                           |
  H – C – OH                   C = O
     |                           |
 HO – C – H                   H – C – H
     |              ⟶           |           + H₂O
  H – C – OH                   H – C – OH
     |                           |
  H – C – OH                   H – C – OH
     |                           |
   CH₂O – P                    CH₂O – P

6-phosphogluconate       2-keto, 3-deoxy, 6-phosphogluconate
```

KDPG Aldolase (19). This enzyme is the cleavage enzyme of the KDPG variant of the RuMP pathway.

```
    COOH
     |
   C = O
     |
  H – C – H
     |                    COOH            CHO
  H – C – OH               |               |
     |         ⟶         C = O     +    H – C – OH
  H – C – OH               |               |
     |                    CH₃            CH₂O – P
   CH₂O – P

   KDPG                 pyruvate      glyceraldehyde
                                      3-phosphate
```

4.2.5.3 Cleavage Phase Enzymes in the FBP Aldolase Variant.
The FBP aldolase variant pathway shown in Figure 4–4 uses the two following enzymes, which are not required in bacteria having the KDPG variant (Figure 4–3).

Phosphofructokinase (15).

$$\begin{array}{c} CH_2OH \\ | \\ C=O \\ | \\ HO-C-H \\ | \\ H-C-OH \\ | \\ H-C-OH \\ | \\ CH_2O-P \end{array} \; + \; ATP \;\longrightarrow\; \begin{array}{c} CH_2O-P \\ | \\ C=O \\ | \\ HO-C-H \\ | \\ H-C-OH \\ | \\ H-C-OH \\ | \\ CH_2O-P \end{array} \; + \; ADP$$

fructose 6-phosphate → fructose 1,6-bisphosphate

Fructose Bisphosphate Aldolase (5). This is the typical "glycolytic" cleavage enzyme. Although the equilibrium constant (M) (about 10^{-4}) favors condensation, at low concentrations of reactants and products, the cleavage reaction will be favored.

$$\begin{array}{c} CH_2O-P \\ | \\ C=O \\ | \\ HO-C-H \\ | \\ H-C-OH \\ | \\ H-C-OH \\ | \\ CH_2O-P \end{array} \;\longleftrightarrow\; \begin{array}{c} CH_2O-P \\ | \\ C=O \\ | \\ CH_2OH \end{array} \; + \; \begin{array}{c} CHO \\ | \\ H-C-OH \\ | \\ CH_2O-P \end{array}$$

fructose 1,6-bisphosphate / dihydroxyacetone phosphate / glyceraldehyde 3-phosphate

4.2.5.4 Rearrangement Phase Enzymes. The enzymes involved in rearrangement reactions are the same as those described above for the RuBP pathway (see Figures 4–1 and 4–2). They include transketolase (8), pentose phosphate epimerase (9), pentose phosphate isomerase (11), transaldolase (12), aldolase (5), and sedoheptulose bisphosphatase (7).

4.2.5.5 Interconversion of Three- and Four-Carbon Compounds. These enzymes are responsible for ensuring that there is a constant supply of C_3 and C_4 precursors for biosynthesis (see Figure 4–7). For convenience, all the possible reactions are noted here although some may not be important in growth by the RuMP pathway and some probably do not occur in any methylotrophs.

4.2 The Ribulose Monophosphate (RuMP) Pathway

Enolase (21).

```
  COOH                              COOH
   |                                 |
  C - O - P   + H₂O  ⟷          H - C - O - P
  ‖                                  |
  CH₂                               CH₂OH
```
phosphoenolpyruvate   2-phosphoglycerate

Phosphoglycerate Mutase (22).

```
      COOH                              COOH
       |                                 |
  H - C - O - P   ⟷              H - C - OH
       |                                 |
      CH₂OH                            CH₂O - P
```
2-phosphoglycerate   3-phosphoglycerate

Phosphoenolpyruvate Synthetase (20). This enzyme catalyzes the direct synthesis of PEP from pyruvate during growth of enteric bacteria on C_3 compounds but it has not been demonstrated so far in methylotrophs. ATP donates the phosphate group, the products being AMP and phosphate:

$$\text{pyruvate} + \text{ATP} \rightarrow \text{PEP} + \text{AMP} + \text{Pi}$$

Phosphoenolpyruvate Carboxylase (24). This enzyme catalyzes the irreversible carboxylation of PEP to oxaloacetate and has been mainly studied in bacteria using the serine pathway, in which it is a major route for the assimilation of cell carbon:

$$\text{PEP} + CO_2 \rightarrow \text{oxaloacetate} + \text{Pi}$$

Pyruvate Carboxylase (25). The function of this enzyme is usually anaplerotic, i.e., it "fills up" the TCA cycle when intermediates are being withdrawn for biosynthesis. In this context many examples of this enzyme are activated by acetyl-CoA, but the carboxylases in methylotrophs are not usually activated in this way:

$$\text{pyruvate} + \text{ATP} + CO_2 \rightarrow \text{oxaloacetate} + \text{ADP} + \text{Pi}$$

Phosphoenolpyruvate Carboxykinase (26). This enzyme catalyzes the reaction by which C_3 compounds are synthesized from C_4 compounds during growth on these compounds, and it is also involved in synthesis of PEP from pyruvate (together with pyruvate carboxylase):

$$\text{oxaloacetate} + \text{ATP} \rightarrow \text{PEP} + \text{ADP} + CO_2$$

Pyruvate Kinase (27). This enzyme catalyzes pyruvate formation from PEP (as in glycolysis):

$$PEP + ADP \rightarrow pyruvate + ATP$$

Pyruvate Phosphate Dikinase (28). This single enzyme catalyzes the direct formation of PEP from pyruvate. It occurs in propionic acid bacteria and some grasses but has not been described in methylotrophs. The hydrolysis of pyrophosphate drives the reaction in the direction of phosphorylation.

$$pyruvate + ATP + Pi \rightarrow PEP + AMP + PPi$$

4.3 THE DIHYDROXYACETONE (DHA) CYCLE OF FORMALDEHYDE ASSIMILATION IN YEASTS

4.3.1 Summary of the DHA Cycle

This pathway, which is also called the xylulose monophosphate pathway, occurs only in methylotrophic yeasts growing on methanol (Figure 4–9).

The DHA pathway is similar to the RuBP and RuMP pathways in principle. The C_1 unit (formaldehyde in this case) is condensed with a phosphorylated pentose sugar, and the only enzyme required exclusively for formaldehyde assimilation in yeast is the formaldehyde-fixing enzyme, dihydroxyacetone (DHA) synthase; all other enzymes are also required for growth on carbohydrates, ethanol, or glycerol. In this pathway, the C_1 acceptor molecule is not a ribulose derivative but is xylulose 5-phosphate, and the product is 2 molecules of triose phosphate. The pathway is thus different from the RuMP pathway in that it consists of a *fixation* phase and *rearrangement* phase, but it differs in not requiring a *cleavage* phase (Figure 4–9).

This assimilatory cycle achieves the synthesis of 1 molecule of triose phosphate from 3 molecules of formaldehyde at the expense of 3 molecules of ATP. For comparison with other pathways the summary equation is best expressed as a route for synthesis of 3-phosphoglycerate:

$$3HCHO + 2ATP + NAD^+ \rightarrow PGA + 2ADP + NADH + H^+ + Pi$$

For production of carbohydrates, PEP, and pyruvate, from dihydroxyacetone phosphate, the "glycolytic" enzymes for oxidation of triose phosphates are also required, together with the "gluconeogenic" enzymes, fructose bisphosphate aldolase and phosphatase. The further metabolism of pyruvate to acetyl-CoA, for lipid biosynthesis, and to keto acids for amino acid biosynthesis, will involve the usual mitochondrial systems.

4.3.2 The Oxidation of Formaldehyde by a Dissimilatory DHA Cycle

Although yeasts contain enzymes catalyzing the direct oxidation of methanol to carbon dioxide, it has been proposed that a cyclic route might also operate in them (Figure 4–10), analogous to that involved in methanol oxidation

4.3 The Dihydroxyacetone (DHA) Cycle of Formaldehyde Assimilation in Yeasts   99

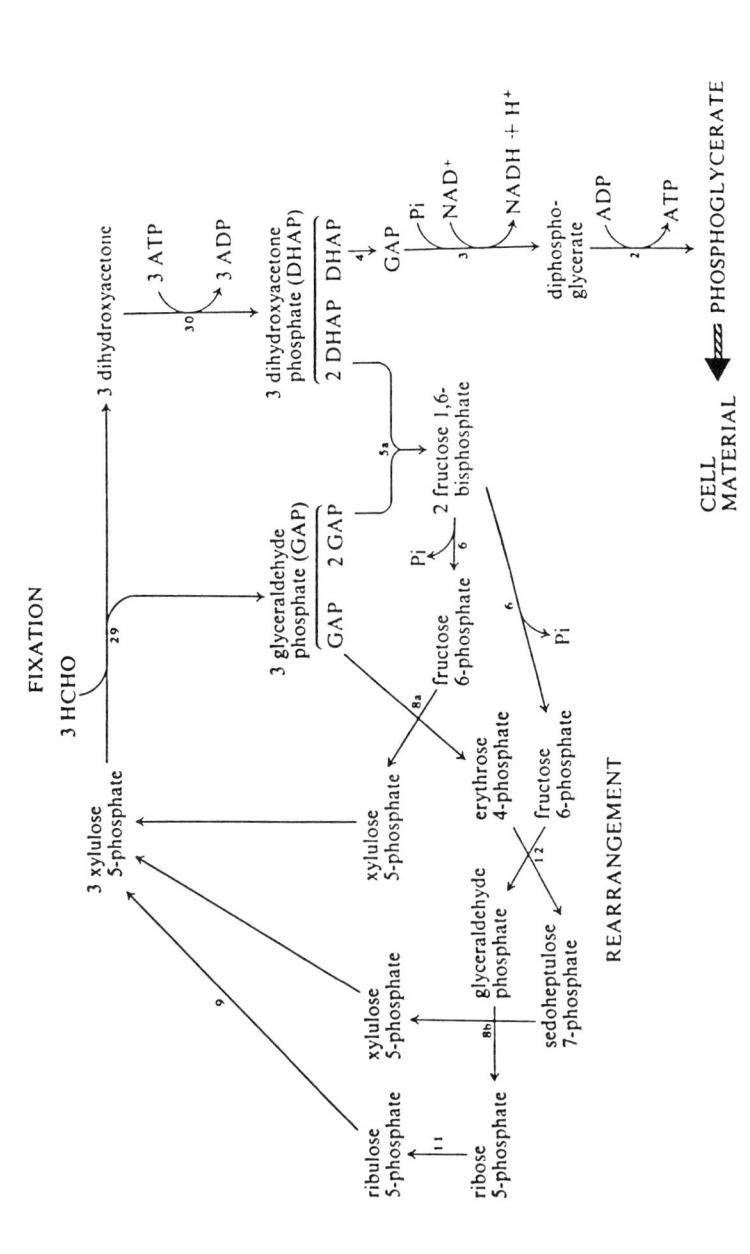

FIGURE 4-9 The dihydroxyacetone (DHA) cycle of formaldehyde assimilation in yeast. The rearrangement part of this cycle involves transaldolase (reaction 12), but an alternative rearrangement of the triose phosphates might involve sedoheptulose bisphosphate aldolase and phosphatase instead of transaldolase. The enzyme numbers correspond to those listed in the text: (2) phosphoglycerate kinase; (3) glyceraldehyde phosphate dehydrogenase; (4) triose phosphate isomerase; (5a) fructose phosphate aldolase; (6) fructose bisphosphatase; (8) transketolase; (9) pentose phosphate epimerase; (11) pentose phosphate isomerase; (12) transaldolase; (29) dihydroxyacetone synthase; (30) triokinase. Reproduced with permission from Anthony (1982).

100 Assimilation of Carbon by Methylotrophs

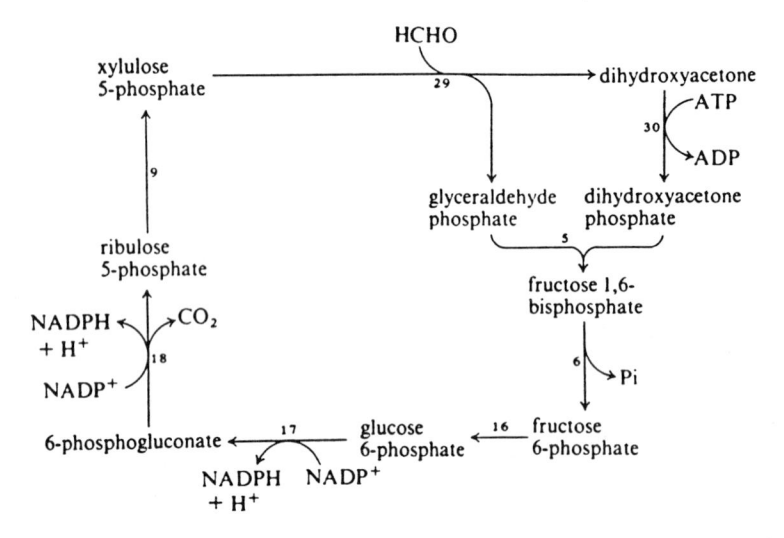

FIGURE 4-10 The dissimilatory DHA cycle of formaldehyde oxidation. The summary equation is: HCHO + 2NADP⁺ + ATP → CO_2 + 2NADPH + 2H⁺ + ADP + Pi. The enzyme numbers correspond to those in the text: (5) fructose bisphosphate aldolase; (6) fructose bisphosphatase; (9) pentose phosphate epimerase; (16) glucose phosphate isomerase; (17) glucose 6-phosphate dehydrogenase; (23) 6-phosphogluconate dehydrogenase; (29) dihydroxyacetone synthase; (3) triokinase. Reproduced with permission from Anthony (1982).

by some obligate methylotrophs (see Figure 4–8). Besides the enzymes for formaldehyde assimilation, three other enzymes are required for this dissimilatory cycle: hexose phosphate isomerase and the dehydrogenases for glucose phosphate and 6-phosphogluconate. Because these enzymes are specific for NADPH in yeast, it is probable that the primary role of this cycle is to produce NADPH for biosynthesis, and that it has no role in the provision of ATP, which is by the usual mitochondrial process. The dissimilatory cycle is regulated by way of its two dehydrogenases, which are inhibited by the "end-product" of the cycle, NADPH.

4.3.3 The Reactions of the DHA Cycle
Figure 4–9 shows both the fixation and rearrangement phases of the assimilatory DHA cycle. Only the enzymes of the *fixation* phase are described here; the rearrangement enzymes of the cycle are the same as those involved in the RuMP pathway and so are not repeated here.

Dihydroxyacetone Synthase (29). This synthase is a transketolase, transferring a glycolaldehyde moiety from xylulose 5-phosphate to formaldehyde and yielding dihydroxyacetone:

$$\text{HCHO}^* + \begin{array}{c} \text{CH}_2\text{OH} \\ | \\ \text{C} = \text{O} \\ | \\ \text{HO} - \text{C} - \text{H} \\ | \\ \text{H} - \text{C} - \text{OH} \\ | \\ \text{CH}_2\text{O} - \text{P} \end{array} \longrightarrow \begin{array}{c} \text{CHO} \\ | \\ \text{H} - \text{C} - \text{OH} \\ | \\ \text{CH}_2\text{O} - \text{P} \end{array} + \begin{array}{c} \text{CH}_2\text{OH} \\ | \\ \text{C} = \text{O} \\ | \\ ^*\text{CH}_2\text{OH} \end{array}$$

xylulose 5-phosphate   glyceraldehyde 3-phosphate   dihydroxyacetone

The "classical" transketolase is synthesized on all substrates tested but the DHA synthase is induced to high levels only during growth on methanol. It is located in the peroxisomes together with the enzymes for the oxidation of methanol to formaldehyde (see Harder and Veenhuis 1988). The DHA produced by the synthase passes out of the peroxisomes into the cytoplasm where it is phosphorylated by the ATP-requiring triokinase. The rate of transport of the substrate for the synthase, xylulose phosphate, into the peroxisomes may be important in regulating the distribution of formaldehyde between its assimilation by the DHA cycle and its oxidation for provision of ATP. Regeneration of xylulose phosphate requires ATP (for the triokinase reaction) whose production will depend on the energy status of the cell. Metabolic control may act in such a way that during energy limitation, the rate of supply of xylulose phosphate to the DHA synthase is diminished, thereby resulting in an increase in the amount of formaldehyde available for oxidation.

Triokinase (30).

$$\begin{array}{c} \text{CH}_2\text{OH} \\ | \\ \text{C} = \text{O} \\ | \\ \text{CH}_2\text{OH} \end{array} + \text{ATP} \longrightarrow \begin{array}{c} \text{CH}_2\text{O} - \text{P} \\ | \\ \text{C} = \text{O} \\ | \\ \text{CH}_2\text{OH} \end{array} + \text{ADP}$$

dihydroxyacetone   dihydroxyacetone phosphate

This enzyme is much more active with dihydroxyacetone than with other trioses, and is only induced to high specific activities during growth with methanol or with glycerol.

4.4 THE SERINE PATHWAY OF FORMALDEHYDE ASSIMILATION

This pathway differs from all other assimilation pathways in methylotrophs in that its intermediates are carboxylic acids and amino acids rather than carbohydrates. As seen in Figure 4–11, at least one-third of the carbon

102 Assimilation of Carbon by Methylotrophs

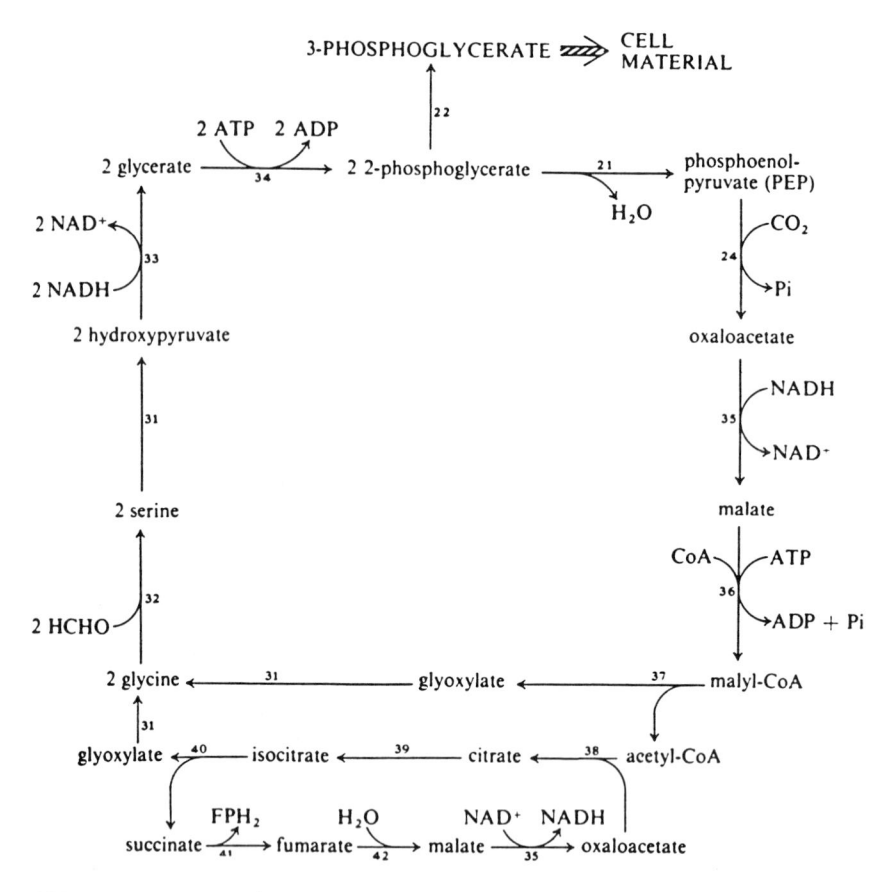

```
                          3-PHOSPHOGLYCERATE ⟹ CELL MATERIAL
                                   ↑
                                  |22
         2 ATP  2 ADP              |
  2 glycerate ──34──→ 2 2-phosphoglycerate ──21──→ phosphoenol-
       ↑                              H₂O         pyruvate (PEP)
   2 NAD⁺                                              │
       ↘33                                             │╭CO₂
   2 NADH                                             24
       │                                               │↘Pi
  2 hydroxypyruvate                                oxaloacetate
       ↑                                               │
       │31                                             │╭NADH
       │                                              35
  2 serine                                             │↘NAD⁻
       ↑                                            malate
       │32                                   CoA↘     │╭ATP
  2 HCHO                                           36
                                             ADP+Pi↙  │
  2 glycine ←──31────── glyoxylate ←──37── malyl-CoA
       ↑
      │31
  glyoxylate ←──40── isocitrate ←──39── citrate ←──38── acetyl-CoA
       │
       │   FPH₂        H₂O         NAD⁺ NADH
       ↓    ↗           ↗            ↗
    succinate ──41──→ fumarate ──42──→ malate ──35──→ oxaloacetate
```

FIGURE 4–11 The serine pathway of formaldehyde assimilation (*icl⁺* variant). The *icl⁻* variant differs from this in lacking a measurable malate thiokinase and in having an alternative route for oxidation of acetyl-CoA to glyoxylate not involving isocitrate lyase. Precursors can be removed from the cycle at the level of oxaloacetate or of succinate (see Figure 4–12). The enzyme numbers correspond to those in the text: (21) enolase; (22) phosphoglycerate mutase; (24) PEP carboxylase; (31) serine-glycine aminotransferase; (32) serine transhydroxymethylase; (33) hydroxypyruvate reductase; (34) glycerate kinase; (35) malate dehydrogenase; (36) malate thiokinase; (37) malyl-CoA lyase; (38) citrate synthase; (39) aconitase; (40) isocitrate lyase; (41) succinate dehydrogenase; (42) fumarase. Reproduced with permission from Anthony (1982).

assimilated is at the level of carbon dioxide. In this figure the serine pathway is shown as it operates to synthesize phosphoglycerate from 2 molecules of formaldehyde plus 1 of carbon dioxide. Two molecules of formaldehyde plus 2 of glyoxylate give 2 molecules of phosphoglycerate, 1 of which is assimilated into cell material. The second is carboxylated, yielding, even-

4.4 The Serine Pathway of Formaldehyde Assimilation

tually, malyl-CoA. In the final part of the pathway, this is cleaved to glyoxylate plus acetyl-CoA whose oxidation to a second molecule of glyoxylate completes the cycle.

The route for the oxidation of acetyl-CoA to glyoxylate as shown in Figure 4–11 is the same as that occurring during growth of bacteria on acetate or ethanol, and it involves isocitrate lyase (this is called the icl^+ variant of the pathway). This variant of the serine pathway is not very common; the alternative (icl^-) variant occurs in the pink-pigmented facultative methylotrophs and is generally more common, but the route for this oxidative part of the cycle has not yet been elucidated (see Anthony 1982).

The summary equation for the synthesis of 3-phosphoglycerate by the icl^+ serine pathway is as follows (FPH_2 is reduced succinate dehydrogenase):

$$CO_2 + 2HCHO + 2NADH + 3ATP \rightarrow$$
$$PGA + 2NAD^+ + 3ADP + 2Pi + FPH_2$$

The serine pathway must also be able to provide precursors for the biosynthesis of all cell components from the C_1 substrate. Oxaloacetate is produced in the cycle itself and so will be "withdrawn" at a later stage than indicated in Figure 4–11. This is shown in Figure 4–12A together with the route for succinate biosynthesis (Figure 4–12B). For the biosynthesis of carbohydrates from phosphoglycerate, the usual gluconeogenic enzymes are also required and pyruvate will be produced from PEP by way of pyruvate kinase. The usual source of acetyl-CoA for lipid biosynthesis is by the malyl-CoA lyase reaction, and this is the only route in those bacteria such as obligate methylotrophs and restricted facultative methylotrophs (e.g., *Hyphomicrobia*) which lack pyruvate dehydrogenase. It should be noted that methylotrophs differ from almost all other aerobic bacteria in not requiring

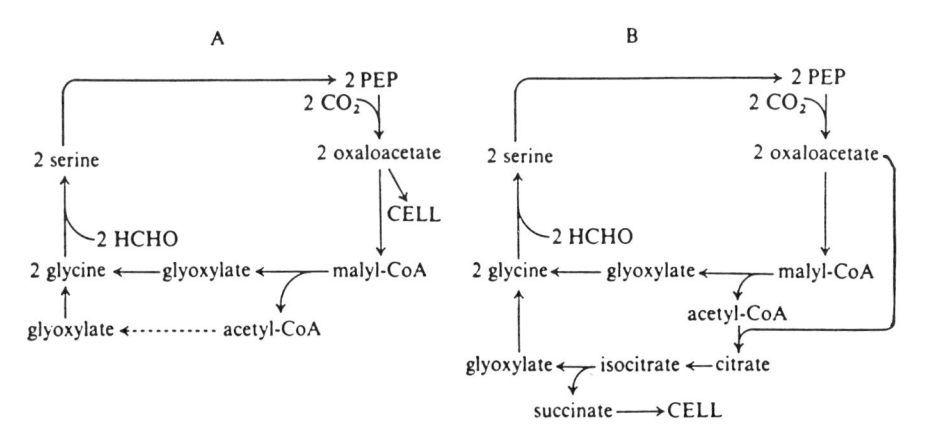

FIGURE 4–12 The serine pathway of formaldehyde assimilation operating to synthesize (A) oxaloacetate and (B) succinate. Reproduced with permission from Anthony (1982).

4.4.1 Reactions of the Serine Pathway

4.4.1.1 The Synthesis of Oxaloacetate.
Serine Transhydroxymethylase (32).

$$\text{HCHO} + \begin{array}{c}\text{CH}_2\text{NH}_2\\|\\\text{COOH}\end{array} \longleftrightarrow \begin{array}{c}\text{CH}_2\text{OH}\\|\\\text{CHNH}_2\\|\\\text{COOH}\end{array}$$

$$\text{glycine} \qquad\qquad \text{serine}$$

Tetrahydrofolate is required for this aldol condensation during which it acts as a "protective" carrier of formaldehyde to the active site. Facultative methylotrophs usually contain two species of this enzyme; one functions specifically in the serine pathway, and the other provides glycine from serine during growth on multicarbon compounds. The "methylotrophic enzyme" is activated by glyoxylate, a characteristic presumably important in regulation of the serine pathway.

Serine-Glyoxylate Aminotransferase (31). This specific aminotransferase, which is induced during methylotrophic growth, catalyzes two functions in the serine pathway. In a single reaction, it synthesizes glycine from glyoxylate and also produces hydroxypyruvate from serine:

$$\begin{array}{c}\text{CH}_2\text{OH}\\|\\\text{CHNH}_2\\|\\\text{COOH}\end{array} + \begin{array}{c}\text{CHO}\\|\\\text{COOH}\end{array} \longleftrightarrow \begin{array}{c}\text{CH}_2\text{OH}\\|\\\text{C}=\text{O}\\|\\\text{COOH}\end{array} + \begin{array}{c}\text{CH}_2\text{NH}_2\\|\\\text{COOH}\end{array}$$

$$\text{serine} \qquad \text{glyoxylate} \qquad \text{hydroxypyruvate} \quad \text{glycine}$$

Hydroxypyruvate Reductase (33).

$$\begin{array}{c}\text{CH}_2\text{OH}\\|\\\text{C}=\text{O}\\|\\\text{COOH}\end{array} + \text{NADH} + \text{H}^+ \longleftrightarrow \begin{array}{c}\text{CH}_2\text{OH}\\|\\\text{CHOH}\\|\\\text{COOH}\end{array} + \text{NAD}^+$$

$$\text{hydroxypyruvate} \qquad\qquad \text{glycerate}$$

Although this inducible enzyme plays a specific role in methylotrophs growing by way of the serine pathway, it also occurs in some whose main pathway is the RuBP or RuMP pathway, in which cases the reductase is constitutively formed.

4.4 The Serine Pathway of Formaldehyde Assimilation

Glycerate Kinase (34). This enzyme catalyzes the phosphorylation of glycerate to 2-phosphoglycerate instead of the more usual 3-phosphoglycerate:

$$\begin{array}{c} CH_2OH \\ | \\ H-C-OH \\ | \\ COOH \end{array} + ATP \longrightarrow \begin{array}{c} CH_2OH \\ | \\ H-C-OP \\ | \\ COOH \end{array} + ADP$$

glycerate   2-phosphoglycerate

Phosphoglycerate Mutase (22).

$$\begin{array}{c} CH_2OH \\ | \\ H-C-OP \\ | \\ COOH \end{array} \longleftrightarrow \begin{array}{c} CH_2OP \\ | \\ CHOH \\ | \\ COOH \end{array}$$

2-phosphoglycerate   3-phosphoglycerate

This inducible mutase is the first enzyme specifically involved in gluconeogenesis during growth on C_1 compounds by the serine pathway.

Phosphoenolpyruvate Carboxylase (24). This inducible carboxylase is the only enzyme of importance in the formation of C_4 compounds from C_3 compounds in methylotrophs growing by the serine pathway:

$$\begin{array}{c} CH_2 \\ \| \\ C-OP \\ | \\ COOH \end{array} + CO_2 \longrightarrow \begin{array}{c} CH_2COOH \\ | \\ C=O \\ | \\ COOH \end{array} + Pi$$

PEP   oxaloacetate

4.4.1.2 Formation of Acetyl-CoA and Glyoxylate from Oxaloacetate. Oxaloacetate is reduced by malate dehydrogenase to malate from which malyl-CoA is produced; this is then cleaved by a lyase to acetyl-CoA and glyoxylate. In bacteria with the *icl⁻* variant, no malate thiokinase has been described and no alternative coenzyme A transferases have been found.

Malate Thiokinase (Malyl-CoA Synthetase) (36). This reaction is analogous to the succinate thiokinase reaction:

$$\begin{array}{c} HOCHCOOH \\ | \\ CH_2COOH \end{array} + \text{Coenzyme A} + ATP \longrightarrow \begin{array}{c} HOCHCOOH \\ | \\ CH_2CO-CoA \end{array} + ADP + Pi$$

malate   4-malyl-CoA

Malyl-CoA Lyase (37).

$$\text{HOCHCOOH}\;|\;\text{CH}_2\text{CO}-\text{CoA} \longleftrightarrow \text{CHOCOOH} + \text{CH}_3\text{CO}-\text{CoA}$$

4-malyl-CoA   glyoxylate   acetyl-CoA

This key cleavage enzyme of the serine pathway has been found in all methylotrophs having this pathway, but it is also present at low levels in some bacteria having the RuMP pathway. In these cases it probably supplies glyoxylate and hence glycine for porphyrin and protein biosynthesis, as suggested for *Rhodopseudomonas sphaeroides* growing on malate plus glutamate.

4.4.1.3 Oxidation of Acetyl-CoA to Glyoxylate in *icl*+ Bacteria. This route involves isocitrate lyase plus some of the enzymes of the TCA cycle as shown in Figure 4–11. In methylotrophic growth, they only function in biosynthesis and this should be considered when investigating their regulation compared with that in other aerobic bacteria.

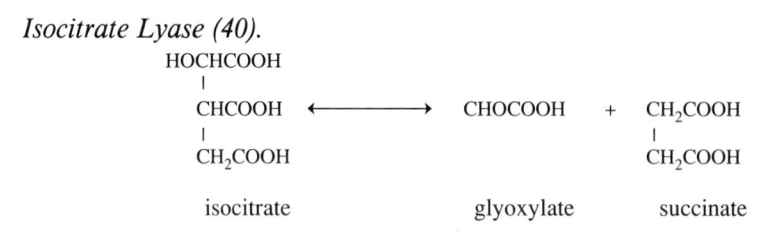

Isocitrate Lyase (40).

$$\text{HOCHCOOH}\;|\;\text{CHCOOH}\;|\;\text{CH}_2\text{COOH} \longleftrightarrow \text{CHOCOOH} + \text{CH}_2\text{COOH}\;|\;\text{CH}_2\text{COOH}$$

isocitrate   glyoxylate   succinate

Before the importance of this enzyme during growth of some methylotrophs on C_1 compounds was understood, its sole function was thought to be in the glyoxylate cycle during growth on compounds assimilated exclusively by way of acetyl-CoA. Because methylotrophs having the *icl*+-serine pathway are all able to grow on such compounds, it might be assumed that they would possess a single isocitrate lyase. This is not the case, however; *Pseudomonas* MA possesses one enzyme induced during growth on C_1 compounds and a second enzyme induced on acetate.

4.4.1.4 Oxidation of Acetyl-CoA to Glyoxylate in *icl*− Bacteria. The route for this is not known. Much of the evidence is consistent with the operation of an isocitrate lyase but the presence of this enzyme has not been demonstrated. *Pseudomonas* AM1 has no isocitrate lyase during growth on either C_1 compounds or C_2 compounds, and mutant evidence indicates that the same enzymes are involved in the oxidation of acetyl-CoA to glyoxylate during growth on both classes of compound.

It has proved impossible to verify an earlier proposal that a homologous pathway operates in this oxidation process, having homoisocitrate lyase as an alternative to isocitrate lyase.

A second alternative, for which there is rather limited evidence, and which has not been independently confirmed, is a pathway which uses the coenzyme B_{12}-dependent enzymes more commonly found in fermentative bacteria and which are only present in very small amounts (if at all) in methylotrophs. The intermediates in this pathway include glutamate, methylaspartate, mesaconyl-CoA, and propionyl-CoA.

It is probable that elucidation of the final part of the icl^--serine pathway will eventually be achieved only by NMR studies of whole cells of wild-type bacteria and of mutant bacteria which have been shown to lack the enzymes involved in oxidation of acetyl-CoA to glyoxylate.

4.5 DISTRIBUTION AND OCCURRENCE OF THE ASSIMILATION PATHWAYS

The distribution of assimilation pathways among methylotrophs is summarized in Table 4–1 and discussed below.

The RuBP Pathway. The RuBP pathway occurs in relatively few methylotrophic bacteria, and it is not known which of the variants operates in them. The first methylotroph shown to assimilate its cell carbon by this pathway was *Pseudomonas oxalaticus,* an otherwise typical heterotroph which grows on formate but not on other C_1 compounds. The best known of the methylotrophs that grows by this route on the more reduced substrate, methanol, is *Paracoccus denitrificans.* In this facultative autotroph methanol replaces the more usual substrates of hydrogen plus carbon dioxide. Another important methylotroph using this pathway is the phototroph *Rhodopseudomonas acidophila,* which can grow anaerobically in the light, using methanol as the source of reductant. Growth yield measurements indicate that the assimilation equation approximates to:

$$2CH_3OH + CO_2 \rightarrow 3HCHO \text{ (cell material)} + H_2O$$

In this process carbon is assimilated as carbon dioxide, and the NAD(P)H for the reductive steps of the RuBP pathway is provided by the oxidation of methanol to carbon dioxide. The first reaction in this process (catalyzed by methanol dehydrogenase) must be coupled to the reduction of $NAD(P)^+$ by reversed electron transport. *Rhodopseudomonas acidophila* can also grow aerobically on methanol in the dark.

Other well-studied bacteria using the RuBP pathway are *Thiobacillus novellus* and *T. versutus* (previously *Thiobacillus* A2).

It is unlikely that any of these bacteria will be of major importance in biotechnology because their growth yields are low and any useful "meth-

TABLE 4-1 Distribution of Assimilation Pathways in Methylotrophs

1. *RuBP Pathway* (Figures 4-1 and 4-2)
   Facultative autotrophs; e.g., *Paracoccus denitrificans*, *Thiobacillus* sp.
   Photosynthetic bacteria; e.g., *Rhodopseudomonas acidophila*
2. *RuMP Pathway—KDPG aldolase/transaldolase variant* (Figure 4-3)
   Type I methanotrophs; e.g., *Methylococcus capsulatus*
   Obligate methylotrophs unable to use methane; e.g., *Methylophilus methylotrophus*
3. *RuMP Pathway—FBP aldolase/sedoheptulose bisphosphatase variant* (Figure 4-4)
   Facultative methylotrophs unable to use methane; e.g., *Arthrobacter* P1, *Bacillus* sp., *Acetobacter methanolicus* MB58
4. *DHA Cycle* (Figure 4-9)
   Yeasts: mainly *Hansenula, Candida, Pichia,* and *Torulopsis*
5. *Serine Pathway*—icl⁺ *variant* (Figure 4-11)
   Facultative methylotrophs unable to use methane or methanol; e.g., *Pseudomonas* MA, *Pseudomonas aminovorans*
6. *Serine Pathway*—icl⁻ *variant* (Figure 4-11)
   a. Type II methanotrophs; e.g., *Methylosinus* sp.
   b. Facultative methylotrophs unable to use methane
      i. Pink-pigmented facultative methylotrophs; e.g., *Methylobacterium extorquens* AM1 (uses methanol or methylated amines)
      ii. *Hyphomicrobium* (uses methanol or methylated amines)

ylotrophic enzymes" are also found in the RuMP bacteria. The RuBP bacteria are proving most useful in the study of the molecular biology of methylotrophs.

The RuMP Pathway. The KDPG aldolase/transaldolase variant of the RuMP pathway operates in those methanotrophs that have type I membranes and those obligate methylotrophs unable to use methane, the best known of which is the Imperial Chemical Industries "Pruteen" strain *Methylophilus methylotrophus*. This is not surprising because the assimilation of methanol by the RuMP pathway is the most "cost-effective" route in terms of requirement for ATP and NADH, and so these bacteria provide the highest growth yields on methanol.

The FBP aldolase/sedoheptulose bisphosphatase variant is much less common and is found only in those facultative methylotrophs that use methylated amines or methanol but are unable to use methane as carbon substrates. The best-known examples of this group are *Arthrobacter* P1 and the acidophilic methylotroph *Acetobacter methanolicus* MB58.

The DHA Cycle. This pathway only occurs in yeasts.

The Serine Pathway. Although the energy requirements of the serine pathway preclude the use of those bacteria for biotechnological purposes in which growth yields are of major importance (as in SCP production), these bacteria are likely to be important in biotechnology because it may become possible to exploit the pathway for the production of amino acids and carboxylic acids.

Methanotrophs having type II membranes such as *Methylosinus trichosporium* use the serine pathway but the presence or absence of isocitrate lyase has not been reported. In the facultative methanotroph *Methylobacterium organophilum* XX, there is no isocitrate lyase, but this organism appears to have lost its ability to use methane and now appears to be more like *M. extorquens;* it is thus not advisable to generalize to all type II methanotrophs from this example.

The *icl*⁺ variant of this pathway is uncommon and has only been shown to occur in facultative methylotrophs (pseudomonads) that are able to use methylated amines but not methane or methanol, such as *Pseudomonas aminovorans* and *Pseudomonas* MA.

The more common *icl*⁻ variant is found in the pink-pigmented facultative methylotrophs such as *Methylobacterium extorquens* AM1 (previously called *Pseudomonas* AM1) and in restricted facultative methylotrophs such as *Hyphomicrobium*.

REFERENCES

Anthony, C. (1982) *The Biochemistry of Methylotrophs,* Academic Press, London.

Harder, W., and Veenhuis, M. (1988) in *The Yeasts,* 2nd ed., vol. III (Rose, A.H., and Harrison, J.S., eds.), pp. 289–316, Academic Press, London.

CHAPTER 5

Oxidation Pathways in Methylotrophs

J. Stefan Rokem
Israel Goldberg

Microbial cells (methylotrophs) growing on substrates with no carbon-to-carbon bonds have to form these bonds in order to synthesize cell constituents and to oxidize these substrates to obtain the energy required for growth. By changing the redox level of these one-carbon (C_1) substrates, energy can be obtained for maintenance and growth. This chapter describes the various routes (the oxidation pathways) and the enzymes used to obtain the energy required during the growth of both prokaryotic and eukaryotic methylotrophs on C_1 compounds, including a review of the literature up to 1990. Very extensive and comprehensive reviews on this subject, emphasizing bacteria, have been published previously (Anthony 1982 and 1986).

5.1 BACTERIAL OXIDATION OF C_1 COMPOUNDS

Bacteria utilizing C_1 compounds need to oxidize them first in order to obtain the energy required for growth. The oxidation reaction(s) is coupled to ATP synthesis. Methylotrophic bacteria have multiple pathways for oxidizing the

---

This chapter is dedicated to the memory of Amitai Rokem, 1973–1990.

reduced C_1 substrates (e.g., methane, methanol, formaldehyde, methylated amines, and formate) to CO_2. The simplest route is by "direct oxidation": methane or methanol are oxidized via formaldehyde to formate and then to CO_2 (Figure 5–1). Bacteria assimilating the C_1 substrate via the serine pathway use this route exclusively, as described in Chapter 4.

The free energy ($G^{0\prime}$) of the four oxidation reactions of C_1 compounds are given in Table 5–1. The free energy generated by these reactions increases

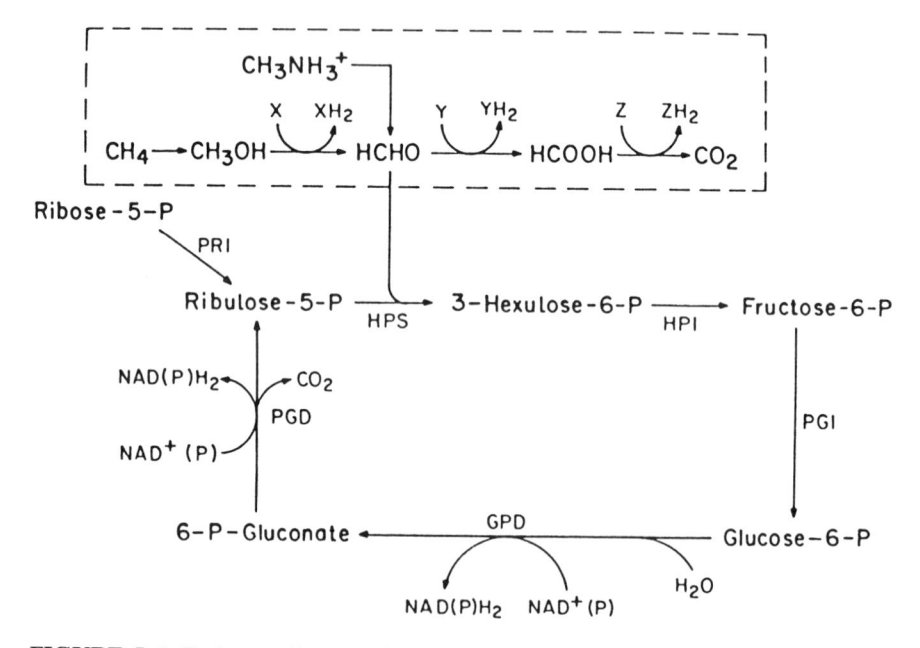

FIGURE 5–1 Pathways for the direct (inside the dashed-line box) and the cyclic oxidation of reduced C_1 compounds to CO_2. Abbreviations: GPD, glucose-6-phosphate dehydrogenase; HPS, hexulose phosphate synthase; PGD, 6-phosphogluconate dehydrogenase; PGI, phosphoglucoisomerase; HPI, hexulose isomerase; PRI, phosphoriboisomerase; X, Y, and Z represent electron acceptors. Reproduced with permission from Goldberg (1985b).

TABLE 5–1 Thermodynamic Constants for the Oxidation Reactions of C_1 Compounds

Reaction	$G^{0\prime}$ (kJ mol^{-1})
$CH_4 + 0.5\ O_2 \rightarrow CH_3OH$	−109.7
$CH_3OH + 0.5\ O_2 \rightarrow HCHO + H_2O$	−188.2
$HCHO + 0.5\ O_2 \rightarrow HCOOH$	−240.0
$HCOOH + 0.5\ O_2 \rightarrow CO_2 + H_2O$	−244.7

Data based on Ribbons et al. (1970).

with the degree of the oxidation of the C_1 substrate. In the first oxidation reaction from methane to methanol, molecular oxygen must be incorporated into the methane molecule at a very low free energy gain. This indicates that the enzyme catalyzing this reaction must be very unusual.

Another route for the oxidation of C_1 compounds is the cyclic pathway. This route is used by bacteria having the assimilatory ribulose monophosphate (RuMP) pathway. In these bacteria, the carbon assimilation pathway is used in part for oxidation purposes (Figure 5-1). In the case of methane and methanol oxidations, the initial oxidative enzymes (i.e., methane monooxygenase and methanol dehydrogenase in methane oxidation; methanol dehydrogenase in methanol oxidation, see below) function in both the direct and the cyclic oxidation pathways. The RuMP bacteria usually have a very low level of formate dehydrogenase activity (Ben-Bassat et al. 1980; Bussineau et al. 1987). This enzyme can be considered specific for the "direct oxidation route" described above.

These oxidation routes were investigated and characterized for a large number of methylotrophic bacteria (mainly Gram-negative) in the 1960s and 1970s, and here we will discuss the most recent developments and findings.

5.1.1 Oxidation of Methane to Methanol

Bacteria able to utilize methane are sometimes called methanotrophs and can be considered a subgroup of the methylotrophs. They contain two different intracellular membrane types, types 1 and 2, which are described and discussed in Chapters 1 and 3.

The first step in methane oxidation is catalyzed by a mixed-function methane monooxygenase (MMO) (methane NAD(P)H: oxygen oxidoreductase [hydroxylating; EC 1.14.13.25]), as shown below:

$$CH_4 + NAD(P)H + H^+ + O_2 \rightarrow H_2O + NAD(P)^+ + CH_3OH$$

One oxygen atom is incorporated into methane to form methanol, and the other oxygen atom is released as water. The common reductant is NADH. The localization of MMO, either soluble (sMMO) or in the particulate (membrane-bound), cell-free fraction (pMMO), was shown to be dependent on the growth conditions, especially on the availability of cupric (Cu^{2+}) ions. Organisms grown at a high ratio of Cu^{2+} to biomass weight primarily contain pMMO, while sMMO predominates in organisms grown at low Cu^{2+} concentrations (Stanley et al. 1983). Both type 1 and type 2 methanotrophs are able to produce soluble and membrane-bound MMO, and they both show a similar intracellular distribution pattern of these enzymes.

Both forms of MMO have been studied and characterized in *Methylosinus trichosporium* OB3b. The soluble form of the enzyme (sMMO) was resolved into three components (Higgins et al. 1981; Fox et al. 1989): a 39.7-kDa NADH reductase containing 1 mol each of FAD and a [2Fe–2S] cluster;

a 15.8-kDa protein factor termed component B with no metals or cofactors; and a 245-kDa hydroxylase, which appears to contain an exo- or hydroxy-bridged binuclear iron cluster. They have been characterized by Lipscomb and his coworkers (Fox et al. 1989), as follows: the hydroxylase catalyzes monooxygenation of both alkanes and alkenes in the absence of the other two components. Component B is not essential for the formation of the activated oxygen species or for electron transfer between the reductase and the hydroxylase components. However, component B affects the efficiency of the reaction and increases the overall specific activity of MMO by up to 150-fold. The soluble MMO of *Methylococcus capsulatus* has also been resolved into three components with similar molecular weights (210 kDa, 15.7 kDa, and 42 kDa; Lund et al. 1985), which seems to have functions similar to those proposed for the sMMO isolated from *Methylosinus trichosporium* OB3b.

It has been more difficult to characterize the membrane-bound pMMO since detergent treatments are needed for the solubilization of the enzyme and enzymatic activity could not be reconstituted after detergent treatments. Recently, however, the nonionic detergent dodecyl-α-D-maltoside (Smith and Dalton 1989) was found to solubilize the membrane-bound pMMO, and with the subsequent addition of egg or soybean lecithin, reconstitution of enzyme activity was possible. The membrane-bound MMO has a more restrictive active site than the soluble form, since its substrate specificity is narrower. Comparison of membrane-bound and solubilized pMMO showed no difference in substrate specificity. Butane is the highest *n*-alkane oxidized by the enzyme (Burrows et al. 1984; Smith and Dalton 1989).

5.1.2 Oxidation of Methanol to Formaldehyde

A NAD$^+$-independent alcohol dehydrogenase catalyzes the oxidation of methanol to formaldehyde in most methylotrophic bacteria studied. This enzyme is usually called methanol dehydrogenase (MDH) (EC 1.1.99.8). It oxidizes a variety of primary alcohols using phenazine methosulfate or phenazine ethosulfate as artificial electron acceptors, and ammonia or methylamine as activators. The natural prosthetic group of MDH is a pyrroloquinoline quinone (PQQ) molecule (Duine and Frank 1981). The reaction catalyzed by MDH is as follows:

$$CH_3OH \xrightarrow[\text{MDH}]{PQQ \quad PQQH_2} HCHO$$

This prosthetic group has also been found in mammalian copper enzymes as well as in other bacterial metalloenzymes involved in oxidation and dehydrogenation reactions (Gallop et al. 1989). The electron acceptor is cytochrome *c*, thus bypassing cytochrome *b*.

A large number of MDH enzymes have been isolated and characterized from many different Gram-negative methylotrophs (Anthony 1986). The MDH enzymes studied are very similar in different bacteria. The PQQ coenzyme has been found in MDH of all Gram-negative facultative and obligate methylotrophs studied. The pH optima for MDH are 9 or higher, however, the enzyme is often stable at pH 4. The enzyme is a dimer containing two identical subunits of about 60-kDa molecular weight. When bacteria are grown on methane or methanol, the MDH is induced and constitutes 5–15% of the cellular soluble proteins. Mutants that lack MDH are not able to grow on methane or methanol (O'Connor and Hanson 1977). This enzyme also catalyzes the oxidation of formaldehyde to formate (Heptinstall and Quayle 1969; Ladner and Zatman 1969). The oxidation rate and the affinity coefficient of the enzyme for both methanol and formaldehyde are very similar. However, the oxidation of formaldehyde by MDH is not affected by inhibitors of methanol oxidation. Mutants lacking MDH are able to oxidize formaldehyde, but not methanol, indicating that MDH must be regulated in vivo to avoid or prevent oxidation of formaldehyde.

Very few Gram-positive methylotrophic bacteria have been isolated and studied. For the methanol-utilizing actinomycete, *Nocardia* sp. 239, a novel NAD-dependent, PQQ-containing MDH has been described (Duine et al. 1984). Recently, a new strain, *Bacillus* sp. C1, was isolated and found to contain a NAD-dependent MDH (Arfman et al. 1989). During growth on methanol, the enzyme constituted a high percentage of the total soluble proteins (up to 22%). The purified enzyme displayed biphasic kinetics with apparent K_m values for methanol of 3.8 and 166 mM, whereas in whole cells, this biphasic effect was not seen. Inhibition studies showed that the electrons from methanol are transferred to the electron transport chain at or above the level of cytochrome *b*. This is in contrast to Gram-negative methylotrophs where cytochrome *c* is the electron acceptor. The transfer of electrons from methanol to cytochrome *b* resulted in a high biomass yield for *Bacillus* sp. C1 when grown on methanol (16 to 18 g cells/mol methanol, see below) (Dijkhuizen et al. 1988).

The redox potential of the methanol-to-formaldehyde oxidation reaction strongly favors the formation of methanol from formaldehyde. Therefore, specially adapted enzymes are required to drive the reaction in the opposite direction. In three Gram-negative methylotrophs (*Methylobacterium extorquens* AM1, formerly *Pseudomonas* AM1, and *Methylophilus methylotrophus*), a modifier protein was shown to regulate MDH activity. The affinity coefficient of MDH to formaldehyde decreased 36- to 64-fold (as measured in vitro by the dye-linked system) and V_{max} was halved in the presence of the modifier protein (Page and Anthony 1986). In the presence of the modifier protein, MDH oxidizes alcohols and acids (such as 1,2-butanediol, 1,3-butanediol, 1,3-propanediol, and 4-hydroxybutyrate) which are not otherwise oxidized by MDH. In the presence of the modifier protein, *M. extorquens* AM1 can grow on 1,2-propanediol and 4-hydroxybutyrate as sole carbon sources. However, it has been argued that the main physiological role of the modifier protein is to regulate MDH activity during the oxidation of methanol (Anthony 1986).

The in vitro activity of the MDH enzyme is always assayed at a relatively high pH, in the presence of NH_4Cl and with artificial electron acceptors. In a recent study it was found that when the purified natural electron acceptor for MDH, cytochrome c, was added to a purified preparation of MDH from *Hyphomicrobium* strain X, very low reaction rates were obtained (Beardmore-Gray et al. 1983). By carrying out all purification procedures of MDH under anaerobic conditions, it was possible to obtain significant MDH activity in the absence of ammonium salts. A low-molecular-weight compound, which is probably oxidized by O_2 in the presence of cytochromes, may be the "natural" activator for MDH (Dijkstra et al. 1988). It was shown that this activator molecule is not an ammonium salt or a divalent cation such as Mg^{2+}, Ca^{2+}, Cu^{2+}, or Mn^{2+} (Dijkstra et al. 1988). Further studies are needed to characterize this low-molecular-weight compound as well as its function in the catalytic reaction of MDH.

The MDH enzyme has been crystallized from *M. methylotrophus* and analyzed at 2Å resolution (Lim et al. 1986). This MDH is a dimer of identical subunits, each having a molecular weight of 62 kDa, including PQQ noncovalently bound to the enzyme. Further characterization of these crystals is needed for a better understanding of the catalytic mechanism of MDH.

5.1.2.1 The Interaction of Methanol Dehydrogenase with Cytochrome c.

It has been shown that in Gram-negative methylotrophs, MDH interacts with the electron transport chain at the level of cytochrome c (Anthony 1986). Two soluble cytochrome c proteins differing in their isoelectric points were found in these organisms. Cytochrome c_L is a large acidic molecule and has a lower isoelectric point than cytochrome c_H, which is a small basic protein. An unusual characteristic of the cytochrome c proteins of methylotrophs is their rapid autoreduction, i.e., the reduction of the heme iron of ferricytochrome c occurs in the absence of reducing agents. This has been one of the sources of difficulties encountered when studying MDH-cytochrome c interactions. The reduction of the heme iron is a first-order intramolecular reaction and involves an electron transfer between a dissociable group (XH) and iron. Stabilization of the resulting radicals occurs simultaneously. The MDH of *M. methylotrophus* can only autoreduce the homologous cytochrome c_L, whereas the MDH of *Methylobacterium extorquens* AM1 can induce autoreduction of both the homologous cytochrome c_L and c_H (Beardmore-Gray et al. 1983).

Studies of the electron transfer rates between MDH and cytochrome c_L have been performed with reconstituted systems. The in vitro electron transfer rates are much lower than those necessary to support growth. This discrepancy has been explained by the very high concentrations of MDH and cytochrome c in methanol-grown bacteria, and it has been suggested that the two proteins completely cover the outer surface of the bacterial membrane (Anthony 1986). Anthony (1986) suggests that when the cell is disrupted for in vitro analysis, environmental changes contribute to the low activity of the proteins in reconstituted systems.

5.1.3 Oxidation of Methylated Amines

Many methylotrophs can utilize methylated amines as their sole carbon and energy source. The methyl groups of methylated amines are all oxidized to formaldehyde as presented schematically for methylamine in Figure 5-1. Different enzymes have been found for the various substrates [$(CH_3)_4N^+$, CH_3N, CH_2NH, CH_3NH_2]. (For more details see Anthony 1982.) The prosthetic group is PQQ as in MDH, but in methylamine dehydrogenase, the PQQ molecule is covalently bound to the enzyme (de Beer et al. 1980). The nature of the electron acceptor for methylamine dehydrogenase is still controversial.

5.1.3.1 Electron Transfer Proteins in Methylamine-Grown Methylotrophs.

During the growth of methylotrophs on methylamine, the soluble cytochrome c_L does not function as an electron acceptor, and therefore other molecules must perform this function. It has been suggested that copper-containing proteins replace the cytochrome c_L in certain methylotrophic bacteria (strain 4025 and *M. extorquens* AM1) (Auton and Anthony 1989; Fukumori and Yamanaka 1987). The obligate methylotrophic strain 4025, able to grow only on methanol and methylamine, is exceptional since maximal growth on both C_1 substrates occurs only at high copper concentrations. Under these conditions, a copper-containing protein is induced that catalyzes an electron transport reaction coupled to methanol or methylamine oxidation. However, it was found that the copper-containing protein is not essential for growth, since at low copper concentration, this protein was not detected. It was concluded that either extremely low levels of the copper protein could mediate electron transport for methylamine dehydrogenase or that some other unknown electron acceptor is present in these bacteria (Auton and Anthony 1989).

The reconstitution of the electron transport system of methylamine oxidation of *M. extorquens* AM1 with highly purified components, showed that oxygen consumption was not affected by the species-specific copper protein previously thought to be responsible for electron transport (Fukumori and Yamanaka 1987). This result indicates that for this organism, methylamine oxidation can occur without the participation of a copper protein. Thus, the nature of the primary electron acceptor for methylamine dehydrogenase in some methylotrophs is still unknown.

5.1.4 Energy Transduction and Growth Yields

The efficiency by which the MDH reaction is coupled to ATP synthesis determines the biomass yield obtained. The mid-point redox potential for methanol/formaldehyde coupling is -0.182 V, which in principle is sufficient for the synthesis of two ATP molecules during the oxidation of one molecule of methanol by oxygen. However, all studies have demonstrated the synthesis of only *one* molecule of ATP from the oxidation of one molecule of methanol to formaldehyde (Anthony 1986).

For methylotrophic bacteria, which in many cases are NADH-limited but not ATP-limited (Anthony 1982 and 1986), an increase in the P/O ratio of this oxidation step would not result in any substantial increase in cell yield. However, with a MDH enzyme yielding NADH, the cell yields would be much higher (see Figure 5–2). In recent reports (Dijkhuizen et al. 1988; Arfman et al. 1989) the yields obtained with Gram-positive cells having a NAD^+-coupled MDH enzyme, were not substantially higher (16 to 18 g cells dry weight/mol of methanol utilized) than those obtained for Gram-negative methylotrophs having the MDH-PQQ enzyme (Rokem et al. 1978). It is possible that the growth conditions of *Bacillus* sp. C1 were not optimized for maximal cell yields. Thermodynamic considerations have been proposed to explain the lower experimental than theoretically expected biomass yields in those cases where NADH limitation is not apparent (Anthony 1986).

The other enzymatic reactions involved in methanol oxidation (formaldehyde and formate dehydrogenation, and the reactions of the cyclic oxidation route) are coupled at high efficiencies to ATP formation and will not be discussed in this context.

5.1.5 Oxidation of Formaldehyde to Formate

The enzymes involved in the oxidation of formaldehyde to formate are not easy to characterize. They include a nonspecific methanol dehydrogenase and various formaldehyde dehydrogenases. The enzyme MDH has been shown to oxidize formaldehyde in vitro (Anthony 1982). However, the activity of the enzyme in vivo has been questioned for two reasons: First it is uncommon for one enzyme to catalyze two successive reactions, and the second, stronger reason is that formaldehyde is a central metabolite in the oxidation and assimilation pathways. Experimental evidence also suggests that MDH does not oxidize formaldehyde in vivo. Mutants of *M. extorquens* AM1 and *Hyphomicrobium* X lacking MDH grow unimpaired on methylamine and oxidize formaldehyde at the same rate as wild-type bacteria (Marison and Attwood 1982). Therefore, MDH is most probably not involved in formaldehyde oxidation in vivo in these two species.

The most common formaldehyde dehydrogenase in methylotrophic bacteria is a NAD^+-dependent enzyme which may or may not require reduced glutathione for full activity (van Dijken et al. 1981). This enzyme is induced during growth on C_1 compounds. Many methylotrophic bacteria show dye-linked formaldehyde dehydrogenase activities, and the enzymes responsible are nonspecific dehydrogenases which can oxidize a variety of aldehydes (Marison and Attwood 1980). As such, they are not induced during methylotrophic growth.

In RuMP pathway bacteria, formaldehyde is oxidized via the dissimilatory hexulose monophosphate pathway to CO_2, and therefore no formaldehyde dehydrogenase activity is required (see below). The formaldehyde dehydrogenase activity in these bacteria is very low or nonexistent. The mechanism of formaldehyde oxidation in bacteria is still an open question, in contrast to the situation in yeast, where formaldehyde oxidation is known to be catalyzed by one enzyme only.

During the growth of methylotrophic cells on methanol or on formaldehyde, the intracellular formaldehyde concentration must be kept very low since this compound is toxic to the cells. Different mechanisms are utilized by various bacteria to maintain low intracellular formaldehyde concentrations. In one species, compartmentalization of formaldehyde metabolism occurs: the MDH and cytochrome c_L are attached to the cell membrane towards the periplasmic side so that the intermediary formaldehyde is generated on the outer side of the cytoplasmic membrane (Carver and Jones 1983). In a methane-grown *Methylococcus capsulatus* strain (Ferenci et al. 1974) the formaldehyde-assimilating enzyme is attached to the inner side of the cytoplasmic membrane, thereby preventing a high concentration of formaldehyde in the cytoplasm. Another mechanism was found in the methylotroph *Arthrobacter* sp. strain P1. In this strain, formaldehyde exerts a feedback inhibition of the enzymes involved in its formation, thus preventing the accumulation of toxic levels of formaldehyde (Dijkhuizen et al. 1982).

Other mechanisms for lowering the cellular concentration of formaldehyde by its oxidation have also been proposed. One such mechanism, which has not been demonstrated, involves the enzyme N^5, N^{10}-methylene tetrahydrofolate dehydrogenase. The substrate for this enzyme is formed by a nonenzymatic reaction of formaldehyde with tetrahydrofolate. For this mechanism to be functional, an enzyme releasing the formate from formyltetrahydrofolate is essential, but it has as yet not been found. Another possibility is a route involving most of the enzymes of the serine pathway, as suggested for *Pseudomonas* MA (Newaz and Hersh 1975), with either pyruvate kinase and pyruvate dehydrogenase, or pyruvate kinase, malic enzyme, and the decarboxylating enzymes of the tricarboxylic acid (TCA) cycle being responsible for formaldehyde oxidation. However, no experimental proof of this possible route for formaldehyde oxidation has been presented.

5.1.6 Oxidation of Formate to CO_2

Formate dehydrogenases are common in nature and are not only found in methylotrophs. They include a NAD^+-dependent formate dehydrogenase (EC 1.2.1.2), a $NADP^+$-dependent formate dehydrogenase (EC 1.2.1.43), and several formate dehydrogenases participating in the electron transport chain and in the synthesis of formate from CO_2 (EC 1.2.2.1 and 1.2.2.3).

The NAD^+-dependent formate dehydrogenase has been purified from a variety of methylotrophic bacteria. *S*-formylglutathione has been found to be the substrate for the purified NAD^+-dependent formate dehydrogenase isolated from the methylotroph *Achromobacter parvulus* (Egorov et al. 1982). The rate of the enzymatic activity with *S*-formylglutathione was 60% of that with formate as a substrate. In *Achromobacter parvulus* (Egorov et al. 1979) and *Moraxella* sp. C-1 (Asano et al. 1988), the two enzymes were similar, with a molecular weight of each of their two identical subunits of 46 to 48 kDa. The K_m values of these enzymes for formate are relatively high, 15 and 13 mM, respectively.

5.1.7 The Cyclic Oxidation Pathway (the Dissimilatory Ribulose-Monophosphate Pathway)

Colby and Zatman (1975a and 1975b) found low or negligible activities of formaldehyde dehydrogenase and NAD^+-dependent formate dehydrogenase in extracts of several methylamine-grown bacteria. However, these extracts contained high specific activities of phosphoglucoisomerase, glucose-6-phosphate dehydrogenase, and 6-phosphogluconate dehydrogenase. Therefore, Colby and Zatman suggested that these enzymes, together with hexulose phosphate synthase and phosphohexuloisomerase, may participate in a cyclic sequence for the complete oxidation of formaldehyde to CO_2 (Figure 5–1). Since the first two enzymes of the oxidation cycle (i.e., hexulose phosphate synthase and phosphohexuloisomerase) are common to the assimilatory ribulose monophosphate (RuMP) pathway, such a formaldehyde dissimilatory pathway is limited to microorganisms that possess the RuMP assimilatory cycle (Goldberg 1985a).

Two mols of reduced pyridine nucleotides and one mol of CO_2 are formed for each mol of formaldehyde oxidized in both the direct oxidation pathway (via formate) and the cyclic oxidation pathway (Figure 5–1). However, the cyclic oxidation pathway yields both NADPH and NADH, whereas the direct oxidation pathway generates only NADH (Ben-Bassat and Goldberg 1977), with the exception of a $NADP^+$-linked formaldehyde dehydrogenase in *Methylococcus capsulatus* (Bath) (Stirling and Dalton 1978). Both the cyclic oxidation and the direct oxidation pathways operate in methanol-grown *Pseudomonas* C. This was deduced from the following observations (Ben-Bassat and Goldberg 1977):

1. High specific activities of enzymes of the cyclic oxidation pathway and low activities of formaldehyde and formate dehydrogenases are detected in cell extracts.
2. Stimulation of NAD^+- or $NADP^+$-formaldehyde-dependent reduction by D-ribulose-5-phosphate in cell extracts.

3. Oxidation of [^{13}C] formaldehyde to CO_2 by cell extracts requires the presence of NAD^+ (or $NADP^+$) and is greatly increased by the addition of D-ribulose-5-phosphate.
4. Demonstration of the cyclic oxidation pathway by incubating cell extracts, labeled glucose-6-phosphate (1-^{14}C and U-^{14}C), and formaldehyde results in an initial ratio of $^{14}CO_2$ derived from [1-^{14}C] glucose-6-phosphate/[U-^{14}/C] glucose-6-phosphate of 6.8. This ratio subsequently decreases upon a prolonged incubation, owing to interconversion of ribulose-5-phosphate and unlabeled formaldehyde to glucose-6-phosphate.

The key metabolite for assimilation and oxidation is 6-phosphogluconate since it serves as the substrate for two enzymes, namely 6-phosphogluconate dehydrogenase and 6-phosphogluconate dehydrase. Decarboxylation of 6-phosphogluconate to ribulose-5-phosphate directs the flow of carbon into dissimilation via the cyclic oxidation pathway, whereas cleavage to pyruvate and glyceraldehyde-3-phosphate commits the flow into assimilation.

Glucose-6-phosphate dehydrogenase and 6-phosphogluconate dehydrogenase have been purified from several methylotrophs (Goldberg 1985a). In these bacteria, both enzymes have higher affinities for $NADP^+$ than for NAD^+, but the intracellular NAD^+ concentration (2–6 mM) is higher than that of $NADP^+$ (0.3–1.8 mM). It appears that the lower affinities of the enzymes for NAD^+ are compensated by higher intracellular NAD^+ concentrations (Goldberg 1985a).

In *Methylophilus methylotrophus*, the electron transport chain can oxidize both NADH and NADPH. From studies of the thermophile *Methylobacillus flagellatum* KT, it was reported that temperature-sensitive mutants were dependent on the reduction of NAD^+ for growth whereas $NADP^+$ reduction by itself could not support growth (Kletsova et al. 1988). Two isoenzymes of 6-phosphogluconate dehydrogenase with different specificities were described in *M. methylotrophus*: one isoenzyme used both NAD^+ and $NADP^+$ as cofactors and the other isoenzyme used only NAD^+ (Beardsmore et al. 1982); indicating the possibility that one isoenzyme participates in the oxidation pathway and the other in the assimilation pathway. However, it has still not been possible to differentiate the NAD^+- and $NADP^+$-dependent activities of these dehydrogenases in the cell, and it is not clear if they are catalyzed by single or different enzymes in methylotrophs.

Reduced pyridine nucleotides and ATP inhibit the activity of these enzymes in methylotrophs with either NAD^+ or $NADP^+$ as coenzymes (Ben-Bassat and Goldberg 1977; Goldberg 1985a).

The inhibitory effect of ATP on these enzymes might be of physiological importance. It is possible that since glucose-6-phosphate dehydrogenase and 6-phosphogluconate dehydrogenase are consecutive enzymes, small variations in ATP concentrations can cause profound changes in the flux of carbon via the various biochemical pathways. When the concentration of

ATP increases, the NAD⁺-dependent activity of these enzymes is inhibited; thus the continued flux of carbon through the cyclic oxidation pathway will result in a higher production of NADPH (relative to NADH) (Goldberg 1985a).

5.2 OXIDATION OF C_1 COMPOUNDS IN YEASTS

Methylotrophic yeasts were first described in 1969 (Ogata et al. 1969). Several species of yeasts have been described which are able to utilize methanol (see Chapter 2). Methane-utilizing yeasts have been isolated, but their enzymes of methane oxidation have not been studied. In yeasts, methanol is oxidized to CO_2 by the linear pathway shown in Figure 5-1. The first oxidation step is catalyzed by an alcohol oxidase, also known as methanol oxidase. The formaldehyde formed is further oxidized to formate and then to CO_2 by formaldehyde dehydrogenase and formate-dehydrogenase, respectively. Both of these enzymes are present in the cytoplasm.

5.2.1 Oxidation of Methanol to Formaldehyde

The primary oxidation step of methanol to formaldehyde in yeasts occurs in microbodies called peroxisomes. These organelles are induced by methanol and contain the methanol oxidase (see Chapters 2 and 3). The products of this reaction are formaldehyde and hydrogen peroxide. The hydrogen peroxide is reduced by catalase, which is also found within the peroxisomal matrix. The enzymes in the peroxisomes are subject to catabolite repression and to inactivation by glucose, whereas their derepression and induction are mediated by methanol. Different yeast species vary in the control mechanisms they employ for the synthesis of the enzymes involved in methanol oxidation. If the yeast *Hansenula polymorpha* is grown on glucose or methanol, the control of peroxisomal enzyme synthesis is mainly by the repression-derepression mechanisms. When it is grown on alternative carbon sources such as glycerol or ribose, low levels of alcohol oxidase are present in mid-exponential growth (Egli et al. 1980).

The consumption mode of methanol and glucose by *H. polymorpha* in chemostat cultures is dependent upon the growth-limiting substrate. Under carbon-limited conditions, both carbon sources were used simultaneously at steady state, whereas from nitrogen-limited growth, only glucose was utilized. These results indicate that methanol utilization is also controlled by the availability of a nitrogen source (Egli et al. 1982). In *Pichia pastoris*, however, methanol oxidase is only expressed when methanol is the sole carbon source (Tschopp et al. 1987).

The peroxisomes contain key enzymes for methanol metabolism. In addition to the two enzymes involved in methanol oxidation (methanol oxidase and catalase), the first assimilatory enzyme, dihydroxyacetone synthase, is also present within the peroxisomes. Under optimal growth conditions, 80% of the cytoplasmic volume is taken up by peroxisomes. Transfer of cells from a medium where no peroxisomes were present (i.e., glucose), to a medium containing methanol as the sole carbon source, indicates that the peroxisomes that are induced form from preexisting glyoxisomes. In the presence of methanol, the import of substrate-specific enzymes transforms the glyoxisomes to peroxisomes (Zwart et al. 1983). Peroxisomal biogenesis and regulation has been thoroughly described elsewhere (Veenhuis et al. 1983; Harder et al. 1987). The very rapid and sudden changes that occur when cells are transferred from a medium containing glucose to one containing methanol indicate that these inducible enzymes must have strong initiation signals (promoters) to ensure their quick and massive induction. The use of these signals, especially the use of the promoter of methanol oxidase of a methylotrophic yeast, is described in detail in Chapter 13.

5.2.2 Oxidation of Formaldehyde to Formate

Formaldehyde plays a key metabolic role in methylotrophic yeasts, similar to the one that it plays in bacteria containing the RuMP pathway. Formaldehyde is the substrate found at the branchpoint of the oxidation and the assimilation pathways. In bacteria, the first enzymes in formaldehyde assimilation (hexulose phosphate synthase) and oxidation (formaldehyde dehydrogenase) are soluble and are present in the cytoplasm. However, in methylotrophic yeasts, the first enzyme of formaldehyde assimilation (dihydroxyacetone synthase) is present in the peroxisomes, whereas oxidation of formaldehyde by formaldehyde dehydrogenase takes place in the cytosol. It is thought that formaldehyde reacts spontaneously with reduced glutathione, so that the substrate of the NAD^+-formaldehyde dehydrogenase is S-hydroxymethyl glutathione (Gleeson and Sudbery 1988).

If the concentration of methanol in the growth medium is high, it may result in a potentially toxic concentration of formaldehyde inside the cells. Several different mechanisms for reducing formaldehyde concentrations have been observed in yeast. In *Candida boidinii*, it was found that formaldehyde was excreted into the medium at high methanol concentrations. Under these conditions the activity of methanol oxidase and catalase decreased and the glutathione level increased. As a result of the decrease in NADH and ATP concentrations, the inhibition of both formaldehyde and formate dehydrogenase decreased, and therefore formaldehyde did not accumulate in the cells (Ubiivovk and Trotsenko 1986). Another mechanism to lower formaldehyde concentration inside the cells utilizes the enzyme formaldehyde reductase. This enzyme, which converts formaldehyde to methanol, has been found in vivo. When cells are suspended in 30 mM formaldehyde, about 5 mM methanol was found in the medium after 1 hr (Trotsenko et al. 1984). In conclusion, the intracellular concentration of formaldehyde is regulated both by the modulation of the activities of enzymes involved in the primary metabolism of methanol and by the changes in the concentrations of metabolites involved in formaldehyde dissimilation (glutathione, NADH, and ATP). The operation of these detoxification mechanisms (if formaldehyde reaches toxic levels) decreases the biomass yield. The control of dissimilatory and assimilatory pathways is discussed in Chapter 6.

5.2.3 Oxidation of Formate to CO_2

As in bacteria, the formate dehydrogenase of yeasts is dependent on NAD^+. The formate dehydrogenase of *Candida methanolica* is composed of two identical 43-kDa subunits, like other formate dehydrogenases purified from methylotrophic yeasts (Izumi et al. 1989). The yeast formate dehydrogenase is stable when heated at 55°C for 10 min, unlike the bacterial enzyme (Izumi et al. 1989). It has been proposed that the yeast enzyme could be used as a NADH regenerator in bioreactor systems where NADH is needed. Formate dehydrogenase from yeast is appropriate for this purpose because the enzymatic reaction is irreversible, the product (CO_2) is easily removed and an inexpensive substrate (formate) is used.

5.3 CONCLUSIONS AND BIOTECHNOLOGICAL IMPLICATIONS

Both the oxidation and the assimilation pathways and their efficiencies are of great importance for the biomass yield, which in turn is crucial for the industrial production of single cell protein (Goldberg 1985b). Our in-depth knowledge of the pathways and the enzymes involved will enable their efficient utilization for novel and useful purposes as described in Chapter 10. In addition, the identification of the rate-limiting enzymes in the oxidation and the assimilation pathways can contribute to the construction of strains in which these enzymes are expressed at amplified levels. These organisms will probably utilize C_1 compounds in a more efficient way than the wild-type strains.

REFERENCES

Anthony, C. (1982) *The Biochemistry of Methylotrophs*, Academic Press, London.
Anthony, C. (1986) *Adv. Microb. Physiol.* 27, 113–210.
Arfman, N., Watling, E.M., Clement, W., et al. (1989) *Arch. Microbiol.* 152, 280–288.
Asano, Y., Sekigawa, T., Inukai, H., and Nakazawa, A. (1988) *J. Bacteriol.* 170, 3189–3193.
Auton, K.A., and Anthony, C. (1989) *J. Gen. Microbiol.* 135, 1923–1931.

Beardmore-Gray, M., O'Keeffe, D.T., and Anthony, C. (1983) *J. Gen. Microbiol.* 129, 923–933.
Beardsmore, A.J., Aperghis, P.N.G., and Quayle, J.R. (1982) *J. Gen. Microbiol.* 128, 1423–1429.
Ben-Bassat, A., and Goldberg, I. (1977) *Biochim. Biophys. Acta* 497, 586–597.
Ben-Bassat, A., Goldberg, I., and Mateles, R.I. (1980) *J. Gen. Microbiol.* 116, 213–223.
Burrows, K.J., Cornish, A., Scott, D., and Higgins, I.J. (1984) *J. Gen. Microbiol.* 130, 3327–3333.
Bussineau, C.M., Keuer, T.A., Chu, I.M., and Papoutsakis, E.T. (1987) *Appl. Microbiol. Biotechnol.* 26, 61–69.
Carver, M.A., and Jones, C.W. (1983) *FEBS Lett.* 155, 187–191.
Colby, J., and Zatman, L.J. (1975a) *Biochem. J.* 148, 505–511.
Colby, J., and Zatman, L.J. (1975b) *Biochem. J.* 148, 513–520.
De Beer, R., Duine, J.A., Frank, J., and Large, P.J. (1980) *Biochim. Biophys. Acta* 622, 370–374.
Dijkhuizen, L., de Boer, L., Boers, R.H., Harder, W., and Konings, W.N. (1982) *Arch. Microbiol.* 133, 261–266.
Dijkhuizen, L., Artman, N., Attwood, M.M., et al. (1988) *FEMS Microbiol. Lett.* 52, 209–214.
Dijkstra, M., Frank, J., and Duine, J.A. (1988) *FEBS Lett.* 227, 198–202.
Duine, J.A., and Frank, J. (1981) in *Microbial Growth on C_1-Compounds* (Dalton, H., ed.), pp. 31–41, Heyden, London.
Duine, J.A., Frank, J., and Berkhout, M.P.J. (1984) *FEBS Lett.* 168, 217–221.
Egli, T., van Dijken, J.P., Veenhuis, M., Harder, W., and Fiechter, A. (1980) *Arch. Microbiol.* 124, 115–121.
Egli, T., Kappele, O., and Fiechter, A. (1982) *Arch. Microbiol.* 131, 1–7.
Egorov, A.M., Avilova, T.V., and Dikov, M.M., et al. (1979) *Eur. J. Biochem.* 99, 569–576.
Egorov, A.M., Tishkov, V.I., Avilova, T.V., and Popov, V.O. (1982) *Biochem. Biophys. Res. Commun.* 104, 1–5.
Ferenci, T., Strom, T., and Quayle, J.R. (1974) *Biochem. J.* 144, 477–486.
Fox, B.G., Froland, W.A., Dege, J.E., and Lipscomb, J.D. (1989) *J. Biol. Chem.* 264, 10023–10033.
Fukumori, Y., and Yamanaka, T. (1987) *J. Biochem.* 101, 441–445.
Gallop, P.M., Paz, M.A., Fluckiger, R., and Kagan, H.M. (1989) *Trends Biochem.* 14, 343–346.
Gleeson, M.A., and Sudbery, P.E. (1988) *Yeast* 4, 1–15.
Goldberg, I. (1985a) in *Biology of Industrial Microorganisms* (Demain, A.L, and Solomon, N.A., eds.), pp. 223–259, The Benjamin/Cummings Publishing Company, Inc., Menlo Park, CA.
Goldberg, I. (1985b) *Single Cell Protein.* Springer-Verlag, Berlin.
Harder, W., Trotsenko, Y.A., Bystryka, L.V., and Egli, T. (1987) in *Microbial Growth on C_1-Compounds: Proceedings of the 5th International Symposium* (Van Verseveld, H.W., and Duine, J.A., eds.), pp. 139–149. Martinus Nijhoff Publishers, Dordrecht, The Netherlands.
Heptinstall, J., and Quayle, J.R. (1969) *J. Gen. Microbiol.* 55, xvi.
Higgins, I.J., Best, D.J., Hammond, R.C., and Scott D. (1981) *Microbiol. Rev.* 45, 556–590.

Izumi, Y., Kanzaki, H., Morita, S., Futazuka, H., and Yamada, H. (1989) *Eur. J. Biochem.* 182, 333-341.

Kletsova, L.V., Chibisova, E.S., and Tsygankov, Y.D. (1988) *Arch. Microbiol.* 149, 441-446.

Ladner, A., and Zatman, L.J. (1969) *J. Gen. Microbiol.* 55, xvi.

Lim, L.W., Xia, Z.X., Mathews, F.S., and Davidson, V.L. (1986) *J. Mol. Biol.* 191, 141-142.

Lund, J., Woodland, M.P., and Dalton, H. (1985) *Eur. J. Biochem.* 147, 297-305.

Marison, I.W., and Attwood, M.M. (1980) *J. Gen. Microbiol.* 117, 305-313.

Marison, I.W., and Attwood, M.M. (1982) *J. Gen. Microbiol.* 128, 1441-1446.

Newaz, S.S., and Hersh, L.B. (1975) *J. Bacteriol.* 124, 825-833.

O'Connor, M.L., and Hanson, R.S. (1977) *J. Gen. Microbiol.* 101, 327-332.

Ogata, K., Nishikawa, H., and Ohsugi (1969) *Agric. Biol. Chem.* 33, 1519-1522.

Page, M.D., and Anthony, C. (1986) *J. Gen. Microbiol.* 1323, 1553-1563.

Ribbons, D.W., Harrison, J.E., and Wadzinski, A.M. (1970) *Ann. Rev. Microbiol.* 24, 135-140.

Rokem (Rock), J.S., Goldberg, I., and Mateles, R.I. (1978) *Biotech. Bioeng.* 20, 1557-1564.

Smith, D.D.S., and Dalton, H. (1989) *Eur. J. Biochem.* 182, 667-671.

Stanley, S.H., Prior, S.D., Leak, D.J., and Dalton, H. (1983) *Biotech. Lett.* 5, 487-492.

Stirling, D.I., and Dalton, H. (1978) *J. Gen. Microbiol.* 107, 19-29.

Trotsenko, Y.A., Bystrykh, L.V., and Ubiivovk, V.M. (1984) in *Microbial Growth on C_1-Compounds* (Crawford, R.L., and Hanson, R.S., eds.), pp. 118-123. American Society of Microbiology, Washington, DC.

Tschopp, J.F., Brust, P.F., Cregg, J.M., Stillman, C.A., and Gingeras, T.R. (1987) *Nucl. Acids. Res.* 15, 3859-3876.

Ubiivovk, V.M., and Trotsenko, Y.A. (1986) *Mikrobiologiya* 55, 181-185.

Van Dijken, J.P., Harder, W., and Quayle, J.R. (1981) in *Microbial Growth on C_1-Compounds* (Dalton, H., ed.), pp. 191-210. Heyden & Son, London, UK.

Veenhuis, M., Van Dijken, J.P., and Harder, W. (1983) *Adv. Microbiol. Physiol.* 24, 2-81.

Zwart, K.B., Veenhuis, M., Plat, G., and Harder, W. (1983) *Arch. Microbiol.* 136, 26-38.

CHAPTER 6

Regulation of Oxidation and Assimilation of One-Carbon Compounds in Methylotrophic Bacteria

L. Dijkhuizen
I.G. Sokolov

A large number of aerobic and anaerobic bacteria are able to carry out the dissimilation and assimilation of C_1 compounds via diverse pathways (Anthony 1982 and 1986; Fuchs 1986; Large and Bamforth 1988; Heijthuijsen and Hansen 1990). Aerobic bacteria able to grow on C_1 compounds as sole carbon and energy sources are under consideration in this review. The oxidation of these C_1 compounds provides the cells with energy and carbon (as formaldehyde and carbon dioxide) for biosynthetic processes. Methylotrophic bacteria assimilate formaldehyde and/or carbon dioxide via the ribulose monophosphate (RuMP) cycle, the serine pathway, and the Calvin cycle (Anthony 1982). Formaldehyde thus is a key intermediate in the dissimilatory and assimilatory pathways of methylotrophs. This, and the fact that formaldehyde is a very toxic compound, necessitates careful control of the enzyme systems involved in its production and further utilization (Attwood and Quayle 1984).

Regulation of enzyme synthesis is most pronounced in facultative methylotrophs. These organisms generally are able to accurately tune the levels of enzymes required for growth on C_1 compounds to changing environmental conditions. Moreover, the activities of existing enzymes may be affected in various ways. Progress made in recent years in studies on a variety of these control mechanisms in methylotrophic bacteria will be reviewed in the following sections.

6.1 REGULATION OF THE OXIDATION OF C_1 COMPOUNDS IN BACTERIA

The biochemistry of oxidation of C_1 compounds is dealt with in Chapter 5. Most of the methylotrophic bacteria studied are Gram-negative. In recent years, however, a number of Gram-positive strains have been isolated in pure culture. Their characterization has led to the realization that the enzyme systems involved in methanol and methylamine oxidation in these organisms are clearly different from those found in Gram-negative methylotrophs (Dijkhuizen and Arfman 1990).

6.1.1 Transport of C_1 Compounds into the Cells

Most C_1 compounds are uncharged, small molecules (e.g., methane, methanol, formaldehyde), and it is generally assumed that they can permeate freely through bacterial cytoplasmic membranes. Various reports in the recent literature, however, deal with studies on the uptake of these compounds, and in some cases the presence and involvement of active transport systems have been suggested. Since the rate of C_1 compound utilization might be controlled most effectively at the level of its uptake, the current evidence for active transport systems for C_1 compounds (and their regulation) in methylotrophs is reviewed in this section.

Studies of active transport systems for C_1 compounds with whole cells are complex. In all Gram-negative, methanol-utilizing bacteria studied the pyrroloquinoline quinone-(PQQ-dependent), methanol dehydrogenase (MDH) enzyme system involved is located in the periplasmic space (Alefounder and Ferguson 1981; Burton et al. 1983; Kasprzak and Steenkamp 1983; Carver et al. 1984). Methanol transport therefore is not required for methanol utilization. Nevertheless, studies on the dynamics of methanol utilization during growth of *Methylomonas* strain L3 in batch and continuous cultures (Diwan et al. 1983; Chu and Papoutsakis 1987a and 1987b; Agrawal and Flores 1989) and on ^{14}C-methanol uptake by whole cells of *Methylobacillus flagellatum* KT (Borisov and Netrusov 1989) have led to the conclusion that these bacteria possess an active transport system for methanol. Bellion et al. (1983), however, showed that the incorporation of radioactive label from methanol into cells of *Methylobacterium extorquens* AM1 required the presence of MDH activity. This dependence of methanol uptake on MDH activity suggests that the observed incorporation of label from methanol in cells was based on conversion of methanol into formaldehyde, followed by its diffusion (or active uptake) into the cells and its assimilation into cell material.

Methylamine oxidation in bacteria may occur via an amine dehydrogenase (e.g., as in *Methylobacterium extorquens* strain AM1), an amine oxidase (e.g., as in *Arthrobacter* P1), or a methylglutamate dehydrogenase (e.g., as in *Pseudomonas* MA and *Hyphomicrobium* X; Large 1981; Levering et al. 1981). Whereas amine dehydrogenase is a periplasmic enzyme (Burton et al. 1983), the latter two enzyme systems are located in the cytoplasm of the cells. Organisms employing cytoplasmic enzymes for methylamine oxidation apparently also require the presence of a methylamine transport system (Bellion et al. 1980; Bellion and Wayland 1982; Dijkhuizen et al. 1982; Brooke and Attwood 1984). Strong evidence for the presence of such a methylamine transport system in *Arthrobacter* P1 was obtained in studies with cytoplasmic membrane vesicles devoid of enzymes of methylamine oxidation. This transport system has a high affinity for methylamine (apparent K_m = 20–25 μM), and it accumulates methylamine in unmodified form in a process driven by the membrane potential (Dijkhuizen et al. 1982). Interestingly, ^{14}C-methylamine uptake by these vesicles is not inhibited by ammonia, but it is inhibited by ethylamine (competitively; K_i = 75 μM) and other lower primary amines. This indicates that an amine-specific transport system is involved and not the ammonia carrier. Strong noncompetitive inhibition was also observed with formaldehyde (K_i = 1.6 μM). Formaldehyde inhibition of this transport system may serve to accurately control the rate of formaldehyde production, preventing accumulation of toxic formaldehyde under transient state conditions (Levering et al. 1984). *Arthrobacter* P1 is able to use methylamine, ethylamine, and propylamine as both carbon and nitrogen sources for growth, whereas butylamine and benzylamine only function as nitrogen sources (de Boer et al. 1989). Cells grown on glucose plus the above-mentioned amines as a nitrogen source are able to accumulate ^{14}C-methylamine, but at greatly reduced rates when compared to cells grown on methylamine as a carbon source (Table 6–1). Invariably, methylamine uptake by these cells was insensitive to inhibition by ammonia. These data indicate that the various amines induce synthesis of this amine transport system, with the presence of glucose resulting in its partial repression (de Boer et al. 1989).

The possible presence of methylamine transport systems in methylotrophs employing periplasmic methylamine dehydrogenase (*Methylobacterium extorquens* AM1; *Methylobacillus flagellatum* KT; *Paracoccus denitrificans*) has been suggested (Bellion et al. 1983; Holtel and Kleiner 1985; Borisov and Netrusov 1989). These studies, however, were performed with whole cells. It therefore remains possible that methylamine was converted into formaldehyde by the methylamine dehydrogenase on the outside face of the cell membrane, and that formaldehyde uptake and incorporation in

TABLE 6-1 Growth of *Arthrobacter* P1 in Batch Cultures Using Various Aliphatic Primary Amines and Benzylamine as Carbon or Nitrogen Sources

Carbon Source[1] (25 mM)	Nitrogen Source (25 mM)	Doubling Time (hr)	(^{14}C-)Methylamine Uptake Rate	Amine Oxidase[2] Measured on Methylamine	Amine Oxidase[2] Measured on Nitrogen Source	Hexulose-6-Phosphate Synthase[2]	Aldehyde Excretion (mM)	Residual Ammonia (mM)
Methylamine	Ammonia	2.6	568	740	—	742	0	20.5
Ethylamine	Ammonia	2.4	224	780	—	47	0	20.8
Glucose	Ammonia	1.2	0	0	—	0	0	1.5
Glucose	Methylamine	1.5	60	193	193	528	0	0.5
Glucose[3]	Ethylamine	1.6	52	35	42	61	6.4	1.2
Glucose[4]	Propylamine	1.5	92	38	46	15	5.8	0.1
Glucose	Butylamine	1.8	68	62	50	0	5.3	0.1
Glucose	Benzylamine	3.0	40	566	171	45	5.0	0.7

[1] Cells grown on methylamine and ethylamine as a carbon source were harvested at an optical density (433 nm) of 1.2–1.4; cells grown on glucose were harvested at an optical density of 4.0 (mid-exponential growth phases).
[2] The specific activity of the enzymes is given in nmol min^{-1} mg^{-1} of protein.
[3] No acetaldehyde dehydrogenase and isocitrate lyase activity was detected in cells grown on a mixture of glucose and ethylamine.
[4] No propionaldehyde dehydrogenase activity was detected in cells grown on glucose and propylamine.
Reproduced with permission from de Boer et al. (1989).

cell material via cytoplasmic enzymes was measured, rather than methylamine uptake directly.

As outlined above, in many methylotrophs, methanol and methylamine are converted into formaldehyde by the action of periplasmic enzymes. In studies with whole cells of various methylotrophs, incorporation of radioactive label from ^{14}C-formaldehyde into cell material was found to be inhibited by the uncoupler carbonylcyanide p-trifluoromethoxyphenylhydrazone (FCCP). This was observed for instance with the methanotroph *Methylococcus thermophilus* 111n, under conditions where the possible oxidation of formaldehyde by methanol dehydrogenase in the periplasmic space was blocked by EDTA, or by formamide (Pinchuk 1989), and with *Paracoccus denitrificans* (Köstler and Kleiner 1989). Incubation of two *Methylomonas* strains under anaerobic conditions in the presence of FCCP only resulted in formaldehyde uptake following pre-energization of the cells with methanol (Diwan et al. 1983; Bussineau and Papoutsakis 1988). The evidence obtained was taken to suggest that formaldehyde uptake was driven by the proton motive force, controlled by the intracellular ATP level, and strongly inhibited by FCCP. In each of these cases, however, it remains to be confirmed that active transport of formaldehyde was the actual target for inhibition by FCCP. Data on the inhibitory effects of FCCP on the metabolism of whole cells clearly should be treated with great care. Studies of Jones and co-workers on the obligate methylotroph *Methylophilus methylotrophus* (Dawson and Jones 1982; Patchett et al. 1985; Jones et al. 1987), for instance, showed that de-energization of cells with FCCP may result in complicated changes, affecting amongst others the kinetics of the intracellular, interlinked pathways of C_1 compound utilization and intermediary metabolism (Jones et al. 1987).

Arthrobacter P1 is able to grow on formaldehyde as the sole carbon and energy source if formaldehyde is supplied as the growth-limiting compound in chemostat culture (Levering et al. 1986). Whole cells rapidly incorporate ^{14}C-formaldehyde, but this activity is absent in mutants that are deficient in hexulose phosphate synthase (HPS) activity. Moreover, no formaldehyde uptake is observed with cytoplasmic membrane vesicles isolated from formaldehyde-grown cells of the wild-type organism (P.R. Levering and L. Dijkhuizen, unpublished observations). The data thus indicate that, at least in *Arthrobacter* P1, a formaldehyde transport system, separate from HPS activity, is unlikely to be present.

Methane oxidation by whole cells of some methanotrophs (*Methylomonas rubra* 15m, *Methylococcus thermophilus* 111n) is inhibited in the presence of compounds dissipating the proton motive force (protonophores; Pinchuk et al. 1989) or the membrane potential (permeating ions like thiocyanate; Sokolov 1986). Further studies with *Methylococcus thermophilus* 111n showed that protonophores did not inhibit methane monooxygenase (MMO) activity directly (I.G. Sokolov and G.E. Pinchuk, unpublished observation). Although the inhibitory effects of these compounds on methane oxidation may be taken to suggest that methane uptake involves an active transport system, at the moment the possibility cannot be ruled out that the de-energization of the membrane by protonophores negatively affected the rate of methane oxidation.

In conclusion, clear-cut evidence for the involvement of active transport systems in the uptake of C_1 compounds has been obtained only for methylamine, as studied with cytoplasmic membrane vesicles of *Arthrobacter* P1 (Dijkhuizen et al. 1982). A prerequisite for further studies on uptake mechanisms of C_1 compounds therefore is the development and application of similar methods for the isolation of cytoplasmic membrane vesicles of the various methylotrophs under investigation. It is only from this approach, allowing a clear separation of membrane systems from associated enzymes and cytoplasmic enzymes, that unambiguous proof for the presence of active transport systems for the intracellular accumulation of various C_1 compounds may be generated.

6.1.2 Regulation of Enzymes Involved in the Oxidation of C_1 Compounds

6.1.2.1 Methane Oxidation. Methanotrophs are generally only able to utilize C_1 compounds as carbon and energy sources for growth, i.e., they are obligate methylotrophs. Evidence for the existence of facultative methanotrophs is still limited (see Chapter 1), and only one facultative methanotroph has been described thus far (Reed and Dugan 1987). Methane oxidation in one specific organism may involve both soluble (sMMO) and particulate (pMMO) methane monooxygenases (MMO) (Dalton et al. 1984; Dalton and Higgins 1987). Synthesis of the sMMO and pMMO enzymes, which differ in a number of properties, in *Methylococcus capsulatus* (Bath) and *Methylosinus trichosporium* is controlled by the availability of copper ions, with copper deficiency resulting in sMMO synthesis (Stanley et al. 1983; Burrows et al. 1984; Green et al. 1985; Prior and Dalton 1985). Following addition of copper to the medium, sMMO is rapidly inactivated, its synthesis stops, and pMMO synthesis dramatically increases. Compared to sMMO, pMMO allows a higher specific growth rate, efficiency, and affinity towards methane (Leak and Dalton 1986a and 1986b; Joergensen and Degn 1987). MMO activity can also be detected in methanol-grown cells of various methanotrophs (Linton and Vokes 1978; Prior and Dalton 1985; Davis et al. 1987). MMO synthesis in methanol-grown cells of *Methylococcus capsulatus* (Bath) was significantly stimulated by the presence of copper in the growth medium, and only the particulate enzyme was detectable under the conditions tested (Prior and Dalton 1985). Methanol-grown cells of *Methylosinus trichosporium* OB3b on the other hand were able to express both forms of MMO, with sMMO predominating during copper-limited growth while pMMO is expressed fully only in copper-sufficient cultures (Davis et al. 1987). Formaldehyde may act as a repressor of sMMO synthesis in *Methylococcus capsulatus* (Bath) (Dalton et al. 1984). Some methanotrophs (representatives of the genus *Methylomonas*) do not possess sMMO. It has been suggested that the ability to express sMMO is an adaptation process to lower copper concentrations in the environment (Dalton and Leak 1985).

6.1.2.2 Methanol Oxidation in Gram-Negative Bacteria. The available evidence suggests that in pink-pigmented facultative methylotrophs employing the serine pathway for C_1 carbon assimilation, synthesis of methanol dehydrogenase (MDH) proceeds constitutively. Nevertheless, a clear further induction (derepression) of MDH synthesis generally can be observed upon transfer of the organisms into methanol-containing media (Dunstan et al. 1972; O'Connor and Hanson 1977; McNerney and O'Connor 1980; Weaver and Lidstrom 1985). MDH genes may be controlled coordinately with several genes coding for assimilatory enzymes (O'Connor 1981). Strong derepression of MDH synthesis also may be observed under growth-limiting conditions with multicarbon substrates, or when such substrates only allow very low growth rates of the organisms, and during methanol limitation (Weaver and Lidstrom 1985 and 1987; Roitsch and Stolp 1986). The latter also may be observed with obligate and restricted facultative, RuMP-cycle methylotrophs (Roitsch and Stolp 1985; Greenwood and Jones 1986; Jones et al. 1987). In these organisms, MDH appears to be further regulated by dissolved oxygen tension and growth rate. The molecular properties of the underlying regulatory phenomena are not yet known, but rapid progress is being made in the genetic analysis of these systems (for reviews, see de Vries 1986; Goodwin 1990; de Vries et al. 1990; also see Chapter 12).

Methanol-utilizing bacteria that assimilate carbon dioxide via the Calvin cycle are all facultative methylotrophs. Control of MDH synthesis in these organisms appears to be completely different from the mechanisms observed in RuMP and serine pathway organisms. In these bacteria, MDH activity generally cannot be detected during growth on multicarbon substrates (Harms et al. 1985; Weaver and Lidstrom 1985; de Vries et al. 1988; Meijer et al. 1990). Formaldehyde rather than methanol appears to be the inducer for MDH synthesis in *Paracoccus denitrificans* (de Vries et al. 1988). This product induction apparently necessitates additional posttranscriptional regulation of MDH activity, probably to prevent the accumulation of formaldehyde up to an inhibitory concentration. A possible target for the control of MDH activity might be the synthesis or the stable incorporation of the prosthetic group, PQQ, into the enzyme molecule (de Vries et al. 1988).

6.1.2.3 Methylamine Oxidation. The properties of methylamine-oxidizing enzymes in methylotrophic bacteria have been reviewed previously

(Large 1981; Dijkhuizen and Levering 1987). The amine oxidase system in *Arthrobacter* P1 has been studied in detail. This enzyme possesses activity with various aliphatic primary amines, and with benzylamine (van Vliet-Smits et al. 1981; Levering et al. 1984). Growth of the organism on glucose plus various amines as a nitrogen source resulted in decreased levels of amine oxidase compared to growth on methylamine as sole carbon, energy, and nitrogen source (Table 6–1; de Boer et al. 1989). The data indicate that the various amines induce synthesis of amine oxidase, with the presence of glucose resulting in its partial repression (de Boer et al. 1989). Not only the amine transport system (see Section 6.1.1) in *Arthrobacter* P1, but also the amine oxidase is sensitive to inhibition by formaldehyde, preventing accumulation of toxic formaldehyde under transient state conditions (Levering et al. 1984).

6.1.2.4 Formaldehyde Oxidation. Formaldehyde oxidation may either occur via a linear pathway, via formate to carbon dioxide, or via a cyclic pathway, involving enzymes of the RuMP cycle and glucose-6-phosphate plus 6-phosphogluconate dehydrogenases (Zatman 1981; Anthony 1982; also see Chapter 5). The situation with respect to the precise identity of formaldehyde-oxidizing enzymes in methylotrophic bacteria, however, is still largely obscure. The MDH enzyme in Gram-negative bacteria is able to oxidize formaldehyde, but it is unlikely that this is an important activity during growth (Attwood and Quayle 1984). The levels reported for the many different NAD(P)-formaldehyde dehydrogenases (which may or may not require reduced glutathione) and the dye-linked formaldehyde dehydrogenases are often low (Zatman 1981; Dijkhuizen et al. 1990). Matters may be further complicated by the observation of Bussineau et al. (1987) that serial subculturing of a newly isolated RuMP-type methylotroph, *Methylomonas* T15, resulted in considerable loss of NAD-formaldehyde dehydrogenase and NAD-formate dehydrogenase activities. Moreover, the formaldehyde dehydrogenase enzymes studied are mostly nonspecific (general-aldehyde dehydrogenases), and they are not induced by C_1 substrates (Marison and Attwood 1982; Weaver and Lidstrom 1985). The latter is also the case with the NAD-formaldehyde dehydrogenase observed in choline-, betaine-, dimethylglycine-, or sarcosine-grown cells of *Arthrobacter* P1 (Levering et al. 1987). Cells of *Arthrobacter* P1 grown in formaldehyde-limited continuous cultures did not possess this enzyme; instead, high levels of the enzymes of the dissimilatory RuMP cycle were detected under these conditions.

The situation in Gram-positive methylotrophs employing a linear pathway for formaldehyde oxidation also requires further investigations. Duine et al. (1984) reported evidence for the presence of an NAD-dependent aldehyde dehydrogenase in a multienzyme complex with an NAD-requiring MDH and NADH dehydrogenase in the actinomycete *Nocardia* sp. 239. Further studies have revealed the presence of three different (form)aldehyde dehydrogenases in this particular organism. The possible physiological functions of these enzymes in methanol metabolism in *Nocardia* sp. 239 remain to be elucidated. Interestingly, one of these enzymes requires the presence of an unknown low-molecular-weight factor for activity (van Ophem and Duine 1990), as previously reported for a formaldehyde-oxidizing system in *Rhodococcus erythropolis* growing on compounds containing methyl groups (Eggeling and Sahm 1984 and 1985).

The discussion about the involvement and relative importance of the linear pathway versus the cyclic pathway for formaldehyde oxidation in RuMP cycle methylotrophs has often been confusing. In some organisms the extent to which these routes are involved appears to depend on cultivation conditions, as has been reported for *Pseudomonas* C (Samuelov and Goldberg 1982), *Methylomonas* L3 (Chu and Papoutsakis 1985; Bussineau and Papoutsakis 1986), and *Methylomonas* T15 (Bussineau et al. 1987). Mutant studies with *Methylobacillus flagellatum* KT (Kletsova et al. 1988) have provided evidence that at least in this organism the presence of enzymes of the dissimilatory RuMP cycle is indispensable for growth on methanol.

6.1.2.5 Electron Transport Chain. Growth of Gram-negative methylotrophic bacteria on methanol and methylamine requires the synthesis of special *c*-type cytochromes. Electrons from the prosthetic group (PQQ) of MDH are accepted by a low-potential form of cytochrome c_L and are transferred via a high-potential form of cytochrome c_H or azurine, to terminal oxidases of the aa_3- or *o*-type (Anthony 1982; Anthony and Jones 1987; also see Chapter 5). Both *c*-type cytochromes are able to bind carbon monoxide. Synthesis of these *c*-type cytochromes is induced during methylotrophic growth (Tonge et al. 1974; Weaver and Lidstrom 1985; Bosma 1989) and may be controlled coordinately with other C_1 enzymes (O'Connor 1981; Weaver and Lidstrom 1985; also see Chapter 12).

Different respiratory chains are induced during growth of *Paracoccus denitrificans* on either methylamine or methanol (Hussain and Davidson 1985; Davidson and Kumar 1989). During methylamine oxidation, electrons from PQQ-containing methylamine dehydrogenase are accepted by amicyanine, a blue copper protein, and transferred via cytochromes *c*-55li and *c*-550 to the terminal oxidase, whereas cytochromes *c*-533i and *c*-550 transfer electrons from PQQ-MDH. Amicyanine functions as an inducible electron acceptor for methylamine dehydrogenase in other methylotrophs (Lawton and Anthony 1985; Tobari 1984).

The rate of methanol oxidation in *Methylophilus methylotrophus* is controlled by the synthesis of the alternative terminal oxidase (Greenwood and Jones 1986). Under oxygen-limiting (i.e., methanol-excess) conditions, the *o*-type oxidase content was fivefold higher than under methanol-limiting conditions. Apparently, the *o*-type oxidase confers a definite advantage under oxygen-deficient conditions, possibly due to a higher affinity towards oxygen.

The structure of the electron transport chain in methanotrophs is still largely obscure. Soluble cytochrome c_{CO} acts as electron acceptor for MDH (Tonge et al. 1975; Sokolov et al. 1981). Its content in the cells may reach a level of up to 5% of total soluble protein (Gvozdev et al. 1982). Cytochrome aa_3 was found in the membranes of some methanotrophs (Tonge et al. 1974; Sokolov et al. 1981) where it could function as the terminal oxidase in the respiratory chain. Another electron transport chain appears to be involved in electron transfer from NADH to MMO, since pMMO activity is sensitive towards respiratory chain inhibitors. During growth on methane the major part of substrate electrons (60–75%, depending on the yield; Anthony 1982) is consumed in the oxygen-dependent monooxygenation reaction. The electron transport chain involved therefore could be considerably more powerful than the usual one. This coincides with low NADH oxidase activity in methanotrophs.

The significance of the role of MMO as the main terminal electron acceptor was verified by a simulation of this reaction during the growth of methanotrophs on methanol. The efficiency of methanol utilization by *Methylomonas rubra* 15m improved considerably after addition of substrates for the MMO reaction (e.g., ethane or CO). Moreover, *Methylococcus thermophilus* 111n is unable to grow on methanol alone, but grows well on mixtures of methanol plus CO (Sokolov et al. 1979; Sokolov 1988). This indicates that electrons derived in the oxidation of methanol are transferred to oxygen during monooxygenation of the second substrate added (Figure 6–1A). Inhibition of MMO activity in *Methylosinus trichosporium* OB3b by acetylene addition completely blocked methanol assimilation by the organism (Cornish et al. 1984; Sokolov et al. 1989). This failure to grow on methanol when MMO activity is inhibited suggests that the activity and/or energy coupling of the usual respiratory chain is insufficient (Figure 6–1B).

Acetylene was found to stimulate growth on methanol of two other methanotrophs, *Methylomonas rubra* 15m and *Methylococcus capsulatus* 9m, and to prevent formaldehyde accumulation (Figure 6–1C; Sokolov et al. 1989). This suggests the presence (or activation) of energy-coupled, electron transfer systems to oxygen in forms other than MMO. The failure of the usual terminal oxidase to substitute for MMO (as terminal electron acceptor) in some methanotrophs may explain why attempts to isolate MMO-deficient mutants of *Methylomonas albus* met with failure (McPheat et al. 1987).

6.1.3 Modulation of the Activity of C₁ Enzymes by Protein Factors

In recent years several examples of modulation of the activity of C_1-specific enzymes have been reported involving interaction with protein factors. The precise metabolic function of these activator/modifier proteins is still largely

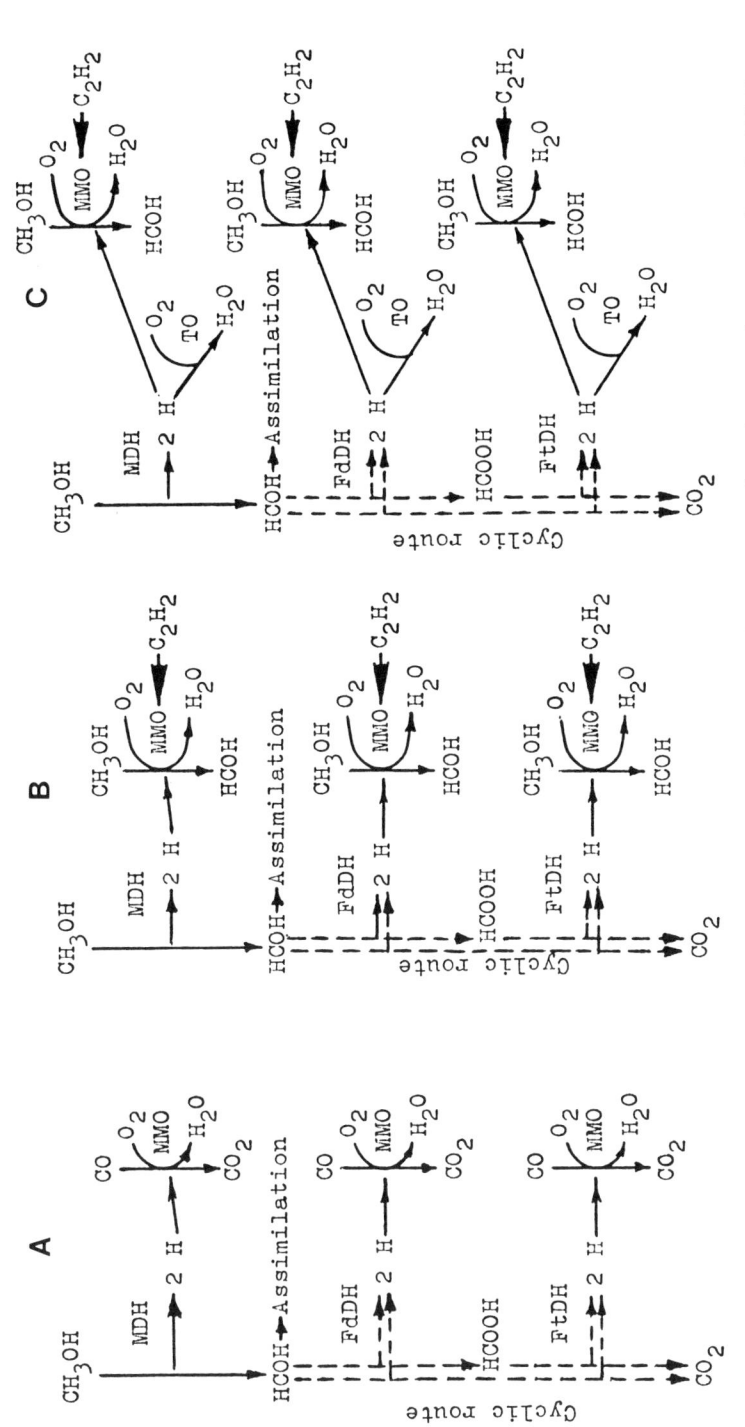

6.1 Regulation of the Oxidation of C₁ Compounds in Bacteria   137

FIGURE 6-1 Three possible routes for terminal electron acceptance during growth of methanotrophs on methanol. (A) Route involving co-oxidation of alternative MMO substrates (e.g., CO). (B) Route used when monooxygenation of methanol by MMO is the only efficient sink for electrons, resulting in formaldehyde formation and inhibition of growth by acetylene. (C) Route used when the usual terminal oxidase is active; acetylene improves growth and blocks formaldehyde formation. Abbreviation: MDH, methanol dehydrogenase; MMO, methane monooxygenase; FdDH, formaldehyde dehydrogenase; FtDH, formate dehydrogenase; TO, normal terminal oxidase; 2 H, reducing equivalents.

obscure but may be of special significance for the kinetics of enzymes converting C_1 compounds.

6.1.3.1 Methane Monooxygenase.

The sMMO of *Methylococcus capsulatus* (Bath) consists of hydroxylase protein A, FAD-containing protein C, and protein B (16 kD), which is necessary for monooxygenase activity (Dalton 1981; also see Chapter 5). The electrons from NADH flow from protein C to protein A, where methane oxidation and oxygen reduction occur (Dalton and Leak 1985; Green and Dalton 1985; Lund et al. 1985). When protein B is absent, proteins A and C still catalyze the NADH-dependent reduction of oxygen, but methane oxidation is blocked. When component B is present but methane is absent, NADH oxidation is blocked thus preventing wasteful oxidation of NADH under these conditions (Green and Dalton 1985; Dalton and Higgins 1987). Thus, the dissociation of protein B modulates the monooxygenase into an oxidase. Formaldehyde enhances this process, which may explain why sMMO-containing cells of *Methylosinus trichosporium* OB3b oxidize methanol without formaldehyde accumulation (Cornish et al. 1984). Formaldehyde as a product of the parallel reactions of methanol oxidation catalyzed by MMO and MDH (see Figure 6–1B) binds protein B and transforms MMO into an oxidase. This results in a block in the formaldehyde-yielding methanol monooxygenation reaction, and MMO proteins act as oxidase-accepting formaldehyde electrons. In the presence of acetylene, the oxidase function of MMO is not expressed, resulting in accumulation of formaldehyde during methanol oxidation by sMMO-containing cells (Cornish et al. 1984). Apparently, pMMO can not be modulated in this way and pMMO-containing cells oxidized methanol with formaldehyde accumulation.

6.1.3.2 Methanol Dehydrogenase.

A protein factor (140 kD, dimer) that modifies the properties of MDH has been isolated from the periplasm of *Methylobacterium extorquens* AM1 and *Methylophilus methylotrophus* (Ford et al. 1985; Page and Anthony 1986; also see Chapter 5). Its interaction with MDH results in a stimulation of methanol oxidation and a decrease in the rate of formaldehyde oxidation by the enzyme. The modifier protein is exclusively found in the periplasm where it is present in low concentrations compared with the MDH protein (Anthony 1989).

In Gram-positive bacteria of the genus *Bacillus*, methanol is oxidized within the cytoplasm by an NAD-dependent MDH (Arfman et al. 1989). Purification of this enzyme, composed of 10 identical subunits of 43 kD each, resulted in loss of 90% of the MDH activity, but not of formaldehyde reductase activity (the reverse reaction). The MDH activity of the purified enzyme could be restored by addition of Mg^{2+} and a 50-kD protein factor.

6.2 REGULATION OF THE ASSIMILATION OF C$_1$ COMPOUNDS IN BACTERIA

Bacteria employing the Calvin cycle or the serine pathway for assimilation of C$_1$ compounds are facultative methylotrophs. These organisms generally clearly regulate the synthesis of the enzymes involved in these pathways. Regulation of carbon assimilation via the Calvin cycle and serine pathway has been dealt with in recent reviews. Thus, the regulation of the Calvin cycle enzymes has been discussed by Tabita (1988) and by Dijkhuizen and Harder (1984 and 1985). The available physiological evidence suggests that synthesis of the key enzymes of the Calvin cycle—ribulose-1,5-bisphosphate carboxylase/oxygenase, phosphoribulokinase, and fructose-bisphosphatase—are controlled by a feedback repression mechanism, as has been observed for biosynthetic pathways in general. In this mechanism, the intracellular level of an intermediate of central metabolism, 3-phosphoglycerate or a closely related metabolite, may function as a signal. Regulation of the serine pathway enzymes has been reviewed recently by Goodwin (1990). In *Methylobacterium extorquens* AM1 and *Methylobacterium organophilum*, a product of methanol oxidation rather than methanol itself is probably the inducer of serine pathway enzymes (O'Connor 1981). In *Methylobacterium extorquens* AM1, four enzymes—serine glyoxylate aminotransferase, hydroxypyruvate reductase, glycerate kinase, and malyl-CoA lyase—appear to be coordinately induced when cells are transferred from succinate to methanol medium; they are repressed during growth on mixtures of succinate and methanol (Dunstan et al. 1972; McNerney and O'Connor 1980; O'Connor 1981). However, the C$_1$-inducible isoenzymes of phosphoenolpyruvate carboxylase and serine transhydroxymethylase are not repressed during growth on succinate plus methanol, and they are induced only after a lag when cells are transferred from succinate to methanol medium. Induction of serine pathway enzymes is also observed following transfer of cells of *Methylobacterium organophilum* from succinate to methanol medium; however, in this organism the enzymes are not strongly repressed during growth on succinate plus methanol (O'Connor and Hanson 1977; Goodwin 1990).

6.2.1 Regulation of the RuMP Cycle of Formaldehyde Assimilation

Most bacteria employing the RuMP cycle of formaldehyde assimilation are obligate or restricted facultative methylotrophs. Although synthesis of the RuMP cycle enzymes in these organisms is constitutive, the level of hexulose-6-phosphate synthase (HPS) may vary strongly. A growth-rate-dependent increase in HPS activity for instance was observed in studies with *Methylomonas* L3 grown in methanol-limited chemostat cultures (Bussineau and Papoutsakis 1986). Further control mechanisms in organisms employing the KDPG-cleavage variant of the RuMP cycle appear to especially affect the activity of two enzymes balancing the flow between dissimilatory and assimilatory pathways, namely glucose-6-phosphate and 6-phosphogluconate dehydrogenases. These enzymes have been purified from various RuMP cycle methylotrophic bacteria and are generally inhibited by reduced pyridine nucleotides and nucleoside triphosphates (Anthony 1982; Beardsmore et al. 1982; Kiriuchin et al. 1988; Kletsova et al. 1988; also see Chapters 4 and 5).

Metabolic studies with facultative RuMP cycle methylotrophs are still scarce. Only a limited number of truly versatile bacteria, able to grow on C_1 compounds and a wide spectrum of multicarbon compounds, have been isolated in pure culture thus far. The available information, however, suggests that these bacteria are highly interesting, both from a fundamental and an applied point of view. These facultative strains are all Gram-positive bacteria, belonging to the genera *Arthrobacter*, *Bacillus*, *Brevibacterium*, *Mycobacterium*, *Nocardia*, and *Amycolatopsis* (Dijkhuizen and Arfman 1990; de Boer et al. 1990a). A brief account of our studies on the regulation of the synthesis of RuMP cycle enzymes in two of these facultative methylotrophs, *Arthrobacter* P1 and *Amycolatopsis methanolica* (*Nocardia* sp. 239), which have been studied in most detail, will be given in the following sections.

6.2.2 Regulation of the RuMP Cycle in *Arthrobacter* P1

Arthrobacter P1 is able to grow on methylated amines but not on methanol. During growth on methylamine, energy is generated by formaldehyde oxidation via the dissimilatory RuMP cycle. Formaldehyde assimilation occurs via the fructose-1,6-bisphosphate aldolase cleavage variant of the RuMP cycle (Levering et al. 1981). Metabolic studies with this organism, involving growth on mixtures of methylamine and acetate, have provided evidence that the synthesis of two RuMP cycle enzymes, HPS and hexulose-6-phosphate isomerase (HPI), is solely controlled via induction by formaldehyde (Levering et al. 1986). As mentioned above, methylamine utilization in this organism involves an amine-specific transport system and a cytoplasmic amine oxidase. Growth of *Arthrobacter* P1 with glucose in the presence of various amines as nitrogen sources resulted, when using ethylamine, propylamine, butylamine, and benzylamine, in accumulation of the corresponding aldehydes. Accumulation of formaldehyde from methylamine, however, was not observed (Table 6–1; de Boer et al. 1989). Excretion of acetaldehyde and propionaldehyde from ethylamine and propylamine was unexpected at first since the organism is able to use these amines as carbon sources. However, enzyme analysis revealed that the presence of glucose caused a complete repression of the synthesis of acetaldehyde dehydrogenase and isocitrate lyase (the enzymes involved in ethylamine metabolism) and propionaldehyde dehydrogenase. Interestingly, glucose/methylamine-grown cells still possessed relatively high levels of HPS, explaining why formaldehyde accumulation did not occur. Growth with ethylamine, propylamine, and benzylamine as nitrogen sources also resulted in synthesis of HPS, most likely via induction by the respective aldehydes produced, although these compounds are not a substrate for the enzyme. These results provide further evidence that the synthesis of HPS in *Arthrobacter* P1 is only regulated via induction by (form)aldehyde (Dijkhuizen and Levering 1987). The mechanisms involved in controlling methylamine metabolism in *Arthrobacter* P1 can be summarized as follows: following incubation of *Arthrobacter* P1 in the presence of methylamine, a toxic level of formaldehyde is prevented by (1) a rapid induction of the synthesis up to high levels of HPS (Levering et al. 1986) and other RuMP cycle enzymes, including a C_1-specific transaldolase isoenzyme involved in the regeneration of RuMP (Levering and Dijkhuizen 1986); and (2) the feedback inhibition of the activities of the methylamine-transport system (section 6.1.1) and amine oxidase (section 6.1.2.3; Dijkhuizen et al. 1982; Levering et al. 1984).

6.2.3 Regulation of the RuMP Cycle in *Nocardia* sp. 239

The precise identity of the enzymes involved in methanol dissimilation in *Amycolatopsis methanolica* (*Nocardia* sp. 239) is still unknown (section 6.1.2.4). The available data (Hazeu et al. 1983; Duine et al. 1984; de Boer et al. 1990a and 1990b) nevertheless indicates that this organism employs a linear pathway via formate to carbon dioxide for the oxidation for formaldehyde. Synthesis of the RuMP cycle enzymes HPS and HPI is under strict control of *A. methanolica*. No activities of these enzymes are detectable in cells grown on glucose or acetate (Figures 6-2 and 6-3; de Boer et al. 1990b). Incubation of these cells in batch cultures with mixtures of glucose or acetate with methanol nevertheless resulted in simultaneous utilization of the two substrates in both experiments. Synthesis of HPS and HPI only occurred after the complete consumption of glucose, and this was followed by a second phase of slower growth on methanol alone, but with a rapid utilization of methanol (Figure 6-2). However, in experiments with acetate-pregrown cells, synthesis of HPS and HPI started immediately (Figure 6-3). The results thus indicate that the regulation of the synthesis of the methanol-dissimilatory and the methanol-assimilatory enzymes in *A. methanolica* differs. Whereas methanol (and/or compounds derived from it) apparently induces the synthesis of the dissimilatory enzymes, the RuMP cycle enzymes are sensitive to (catabolite) repression by glucose, but not by acetate. To investigate whether intracellularly produced formaldehyde, rather than methanol itself, is the inducing signal for HPS and HPI synthesis, the effects of formaldehyde addition to cells of *A. methanolica* growing in batch culture on glucose were investigated (Figure 6-4; de Boer et al. 1990b). To avoid toxic formaldehyde concentrations, this compound was added continuously and at such a rate that its concentration in the culture was maintained at around 0.5 mM. For this purpose, the formaldehyde concentration

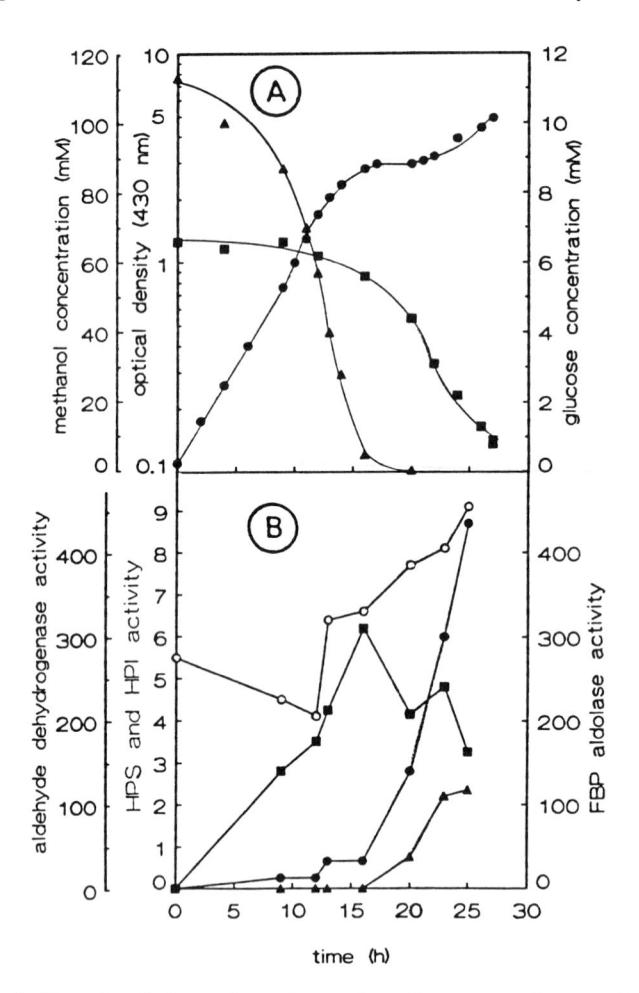

FIGURE 6-2 Growth of *Amycolatopsis methanolica* on a mixture of glucose (11 mM) and methanol (70 mM) in batch culture. Cells were pregrown on glucose. (A) ●, optical density; ■, methanol concentration; ▲, glucose concentration. (B) ●, HPS; ▲, HPI; ■, aldehyde dehydrogenase; ○, fructose-1,6-bisphosphate (FBP) aldolase. HPS and HPI activities are expressed in μmol min^{-1} mg^{-1} of protein and all other enzymes in nmol min^{-1} mg^{-1} of protein. Reproduced with permission from de Boer et al. (1990b).

in the culture was measured every 30 min, and the rate of its addition was adjusted to the rate of formaldehyde consumption. In this formaldehyde-fed batch-culture experiment, the synthesis of HPS and HPI only occurred after depletion of glucose. Although the rate of formaldehyde addition in the latter phase was gradually increased to 2.5 mmol l^{-1} hr^{-1}, no residual formaldehyde could be detected any more in culture supernatants (de Boer et al. 1990b).

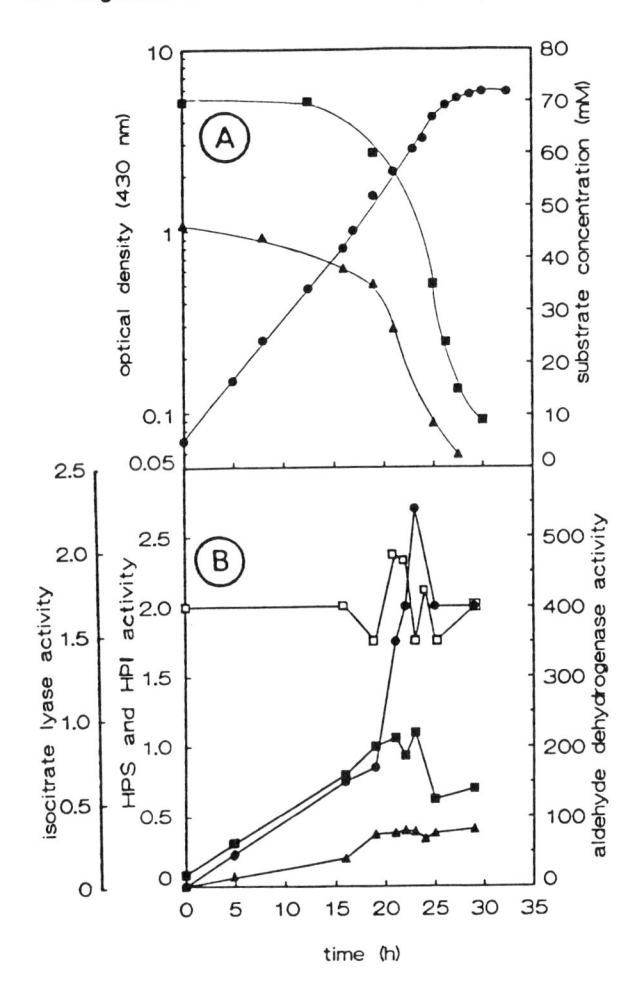

6.2 Regulation of the Assimilation of C₁ Compounds in Bacteria

FIGURE 6-3 Growth of *Amycolatopsis methanolica* on a mixture of acetate (40 mM) and methanol (70 mM) in batch culture. Cells were pregrown on acetate. Legends as in Figure 6-2, except: ▲, acetate concentration; □, isocitrate lyase (expressed in μmol min⁻¹ mg⁻¹ of protein). Reproduced with permission from de Boer et al. (1990b).

Regulation of the synthesis of the RuMP cycle enzymes thus appears to be different in the two facultative methylotrophs investigated. During growth on methylamine or on formaldehyde, *Arthrobacter* P1 does not possess a linear pathway for formaldehyde oxidation, but employs the so-called dissimilatory RuMP cycle of formaldehyde oxidation. In *A. methanolica*, formaldehyde is an intermediate in both the dissimilatory and assimilatory pathways. In *Arthrobacter* P1, synthesis of HPS and HPI is solely controlled via induction by formaldehyde. In *A. methanolica*, the inducing effect of

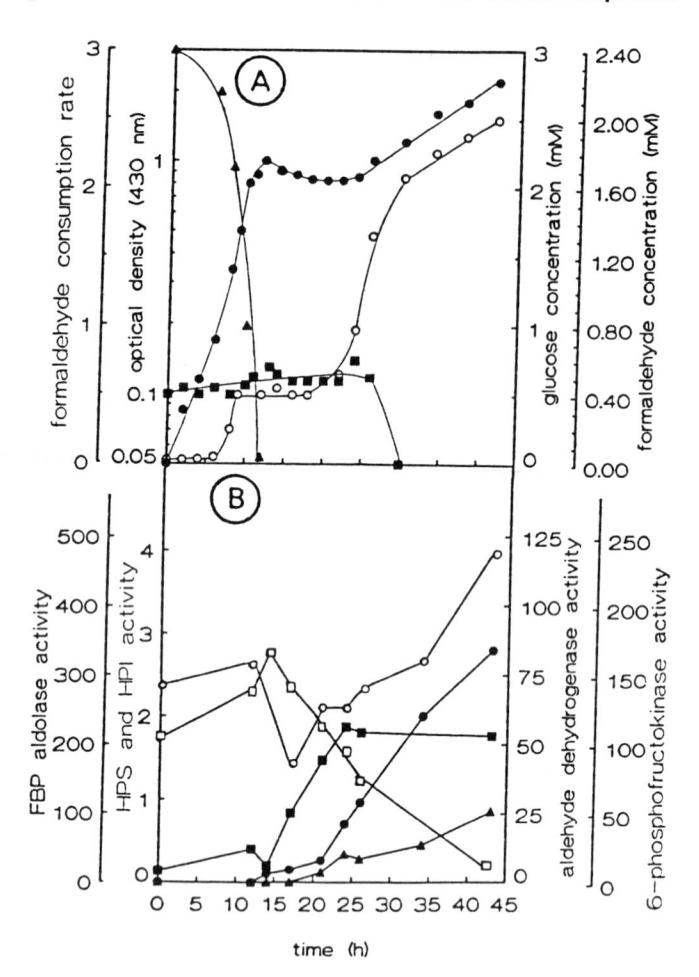

FIGURE 6-4 Growth of *Amycolatopsis methanolica* on a mixture of glucose (3 mM) and formaldehyde in a formaldehyde-fed batch culture. Cells were pregrown on glucose. The formaldehyde supply rate was adjusted to the rate of formaldehyde consumption by the culture to maintain a concentration in the culture of approximately 0.5 mM. (A) ●, optical density; ▲, glucose concentration; ■, formaldehyde concentration; ○, formaldehyde consumption rate (expressed in mmol l^{-1} hr^{-1}). (B) ●, HPS; ▲, HPI; ■, aldehyde dehydrogenase; ○, fructose-1,6-bisphosphate (FBP) aldolase; □, 6-phosphofructokinase. HPS and HPI activities are expressed in μmol min^{-1} mg^{-1} of protein; all other enzymes are expressed in nmol min^{-1} mg^{-1} of protein. Reproduced with permission from de Boer et al. (1990b).

formaldehyde on the synthesis of the RuMP cycle enzymes is overruled in the presence of glucose. This reduces methanol to the status of an ancillary energy source, as is the case in C_1-compound-utilizing autotrophic bacteria (Dijkhuizen and Harder 1984 and 1985). The nature of the control mechanisms involved in RuMP cycle methylotrophs thus may depend on the particular organization mechanism used for C_1 metabolism, namely the linear versus the cyclic pathway for formaldehyde oxidation. In both cases the unperturbed utilization of formaldehyde appears to be ensured, avoiding accumulation of this highly toxic compound.

REFERENCES

Agrawal, P., and Flores, G.E. (1989) *Biotechnol. Bioeng.* 33 (1), 104–114.
Alefounder, P.R., and Ferguson, S.J. (1981) *Biochem. Biophys. Res. Commun.* 98 (3), 778–784.
Anthony, C. (1982) *The Biochemistry of Methylotrophs*, Academic Press, London.
Anthony, C. (1986) *Adv. Microb. Physiol.* 27, 113–210.
Anthony, C. (1989) in *Microbial Growth on C_1 Compounds* Abstr. 6th Int. Symp., Göttingen, FRG.
Anthony, C., and Jones, C.W. (1987) in *Microbial Growth on C_1 Compounds* Proc. 5th Int. Symp. (van Verseveld, H.W., and Duine, J.A., eds.), pp. 195–202, Martinus Nijhoff Publishers, Dordrecht, The Netherlands.
Arfman, N., Watling, E.M., Clement, W., et al. (1989) *Arch. Microbiol.* 152 (3), 280–288.
Attwood, M.M., and Quayle, J.R. (1984) in *Microbial Growth on C_1 Compounds* Proc. 4th Int. Symp. (Crawford, R.L., and Hanson, R.S., eds.), pp. 315–323, American Society for Microbiology, Washington DC.
Beardsmore, A.F., Aperghis, P.N.G., and Quayle, J.R. (1982) *J. Gen. Microbiol.* 128 (7), 1423–1439.
Bellion, E., and Wayland, L. (1982) *J. Bacteriol.* 149 (1), 395–398.
Bellion, E., Ali Khan, M.Y., and Romano, M. (1980) *J. Bacteriol.* 142 (3), 786–790.
Bellion, E., Kent, M.E., Aud, J.C., Ali Khan, M.Y., and Bolbot, J.A. (1983) *J. Bacteriol.* 154 (3), 1168–1173.
Borisov, N., and Netrusov, A. (1989) *Arch. Microbiol.* 152 (2), 201–205.
Bosma, G. (1989) Ph.D. thesis, Free University, Amsterdam.
Brooke, A.G., and Attwood, M.M. (1984) *J. Gen. Microbiol.* 130 (3), 459–463.
Burrows, K.J., Cornish, A., Scott, D., and Higgins, I.J. (1984) *J. Gen. Microbiol.* 130 (12), 3327–3333.
Burton, S.M., Byrom, D., Carver, M., Jones, G.D.D., and Jones, C.W. (1983) *FEMS Microbiol. Lett.* 17 (2), 185–190.
Bussineau, C.M., and Papoutsakis, E.T. (1986) *Appl. Microbiol. Biotechnol.* 24 (6), 435–442.
Bussineau, C.M., and Papoutsakis, E.T. (1988) *Arch. Microbiol.* 149 (3), 214–219.
Bussineau, C.M., Keuer, T.A., Chu, I.-M., and Papoutsakis, E.T. (1987) *Appl. Microbiol. Biotechnol.* 26 (1), 61–69.
Carver, M.A., Humphrey, K.M., Patchett, R.A., and Jones, C.W. (1984) *Eur. J. Biochem.* 138 (3), 611–615.
Chu, I.-M., and Papoutsakis, E.T. (1985) *Biotechnol. Lett.* 7 (1), 15–20.
Chu, I.-M., and Papoutsakis, E.T. (1987a) *Biotechnol. Bioeng.* 29 (1), 55–64.
Chu, I.-M., and Papoutsakis, E.T. (1987b) *Biotechnol. Bioeng.* 29 (1), 65–71.

Cornish, A., Niccholls, K.M., Scott, D., et al. (1984) *J. Gen. Microbiol.* 130 (10), 2565-2575.
Dalton, H. (1981) in *Microbial Growth on C_1 Compounds* Proc. 3rd Int. Symp. (Dalton H., ed.), pp. 1-10, Heyden, London.
Dalton, H., and Higgins, I.J. (1987) in *Microbial Growth on C_1 Compounds* Proc. 5th Int. Symp (van Verseveld, H.W., and Duine, J.A., eds.), pp. 89-94, Martinus Nijhoff Publishers, Dordrecht, The Netherlands.
Dalton, H., and Leak, D.J. (1985) in *Microbial Gas Metabolism* (Poole, R.K., and Dow, C.S., eds.), pp. 173-200, SGM Special Publication no. 14, Academic Press, London.
Dalton, H., Prior, S.D., Leak, D.J., and Stanley, S.H. (1984) in *Microbial Growth on C_1 Compounds* Proc. 4th Int. Symp. (Crawford, R.L., and Hanson, R.S., eds.), pp. 75-82, American Society for Microbiology, Washington DC.
Davidson, V.L., and Kumar, M.A. (1989) *FEBS Lett.* 245 (1,2), 271-273.
Davis, K.J., Cornish, A., and Higgins, I.J. (1987) *J. Gen. Microbiol.* 133 (2), 291-297.
Dawson, M.J., and Jones, C.W. (1982) *Arch. Microbiol.* 133 (1), 55-61.
de Boer, L., Brouwer, J.W., van Hassel, C.W., Levering, P.R., and Dijkhuizen, L. (1989) *Antonie van Leeuwenhoek* 56 (3), 221-232.
de Boer, L., Dijkhuizen, L., Grobben, G., et al. (1990a) *Int. J. Syst. Bacteriol.* 40 (2), 194-204.
de Boer, L., Euverink, G.J., van der Vlag, J., and Dijkhuizen, L. (1990b) *Arch. Microbiol.* 153 (4), 337-343.
de Vries, G.E. (1986) *FEMS Microbiol. Rev.* 39 (3), 235-258.
de Vries, G.E., Harms, N., Maurer, K., Papendrecht, A., and Stouthamer, A.H. (1988) *J. Bacteriol.* 170 (8), 3731-3737.
de Vries, G.E., Kües, U., and Stahl, U. (1990) *FEMS Microbiol. Rev.* 75, 57-102.
Dijkhuizen, L., and Arfman, N. (1990) in *Microbial Growth on C_1 Compounds* Proc. 6th Int. Symp., *FEMS Microbiol. Rev.* 87, 215-220.
Dijkhuizen, L., and Harder, W. (1984) *Antonie van Leeuwenhoek* 50 (5/6), 473-487.
Dijkhuizen, L., and Harder, W. (1985) in *Comprehensive Biotechnology*, vol. 1 (Moo-Young, M., ed.), pp. 409-423, Pergamon Press, Oxford, UK.
Dijkhuizen, L., and Levering, P.R. (1987) in *Microbial Growth on C_1 Compounds* Proc. 5th Int. Symp. (van Verseveld, H.W., and Duine, J.A., eds.), pp. 95-104, Martinus Nijhoff Publishers, Dordrecht, The Netherlands.
Dijkhuizen, L., de Boer, L., Boers, R.H., Harder, W., and Konings, W.N. (1982) *Arch. Microbiol.* 133 (4), 261-266.
Dijkhuizen, L., Levering, P.R., and de Vries, G.E. (1990) in *The Methane and Methanol Utilizers*, Biotechnology Handbook (Dalton, H., and Murrell, J.C., eds.), (in press) Plenum Publishing Corporation, New York.
Diwan, A.R., Chu, I.-M., and Papoutsakis, E.T. (1983) *Biotechnol. Lett.* 5 (9), 579-584.
Duine, J.A., Frank, J., and Berkhout, M.P.J. (1984) *FEBS Lett.* 168 (2), 217-221.
Dunstan, P.M., Anthony, C., and Drabble, W.T. (1972) *Biochem. J.* 128 (1), 107-115.
Eggeling, L., and Sahm, H. (1984) *FEMS Microbiol. Lett.* 25 (3), 253-257.
Eggeling, L., and Sahm, H. (1985) *Eur. J. Biochem.* 150 (1), 129-134.
Ford, S., Page, M.D., and Anthony, C. (1985) *J. Gen. Microbiol.* 131 (9), 2173-2182.
Fuchs, G. (1986) *FEMS Microbiol. Rev.* 39 (3), 181-213.

Goodwin, P.M. (1990) in *Advances in Autotrophic Microbiology and One-Carbon Metabolism*, vol. 1 (Codd, G.A., Dijkhuizen, L., and Tabita, F.R., eds.), pp. 153–172, Kluwer Academic Publishers, Dordrecht, The Netherlands.
Green, J., and Dalton, H. (1985) *J. Biol. Chem.* 260 (29), 15795–15801.
Green, J., Prior, S.D., and Dalton, H. (1985) *Eur. J. Biochem.* 153 (1), 137–144.
Greenwood, J.A., and Jones, C.W. (1986) *J. Gen. Microbiol.* 132 (5), 1247–1256.
Gvozdev, R.I., Nikonova, E.L., Pilyashenko-Novokhatniy, A.I., et al. (1982) *Biochimiya* (in Russian) 47 (7), 1118–1124.
Harms, N., de Vries, G.E., Maurer, K., Veltkamp, E., and Stouthamer, A.H. (1985) *J. Bacteriol.* 164 (3), 1064–1070.
Hazeu, W., de Bruyn, J.C., and van Dijken, J.P. (1983) *Arch. Microbiol.* 135 (2), 205–210.
Heijthuijsen, J.H.F.G., and Hansen, T.A. (1990) in *Advances in Autotrophic Microbiology and One-Carbon Metabolism*, vol. 1 (Codd, G.A., Dijkhuizen, L., and Tabita, F.R., eds.), pp. 173–204, Kluwer Academic Publishers, Dordrecht, The Netherlands.
Holtel, A., and Kleiner, D. (1985) *Arch. Microbiol.* 142 (3), 285–288.
Hussain, M., and Davidson, V.L. (1985) *J. Biol. Chem.* 260 (27), 14626–14629.
Joergensen, L., and Degn, H. (1987) *Biotechnol. Lett.* 9 (1), 71–76.
Jones, C.W., Greenwood, J.A., Burton, S.M., Santos, H., and Turner, D.L. (1987) *J. Gen. Microbiol.* 133 (6), 1511–1519.
Kasprzak, A.A., and Steenkamp, D.J. (1983) *J. Bacteriol.* 156 (1), 348–353.
Kiriuchin, M.Y., Kletsova, L.V., Chistoserdov, A.Y., and Tsygankov, Y.D. (1988) *FEMS Microbiol. Lett.* 52 (3) 199–204.
Kletsova, L.V., Chibisova, E.S., and Tsygankov, Y.D. (1988) *Arch. Microbiol.* 149 (5), 441–446.
Köstler, M., and Kleiner, D. (1989) *FEMS Microbiol. Lett.* 65 (1), 1–4.
Large, P.J. (1981) in *Microbial Growth on C_1 Compounds* Proc. 3rd Int. Symp. (Dalton, H., ed.), pp. 55–69, Heyden, London.
Large, P.J., and Bamforth, C.W. (1988) *Methylotrophy and Biotechnology*, Longman, Harlow, UK.
Lawton, S.A., and Anthony, C. (1985) *J. Gen. Microbiol.* 131 (9), 2165–2171.
Leak, D.J., and Dalton, H. (1986a) *Appl. Microbiol. Biotechnol.* 23 (6), 470–476.
Leak, D.J., and Dalton, H. (1986b) *Appl. Microbiol. Biotechnol.* 23 (6), 477–481.
Levering, P.R., and Dijkhuizen, L. (1986) *Arch. Microbiol.* 144 (2), 116–123.
Levering, P.R., van Dijken, J.P., Veenhuis, M., and Harder, W. (1981) *Arch. Microbiol.* 129 (1), 72–80.
Levering, P.R., Dijkhuizen, L., and Harder, W. (1984) *Arch. Microbiol.* 139 (2), 188–195.
Levering, P.R., Croes, L.M., and Dijkhuizen, L. (1986) *Arch. Microbiol.* 144 (3), 272–278.
Levering, P.R., Tiesma, L., Woldendorp, J.P., Steensma, M., and Dijkhuizen, L. (1987) *Arch. Microbiol.* 146 (4), 346–352.
Linton, J.D., and Vokes, J. (1978) *FEMS Microbiol. Lett.* 4 (1), 125–128.
Lund, J., Woodland, M.P., and Dalton, H. (1985) *Eur. J. Biochem.* 147 (2), 297–305.
Marison, I.W., and Attwood, M.M. (1982) *J. Gen. Microbiol.* 128 (7), 1441–1446.
McNerney, T., and O'Connor, M.L. (1980) *Appl. Environ. Microbiol.* 40 (2), 370–375.

McPheat, W.L., Mann, N.H., and Dalton, H. (1987) *Arch. Microbiol.* 148 (1), 40–43.
Meijer, W.G., Croes, L.M., Jenni, B., Lehmicke, L.G., Lidstrom, M.E., and Dijkhuizen, L. (1990) *Arch. Microbiol.* 153 (4), 360–367.
O'Connor, M.L. (1981) in *Microbial Growth on C_1 Compounds* Proc. 3rd Int. Symp. (Dalton, H., ed.), pp. 294–300, Heyden, London.
O'Connor, M.L., and Hanson, R.S. (1977) *J. Gen. Microbiol.* 101 (2), 327–332.
Page, M.D., and Anthony, C. (1986) *J. Gen. Microbiol.* 132 (6), 1553–1563.
Patchett, R.A., Quilter, J.A., and Jones, C.W. (1985) *Arch. Microbiol.* 141 (1), 95–102.
Pinchuk, G.E. (1989) in *Microbial Growth on C_1 Compounds* Abstr. 6th Int. Symp., Göttingen, FRG.
Pinchuk, G.E., Sokolov, I.G., and Malashenko, Y.R. (1989) in *Microbial Growth on C_1 Compounds* Abstr. 6th Int. Symp., Göttingen, FRG.
Prior, S.D., and Dalton, H. (1985) *J. Gen. Microbiol.* 131 (1), 155–163.
Reed, W.M., and Dugan, P.R. (1987) *J. Gen. Microbiol.* 133 (5), 1389–1395.
Roitsch, T., and Stolp, H. (1985) *Arch. Microbiol.* 142 (1), 34–39.
Roitsch, T., and Stolp, H. (1986) *Arch. Microbiol.* 144 (3), 245–247.
Samuelov, N., and Goldberg, I. (1982) *Biotechnol. Bioeng.* 24 (3), 731–736.
Sokolov, I.G. (1986) *Microbiologiya* (in Russian) 55 (5), 715–722.
Sokolov, I.G. (1988) *Microbiologiya* (in Russian) 57 (5), 716–723.
Sokolov, I.G., Malashenko, Y.R., Karpenko, V.I., and Kryshtab, T.P. (1979) *Ukr. Biochim. Zh.* (in Russian) 51 (4), 393–399.
Sokolov, I.G., Malashenko, Y.R., and Romanovskaya, V.A. (1981) *Microbiology* (English translation) 50 (1), 7–13.
Sokolov, I.G., Stolyar, S.M., Kryshtab, T.P., and Pinchuk, G.E. (1989) *Microbiol. Zh.* (in Russian) 51 (2), 38–46.
Stanley, S.H., Prior, S.D., Leak, D.J., and Dalton, H. (1983) *Biotechnol. Lett.* 5 (7), 487–492.
Tabita, F.R. (1988) *Microbiol. Rev.* 52 (2), 155–189.
Tobari, J. (1984) in *Microbial Growth on C_1 Compounds* Proc. 4th Int. Symp. (Crawford, R.L., and Hanson, R.S., eds.), pp. 106–112, American Society for Microbiology, Washington DC.
Tonge, G.M., Knowles, C.J., Harrison, D.E.F., and Higgins, I.J. (1974) *FEBS Lett.* 44 (1), 106–110.
Tonge, G.M., Harrison, D.E.F., Knowles, C.J., and Higgins, I.J. (1975) *FEBS Lett.* 58 (1), 293–299.
van Ophem, P.W., and Duine, J.A. (1990) *Arch. Biochem. Biophys.* 282 (2), 248–253.
van Vliet-Smits, M., Harder, W., and van Dijken, J.P. (1981) *FEMS Microbiol. Lett.* 11 (1), 31–35.
Weaver, C.A., and Lidstrom, M.E. (1985) *J. Gen. Microbiol.* 131 (9), 2183–2197.
Weaver, C.A., and Lidstrom, M.E. (1987) *J. Gen. Microbiol.* 133 (7), 1721–1731.
Zatman, L.J. (1981) in *Microbial Growth on C_1 Compounds* Proc. 3rd Int. Symp. (Dalton, H., ed.), pp. 42–54, Heyden, London.

CHAPTER 7

Growth Yields, Productivities, and Maintenance Energies of Methylotrophs

L.E. Erickson
P. Tuitemwong

Methylotrophs are important microorganisms; they can be used to produce useful products such as single cell protein and methane, and they can be used to biodegrade waste products. The bioenergetics of methylotrophs have been considered by a number of earlier writers (Van Dijken and Harder 1975; Anthony 1978; Anthony 1982; Lee et al. 1984; Leak and Dalton 1986; Anthony and Jones 1987; Erickson and Fung 1988; Large and Bamforth 1988). A unified treatment of the available experimental data and a summary of the results will be presented in this chapter. The energy and composition regularities employed by Minkevich and Eroshin (1973) are used in the analysis of the data; this allows bioenergetic data from different substrates to be compared.

Although the research described in this chapter has been funded in part by the United States Environmental Protection Agency under assistance agreement number R-815709 to the Hazardous Substance Research Center for U.S. EPA Regions 7 and 8 headquartered at Kansas State University, it has not been subjected to the Agency's peer and administrative review and therefore may not necessarily reflect the views of the agency and no official endorsement should be inferred. The Kansas State University Center for Hazardous Substance Research provided partial support.

7.1 BASIC BIOENERGETIC CONCEPTS

Minkevich and Eroshin (1973), Erickson et al. (1978a), Roels (1983), Erickson (1987), and Erickson and Fung (1988) have developed and described the basic bioenergetic concepts that are utilized in this review.

7.1.1 Regularities

The history of the energy regularity has been reviewed by Erickson (1987). For a wide variety of organic compounds, the heat of combustion per equivalent of oxygen utilized is relatively constant and equal to approximately -110 kJ/equivalent of oxygen (or per equivalent of available electrons). The Gibbs free energy of combustion per equivalent of available electrons does not differ greatly from the heat of combustion for many of the organic compounds of interest. This regularity has a smaller variance for larger molecules than for small molecules. For example, methane has a free energy of combustion of -102.3 kJ/eq under standard conditions. It is this smaller absolute value which allows microorganisms to convert a large number of organic substrates to methane under anaerobic conditions. Table 7–1 gives the heat of combustion Q and free energy of combustion g for several substrates of importance and for microbial biomass.

Minkevich and Eroshin (1973), Minkevich et al. (1977), and Erickson (1980) have reviewed the literature with respect to the elemental composition of biomass. The value 0.462 is a reasonable average value for the weight fraction of carbon in microbial biomass. For microbial biomass, the reductance degree (the equivalents of available electrons per g mole of carbon) has an average value of 4.291; that is, 4.291 equivalents of oxygen are required per g mole of carbon to oxidize microbial biomass to carbon dioxide, water, and ammonia.

TABLE 7–1 Heat of Combustion and Gibbs Free Energy of Combustion in kJ per Equivalent of Available Electrons

Substance	$-(Heat\ of\ Combustion,\ Q)$	$-(Free\ Energy\ of\ Combustion,\ g)$
Methane	111	102
Methanol	121	117
Formaldehyde	135	131
Formic acid	135	138
Biomass	113	113

Data adapted from Roels (1983) and Weast (1975).

7.1.2 Balances and Yields for Aerobic Growth

Consider the following balance equation for aerobic growth and product formation (Erickson 1987):

$$CH_mO_k + aNH_3 + bO_2 = y_cCH_pO_nN_q + zCH_rO_sN_t + cH_2O + dCO_2 \quad (7.1)$$

where CH_mO_k denotes the elemental composition of the organic substrate, $CH_pO_nN_q$ is the elemental composition of the biomass, and $CH_rO_sN_t$ is the elemental composition of the extracellular product. The carbon balance is (Erickson et al. 1978a):

$$y_c + z + d = 1.0 \quad (7.2)$$

where y_c, z, and d give the fraction of organic substrate carbon converted to biomass, extracellular product, and CO_2, respectively. The available electron balance may be written in the form (Erickson et al. 1978a)

$$\frac{4b}{\gamma_s} + y_c\frac{\gamma_b}{\gamma_s} + z\frac{\gamma_p}{\gamma_s} = 1.0 \quad (7.3)$$

provided the physiological dead state is used to define the reductance degree, γ. When the valences $C = 4$, $H = 1$, $O = -2$, and $N = -3$ are used, NH_3, CO_2, and H_2O have zero available electrons and the equations (Erickson et al. 1978a)

$$\gamma_s = 4 + m - 2k \quad (7.4)$$

$$\gamma_b = 4 + p - 2n - 3q \quad (7.5)$$

$$\gamma_p = 4 + r - 2s - 3t \quad (7.6)$$

give the reductance degree of the substrate, biomass, and product, respectively, based on their elemental composition. The regularity value of $\gamma_b = 4.291$ has been used as an average value for equation 7.5 when biomass elemental composition is not available (Erickson 1980).

The available electron balance may be written in the form:

$$\epsilon + \eta + \xi_p = 1.0 \quad (7.7)$$

by defining $\epsilon = 4b/\gamma_s$ as the fraction of available electrons in the organic substrate which is transferred to oxygen, $\eta = y_c\gamma_b/\gamma_s$ as the biomass yield in available electron units, and $\xi_p = z\gamma_p/\gamma_s$ as the product yield in available electron units.

The enthalpy and free energy are quantities which are defined relative to a selected dead state or zero energy state. If the physiological dead state is defined such that the ammonia, water, and carbon dioxide have zero energy level, then energetic yields analogous to the available electron yields can be written for enthalpy.

$$\eta_c = \frac{y_c \gamma_b Q_b}{\gamma_s Q_s} \tag{7.8}$$

$$\xi_{pc} = \frac{z \gamma_p Q_p}{\gamma_s Q_s} \tag{7.9}$$

where Q refers to the enthalpy per equivalent of available electrons. For free energy

$$\eta_{th} = \frac{y_c \gamma_b g_b}{\gamma_s g_s} \tag{7.10}$$

$$\xi_{th} = \frac{z \gamma_p g_p}{\gamma_s g_s} \tag{7.11}$$

where g refers to the free energy per equivalent of available electrons relative to the selected dead state. If the regularity value is used and $Q_b = Q_p = Q_s = g_b = g_s = g_p = Q_0$ is assumed, then the energetic yields are equal to the available electron yields. For this case, the heat of fermentation associated with Equation 7.1 is directly related to ϵ; that is,

$$\Delta H_F = 4bQ_0 \tag{7.12}$$

where ΔH_F is the heat of fermentation per mol of carbon substrate consumed and Equation 7.12 makes use of Thornton's rule (Patel and Erickson 1981). Some representative values of Q_s are given in Table 7–1.

7.1.3 Growth Kinetics, Balances, and Yields for Anaerobic Growth

The bioenergetics of anaerobic growth and product formation are described in general form by Erickson (1988a); the treatment which follows is very similar to that presented previously.

Anaerobic growth and product formation in which the ATP formed is the result of substrate phosphorylation has been extensively investigated (Erickson and Fung 1988). One of the unique aspects of this fermentation is that biomass growth, substrate consumption, and product formation are very closely coupled (Roels 1983; Oner et al. 1984; Erickson 1988a). When no electron acceptors are present, the available electrons initially present in the substrate are distributed to the product and the biomass, with the distribution controlled by the bioenergetics of the process. Consider the anaerobic production of a simple product, such as methane, by anaerobic fermentation. The chemical balance equation is (Erickson 1988a)

7.1 Basic Bioenergetic Concepts

$$CH_mO_l + aNH_3 = y_c CH_pO_nN_q + zCH_rO_sN_t + cH_2O + dCO_2 \quad (7.13)$$

where CH_mO_l, $CH_pO_nN_q$, and $CH_rO_sN_t$ give the elemental compositions of the carbon, hydrogen, oxygen, and nitrogen in the substrate, biomass, and extracellular product, respectively. Using the valences C = 4, H = 1, O = -2, and N = -3, the available electron balance is (Minkevich and Eroshin 1973; Erickson et al. 1978a)

$$y_c \frac{\gamma_b}{\gamma_s} + z \frac{\gamma_p}{\gamma_s} = 1.0 \quad (7.14)$$

or

$$\eta + \xi_p = 1.0 \quad (7.15)$$

where γ is the reductance degree (see Equations 7.4, 7.5, and 7.6).

Luedeking and Piret (1959a and 1959b) reported that product formation kinetics and growth kinetics were closely related; product formation was modeled with two parameters, a growth-associated term and a maintenance-associated term; that is,

$$\frac{dP}{dt} = \alpha \frac{dX}{dt} + \beta X \quad (7.16)$$

where X is biomass concentration and P is extracellular product concentration. Growth and product formation are dependent upon the availability of substrate; the Monod model (Erickson 1988a)

$$\frac{1}{X}\frac{dX}{dt} = \mu = \frac{\mu_{max} S}{K_s + S} \quad (7.17)$$

is commonly used to relate growth rate to substrate concentration. Here S is substrate concentration, μ is specific growth rate, μ_{max} is maximum specific growth rate, and K_s is the saturation constant. The organism uses substrate for growth and maintenance according to the linear model for substrate consumption (Pirt 1975; Roels 1983; Oner et al. 1984; Erickson 1988a); that is,

$$-\frac{dS}{dt} = \frac{1}{Y_{x/s}^{max}} \frac{dX}{dt} + m_s X \quad (7.18)$$

where $Y_{x/s}^{max}$ is the true growth yield corrected for maintenance and m_s is the maintenance coefficient. In this process, product formation occurs. Thus, the product formation kinetic parameters and the bioenergetic parameters are related. Oner et al. (1984) show that Equation 7.18 may be written as (Erickson 1988a)

$$\frac{\mu}{\eta} = \frac{\mu}{\eta_{max}} + m_e \tag{7.19}$$

and Equation 7.16 may be written in the form

$$\frac{\mu}{\eta}\xi_p = \frac{\mu}{\eta}(1-\eta) = \alpha_e\mu + \beta_e \tag{7.20}$$

where

$$m_e = \beta_e = m_s\frac{\sigma_s\gamma_s}{\sigma_b\gamma_b} = \beta\frac{\sigma_p\gamma_p}{\sigma_b\gamma_b} \tag{7.21}$$

$$\frac{1}{\eta_{max}} = \alpha_e + 1 = \frac{\sigma_s\gamma_s}{\sigma_b\gamma_b Y_{x/s}^{max}} = \alpha\frac{\sigma_p\gamma_p}{\sigma_b\gamma_b} + 1 \tag{7.22}$$

The true growth yield must satisfy the relationship (Erickson 1988a)

$$\eta_{max} = \frac{(\sigma_b\gamma_b/12)\ Y_{ATP}^{max}}{(\sigma_b\gamma_b/12)\ Y_{ATP}^{max} + \delta} \tag{7.23}$$

where Y_{ATP}^{max} is the true growth yield based on ATP, δ is the equivalents of available electrons transferred to products to produce 1 mol of ATP from ADP, σ_b is the weight fraction carbon in biomass, and γ_b is the reductance degree of the biomass. Knowledge of the theoretical maximum yield with respect to ATP and the value of δ for a particular fermentation allows one to estimate the acceptable range of values of the growth-associated kinetic term in the product formation model. For $\delta = 12$, equivalents of available electrons per mole of ATP generated, $Y_{ATP}^{max} = 28.8$ g cells per mol ATP, $\sigma_b = 0.462$, and $\gamma_b = 4.291$, $\alpha_e \geq 2.52$.

In summary, there are four independent kinetic parameters (μ_{max}, K_s, η_{max}, and m_e) needed to describe growth and product formation in fermentations with one chemical balance equation and one product that contains available electrons. The estimation of the parameters for such a system is considered in detail in several chapters of Erickson and Fung (1988).

Anthony (1982) and Large and Bamforth (1988) have reviewed the theoretical ATP requirements for formation of biomass from several different substrates for methylotrophs. Figure 7-1, which is based on Equation 7.23, shows how the biomass true growth yield η_{max} varies with δ, the number of equivalents of available electrons transferred to products to produce 1 mol of ATP from ADP. The growth yield based on ATP corrected for maintenance is the parameter varied. The appropriate value of δ for a particular growth process depends on the substrate and products as well as the microorganism. For many of the ATP-producing processes associated with an-

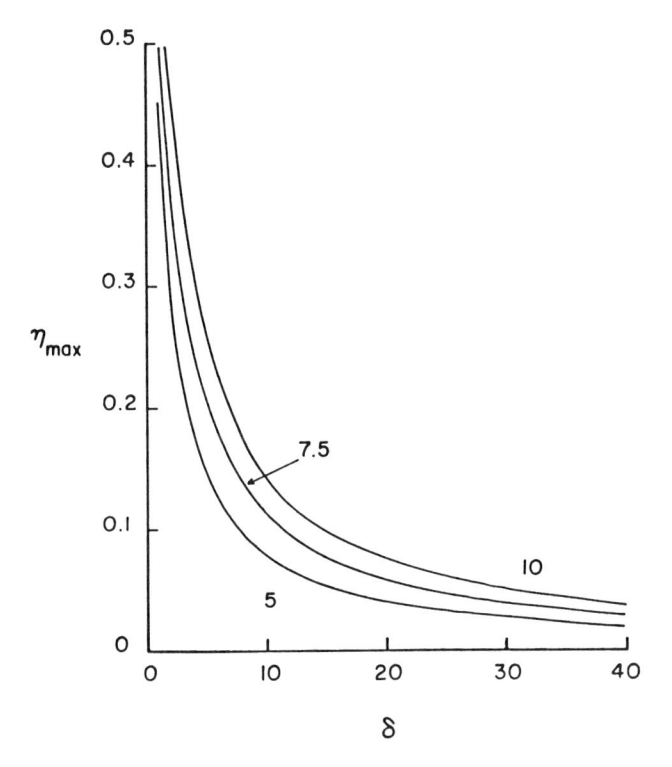

[Figure: Plot of η_{max} vs δ with three curves labeled 5, 7.5, and 10.]

FIGURE 7-1 Variation of the true available electron yield η_{max} with δ (the equivalents of available electrons transferred to products to produce 1 g mol ATP from ADP). The parameter varied in the three curves is Y_{ATP}^{max} in g cells dry weight per mole ATP.

aerobic methane production, the values fall in the range from 12 to 24 equivalents of available electrons to produce 1 mol of ATP from ADP.

For anaerobic processes, the free energy yield associated with the growth process has been defined as follows (Erickson and Oner 1983; Oner et al. 1988)

$$\eta_{th} = \frac{g_b \eta}{g_s - g_p \xi_p} \qquad (7.24)$$

where g is the free energy per equivalent of available electrons relative to the dead state associated with combustion.

7.1.4 Statistical Estimation of Yield Parameters

The statistical estimation of kinetic and yield parameters is treated in detail by Erickson et al. (1988). Erickson and Yang and co-workers have developed the covariate adjustment method to estimate yield and maintenance parameters simultaneously using all of the available experimental data. These methods have been employed to estimate the true growth yield and the maintenance coefficient for several sets of data in which methylotrophic organisms are growing aerobically on several different substrates. Based on the yield model of Pirt (1975), Lee et al. (1984) employ the available electron form of these balances. For the case in which there is no extracellular product, Form I of the balances may be written as follows (Lee et al. 1984):

$$\frac{1}{\eta} = \frac{1}{\eta_{max}} + \frac{m_e}{\mu} \qquad (7.25)$$

Two related equations are obtained by using the equality constraints from Equations 7.2 and 7.7 as follows for $z = \xi_p = 0$:

$$\frac{(y_c+d)}{\eta} = \frac{1}{\eta_{max}} + \frac{m_e}{\mu} \qquad (7.26)$$

$$\frac{(\eta+\epsilon)}{\eta} = \frac{1}{\eta_{max}} + \frac{m_e}{\mu} \qquad (7.27)$$

Two forms of Equations 7.25 through 7.27 are used for parameter estimation. In this work, Equations 7.25 through 7.27 are referred to as Form I. Form II is obtained by multiplying each term in Equations 7.25 through 7.27 by μ.

Several different estimates of the parameters from Equations 7.25 through 7.27 can be obtained. By treating the three equations as a multivariate linear model with common parameters, a method of analysis of growth curves which utilizes the technique of analysis of covariance may be applied (Yang et al. 1984; Erickson et al. 1988). To apply this method, Equations 7.25 through 7.27 are linearized as follows:

for Form I

$$Y_{ji} = \psi + m_e X_i \quad \begin{matrix} j = 1, 2, 3 \\ i = 1, 2, \ldots, N \end{matrix} \qquad (7.28)$$

for Form II

$$Y_{ji} = m_e + \psi \mu_i \quad \begin{matrix} j = 4, 5, 6 \\ i = 1, 2, \ldots, N \end{matrix} \qquad (7.29)$$

where

$$Y_{1i} = \frac{1}{\eta_i}, \ Y_{2i} = \frac{(Y_{ci}+d_i)}{\eta_i}, \ Y_{3i} = \frac{(\eta_i+\epsilon_i)}{\eta_i},$$

$$Y_{4i} = \frac{\mu_i}{\eta_i}, \ Y_{5i} = \frac{\mu_i(Y_{ci}+d_i)}{\eta_i},$$

$$Y_{6i} = \frac{\mu_i(\eta_i+\epsilon_i)}{\eta_i}, \ \psi = \frac{1}{\eta_{max}}, \ X_i = \frac{1}{\mu_i}$$

and N is the number of observations for each variable. The three responses Y_{1i}, Y_{2i}, and Y_{3i} contain information from the experimental measurements. The average of these values for Form I

$$\bar{Y} \cdot i = \frac{1}{3} \sum_{j=1}^{3} Y_{ji} \qquad (7.30)$$

may be used to estimate the parameters m_e and ψ; that is

$$\bar{Y} \cdot i = \psi + m_e X_i + \text{error} \qquad (7.31)$$

However, a conditional model, which includes the error structure of the equality constraints,

$$\bar{Y} \cdot i = \psi + m_e X_i + \sum_{k=1}^{q} \alpha_k Z_{ki} + \text{error} \ (1 \leq q \leq 2) \qquad (7.32)$$

may provide better estimates. In Equation 7.32, Z_{ki} are covariates which are selected such that they have zero expected value. For example,

$$Z_{1i} = Y_{1i} - 2Y_{2i} + Y_{3i} \qquad (7.33)$$

$$Z_{2i} = Y_{1i} - Y_{3i} \qquad (7.34)$$

provide a set of covariates with zero expected value. The coefficients, α_k, are estimated in the regression analysis. The maximum likelihood estimate is obtained when the maximum number of covariates is included (two in this case, with two equality constraints). Since an additional degree of freedom is introduced with each covariate that is introduced, the parameter estimates with shortest confidence interval may arise from using zero, one, or two covariates. Each case may be considered and the results may be compared using the shortness of the confidence interval as one of the measures of the quality of the estimate (Lee et al. 1984).

An analogous procedure is used with Form II. By using multiple linear regression methods, estimates of $\psi = 1/\eta_{\max}$ and m_e may be obtained for both Form I and Form II.

7.2 AEROBIC GROWTH AND PRODUCT FORMATION

Methylotrophic microorganisms grow both aerobically and anaerobically. Table 7-2 lists a number of compounds that are biodegraded by methylotrophs and indicates the literature citation and the organism and medium involved. For some of the compounds in the table such as carbon tetrachloride, there is no evidence that the compound can be used as the sole source of carbon and energy for growth; however, because of the importance of methylotrophs in bioremediation, these compounds are included in the table. A number of chlorinated aliphatic hydrocarbons are present as contaminants in soil and ground water at a large number of sites. There is great

TABLE 7-2 Compounds That Can Serve as Substrates for Methylotrophic Microorganisms

Compound	Formula	Reference and Remarks
Methane[1]	CH_4	Ehhalt (1976); methanogenic bacteria Leak and Dalton (1986); *Methylococcus capsulatus*, aerobic, mineral salts medium Sheehan and Johnson (1971); mixed cultures, aerobic, minimal medium Vary and Johnson (1967); mixed cultures, aerobic, minimal medium
Methanol[1]	CH_3OH	Allais and Baratti (1983); thermotolerant yeast, *Hansenula polymorpha* CMB 11, aerobic, mineral salts medium Hamer et al. (1979); mixed cultures, aerobic, minimal medium Goldberg et al. (1976), *Pseudomonas* sp., aerobic, minimal medium Linton and Vokes (1978); mixed cultures, aerobic, mineral salts medium
Formaldehyde	HCHO	Goldberg et al. (1976); *Pseudomonas* sp., aerobic, minimal medium
Formate	HCOOH	Goldberg et al. (1976); *Pseudomonas* sp., aerobic, minimal medium Mell et al. (1982); *Vibrio succinogenes*, anaerobic, complex medium
Carbon monoxide	CO	Meyer (1985); *Pseudomonas* sp., aerobic Vega et al. (1989a); *Peptostreptococcus productus*, anaerobic, complex medium
Chloromethane	CH_3Cl	Hartmans et al. (1986); *Hyphomicrobium* sp., aerobic, minimal medium
Dichloromethane	CH_2Cl_2	Stucki et al. (1981); *Hyphomicrobium* DM2, aerobic, minimal medium LaPat-Polasko et al. (1984); *Pseudomonas* LP, aerobic, minimal medium Henson et al. (1989); mixed cultures, aerobic, minimal medium Brunner et al. (1980); pure culture, strain DM1, aerobic, minimal medium
Trichloromethane (chloroform)[2]	$CHCl_3$	Strand and Shippert (1986); mixed cultures, aerobic soil exposed to natural gas Bouwer and McCarty (1983a); anaerobic Bouwer and McCarty (1985); mixed cultures, aerobic, complex medium

(continued)

7.2 Aerobic Growth and Product Formation

TABLE 7-2 (Continued)

Compound	Formula	Reference and Remarks
Carbon tetrachloride[2]	CCl_4	Henson et al. (1989); mixed cultures, aerobic, minimal medium Parsons and Lage (1985); mixed cultures, anaerobic, complex medium Bouwer and McCarty (1983a, 1983b); anaerobic Bouwer and McCarty (1985); mixed cultures, aerobic, complex medium
Monomethylamine	CH_3NH_2	Brunner et al. (1980); strain DM1, aerobic, minimal medium
Dimethylamine	$(CH_3)_2NH$	Ghisalba and Küenzi (1983b) Walther et al. (1981); *Methanosarcina barkeri*, anaerobic, minimal medium
Dimethyl sulfoxide	$(CH_3)_2SO$	Suylen and Kuenen (1986); mixed cultures (*Hyphomicrobium* EG) and *Thiobacillus* MS1, aerobic, minimal medium de Bont et al. (1981); *Hyphomicrobium* S, aerobic, minimal medium
Trimethyl sulfonium salts	$(CH_3)_3S^+$	Large and Bamforth (1988)
Methyl sulfate	CH_3OSO_3H	Ghisalba and Küenzi (1983a)
Dimethyl sulfate	$(CH_3O)_2SO_2$	Large and Bamforth (1988)
Formamide	$HCONH_2$	Large and Bamforth (1988)
Dimethyl formamide	$HCON(CH_3)_2$	Schär et al. (1986)
Hydrogen cyanide	HCN	Knowles (1976)
Trimethylamine	$(CH_3)_3N$	Barrett and Kwan (1985)
Tetramethyl ammonium salts	$(CH_3)_4N^+$	Ghisalba and Küenzi (1983b)
Trimethylamine N-oxide	$(CH_3)_3NO$	Large and Bamforth (1988)
Methanethiol	CH_3SH	Kanagawa and Mikami (1989); *Thiobacillus thioparus* TK-m, aerobic, complex medium
Dimethyl sulfide	$(CH_3)_2S$	de Bont et al. (1981); *Hyphomicrobium* S, aerobic, minimal medium Kanagawa and Mikami (1989); *Thiobacillus thioparus* TK-m, aerobic, complex medium

(continued)

TABLE 7-2 (Continued)

Compound	Formula	Reference and Remarks
Dimethyl sulfide (continued)	$(CH_3)_2S$	Suylen and Kuenen (1986); *Thiobacillus* MS1 and mixed cultures (*Hyphomicrobium* EG), aerobic, minimal medium
Dimethyl disulfide	$(CH_3)_2S_2$	Kanagawa and Mikami (1989); *Thiobacillus thioparus* TK-m, aerobic, complex medium

[1] Many other references are not included.
[2] These are compounds that are biodegraded by methylotrophs when growth is supported by other substrates.

interest in the use of methanotrophs for the biodegradation of these chlorinated aliphatic hydrocarbons (Wilson and Wilson 1985; Fogel et al. 1986; Strand and Shippert 1986; Erickson and Fan 1988; Henson et al. 1989; Oldenhuis et al. 1989). Methane has been used to support the growth of the culture in these studies; however, Oldenhuis et al. (1989) reported better results with formate as cosubstrate. Investigators interested in biodegradation have not examined growth yield and maintenance requirements of the cultures as extensively as those interested in cell production. However, maintenance requirements are an important consideration in bioremediation because of the need to achieve low concentrations of the contaminants which are subjected to biodegradation. If maintenance requirements are too large, growth cannot be sustained at low concentrations of the growth-limiting substrate. This area of bioenergetics is in need of further research.

The microbial removal of sulfur-containing compounds is also of environmental interest (Kanagawa and Mikami 1989). Methanethiol, dimethyl sulfide, dimethyl disulfide, and hydrogen sulfide can be removed from contaminated air by *Thiobacillus thioparus* TK-m. As shown in Table 7-3, the available electron yields were 0.55 for methanethiol, 0.55 for dimethyl disulfide, and 0.29 for dimethyl sulfide in this study based on the sulfur-containing substrate in a complex medium.

Many of the important yield and maintenance results for methylotrophs growing aerobically have been generated because of interest in producing single cell protein and other useful products. Tables 7-3, 7-4, and 7-5 provide summary information from representative studies of microbial biomass production. A number of investigators have conducted research on aerobic product formation by methylotrophs; however, yield data from these studies is not included here because of the limited amount of information that is available. Large and Bamforth (1988) have reviewed the production of amino acids, nucleotides, and other complex organic chemicals by methylotrophs. Lin et al. (1989) investigated the microbial production of the

7.2 Aerobic Growth and Product Formation

herbicide 5-aminolevulinic acid; however, better yields are required for commercialization of the process.

Of all the potential substrates for single cell protein production, the one of greatest interest is methanol. Methane is another potential substrate; however, the possibility of explosive mixtures of methane and oxygen, the high oxygen and heat transfer requirements, and the low solubility of methane in water have discouraged investigators from attempting to commercialize microbial protein from methane processes. Data on both substrates is presented in Table 7-3. Table 7-4 provides true growth yield and maintenance coefficient estimates for methanol from several experimental studies; the estimates are from Lee et al. (1984). The consistency of the data of Held et al. (1978) is presented by Lee et al. (1984). Where sufficient data is available, the covariate adjustment method is employed, and all of the data from a set of experiments are analyzed simultaneously. The largest true growth yield estimates were obtained from the data of Dostálek and Molin (1975) who cultured *Methylomonas methanolica*. *Candida boidinii* was cultured by Held et al. (1978); *Hansenula polymorpha* was used by Cooney and Makiguchi (1977), Erickson et al. (1978b), and Allais and Baratti (1983); Jara et al. (1983) cultured *Pichia pastoris*. The data in Tables 7-4 and 7-5 is for *Pseudomonas* AM-1 (Tsuchiya 1982). All of the true growth yield estimates in Table 7-4 are less than the theoretical value of 0.63 based on the analysis of Anthony (1978) for $Y_{ATP} = 10.5$ g cell dry weight per mole of ATP and the ribulose monophosphate pathway (Lee et al. 1984).

In Table 7-5, estimates of true growth yields and maintenance coefficients for *Pseudomonas* AM-1 growing on methanol, formaldehyde, and formate are compared. Because of the small number of data points, the shortest confidence intervals are those obtained using the covariate adjustment method when no covariates are included (Lee et al. 1984).

In Figure 7-2, the carbon growth yield, y_c, and the available electron growth yield, η, are presented graphically for formate, formaldehyde, methanol, and methane. The carbon yield appears to increase linearly with the reductance degree, γ_s, up to $\gamma_s = 6$. These results are similar to those presented elsewhere for aerobic microbial growth on all substrates (Erickson 1987). The results in Figure 7-2 indicate that aerobic growth yields on formate and formaldehyde are limited by the availability of energy.

The results in Table 7-5 show yields based on free energy, η_{th}^{max}, as well as yields based on available electrons, η_{max}. The true growth yield based on free energy is similar for formaldehyde and methanol. This is to be expected since the theoretical free energy yields are 0.54 and 0.55, respectively, for methanol and formaldehyde based on the serine pathway yields of 0.64 g/g and 0.52 g/g reported by Anthony (1978). The true growth yield based on available electrons, η_{max}, is about the same for methanol and formate; these experimental values may be compared to theoretical values of 0.56 and 0.57 for methanol and formate, respectively, based on Anthony's values of 0.64 g/g and 0.15 g/g. The ratio of the observed true growth yields to the theo-

Growth Yields, Productivities, and Maintenance Energies of Methylotrophs

TABLE 7-3 Aerobic Growth Yields of Methylotrophs on Various Substrates

Substrate	γ_s	Y_s	y_c	η	Reference and Remarks
Methane	8	0.5–0.69	0.308–0.425	0.165–0.228	Leak and Dalton (1986); *Methylococcus capsulatus*, KNO_3 as N source
		0.67–0.75	0.413–0.462	0.222–0.248	NH_4Cl as N source
		0.574–0.641	0.354–0.395	0.190–0.212	Sheehan and Johnson (1971); mixed culture
		0.75–0.78	0.462–0.480	0.248–0.258	Linton and Vokes (1978); *Methylococcus* sp. NCIB 11083
		0.565	0.348	0.187	Vary and Johnson (1967); mixed culture
		1.00	0.616	0.337	Harwood and Pirt (1972); *Methylococcus capsulatus*
Methanol	6	0.54	0.665	0.476	Goldberg et al. (1976); *Pseudomonas* C.
		0.53	0.653	0.467	*Pseudomonas methylotropha*
		0.49	0.604	0.432	*Methylomonas methanolica*
		0.38	0.468	0.335	*Pseudomonas* 1, *Pseudomonas* 135
		0.30	0.370	0.264	*Pseudomonas* AM-1
		0.41	0.505	0.361	Vary and Johnson (1967); *Pseudomonas* M-27, *Pseudomonas rosea*
Formaldehyde	4	0.32	0.370	0.396	Goldberg et al. (1976); *Pseudomonas* 1
		0.27	0.312	0.334	*Pseudomonas* 135
		0.24	0.277	0.297	*Pseudomonas* AM-1, *P. rosea*
		0.28	0.323	0.347	*Pseudomonas* M-27

Formate	2	0.140	0.248	0.532	Goldberg et al. (1976); *Pseudomonas* 1
		0.150	0.266	0.570	*Pseudomonas* 135
		0.104	0.184	0.395	*Pseudomonas* AM-1
		0.097	0.172	0.368	*Pseudomonas* M-27
		0.073	0.129	0.227	*P. rosea*
Methylamine	6	0.35	0.418	0.299	Goldberg et al. (1976); *Pseudomonas* 135
		0.30	0.358	0.256	*Pseudomonas* AM-1
		0.37	0.442	0.316	*P. rosea*
Methanethiol	6	0.417	0.770	0.551	Kanagawa and Mikami (1989); *Thiobacillus thioparus* TK-m
Dimethyl sulfide	6	0.338	0.404	0.289	Kanagawa and Mikami (1989); *Thiobacillus thioparus* TK-m
Dimethyl disulfide	5	0.354	0.641	0.550	Kanagawa and Mikami (1989); *Thiobacillus thioparus* TK-m
Chloromethane	6	0.429[1]	0.833	0.596	Hartmans et al. (1986); *Hyphomicrobium* MC1
Dichloromethane	4	0.176	0.605	0.650	Brunner et al. (1980); strain DM1

[1] This value was obtained from the yield of 5 g protein (mol C)$^{-1}$.

TABLE 7-4 Estimates of True Growth Yield, η_{max}, and of Maintenance Coefficient, m_e, for Aerobic Growth on Methanol

Data Used; Source	Covariates Included	Form	η_{max} Point Estimate	η_{max} 95% Confidence Interval	$m_e(h^{-1})$ Point Estimate	$m_e(h^{-1})$ 95% Confidence Interval
All, So = 10 g/L; Held et al. (1978)	None	I	0.327	0.294, 0.369	0.023	0.007, 0.040
		II	0.290	0.260, 0.330	−0.009	−0.053, 0.036
All, So = 30 g/L; Held et al. (1978)	None	I	0.295	0.271, 0.323	0.011	0.006, 0.028
		II	0.249	0.225, 0.279	−0.041	−0.083, 0.002
1/η; Allais and Baratti (1983)	None	I	0.329	0.244, 0.506	0.021	0.087, 0.130
		II	0.280	0.183, 0.596	−0.053	−0.355, 0.249
All; Erickson et al. (1978b)	Z_1	I	0.377	0.286, 0.551	0.014	0.096, 0.124
		II	0.370	0.273, 0.573	0.008	−0.122, 0.137
1/η; Jara et al. (1983)	None	I	0.374	0.367, 0.382	0.010	0.007, 0.013
		II	0.374	0.363, 0.387	0.010	−0.001, 0.018
1/η; Cooney and Makiguchi (1977)	None	I	0.259	0.231, 0.293	−0.019	0.077, 0.039
		II	0.254	0.220, 0.299	−0.030	−0.120, 0.060
All; Dostalek and Molin (1975)	Z_2	I	0.419	0.395, 0.446	0.066	0.007, 0.125
		II	0.452	0.413, 0.498	0.128	0.041, 0.215
All; Tsuchiya et al. (1982)	None	I	0.301	0.269, 0.339	−0.015	−0.057, 0.027
		II	0.299	0.260, 0.351	−0.017	−0.080, 0.046

Adapted from Lee et al. (1984).

7.2 Aerobic Growth and Product Formation

TABLE 7-5 Comparison of True Growth Yield Based on Available Electrons, η_{max}, and Free Energy, η_{max}, and η_{th}^{max}, and Maintenance Coefficient Based on Available Electrons, m_e, and Free Energy, m_{th}, for *Pseudomonas* AM-1 Growing on Methanol, Formaldehyde, and Formate

Substrate	Form	η_{max} Point Estimate	η_{max} 95% Confidence Interval	η_{th}^{max} Point Estimate	η_{th}^{max} 95% Confidence Interval	$m_e (h^{-1})$ Point Estimate	$m_e (h^{-1})$ 95% Confidence Interval	$m_{th} (h^{-1})$ Point Estimate	$m_{th} (h^{-1})$ 95% Confidence Interval
Methanol	I	0.301	0.269, 0.339	0.290	0.260, 0.327	−0.015	−0.057, 0.027	−0.015	−0.055, 0.025
	II	0.299	0.260, 0.351	0.288	0.251, 0.338	−0.017	−0.080, 0.046	−0.016	−0.077, 0.045
Formaldehyde	I	0.337	0.277, 0.397	0.292	0.240, 0.344	0.006	−0.031, 0.042	0.005	−0.027, 0.036
	II	0.333	0.279, 0.413	0.288	0.242, 0.358	0.004	−0.038, 0.045	0.003	−0.033, 0.039
Formate	I	0.311	0.235, 0.459	0.255	0.193, 0.376	0.031	−0.073, 0.135	0.025	−0.060, 0.111
	II	0.306	0.242, 0.417	0.251	0.198, 0.342	0.025	−0.073, 0.124	0.021	−0.060, 0.101

Data adapted from Tsuchiya et al. (1982); results are from Lee et al. (1984).

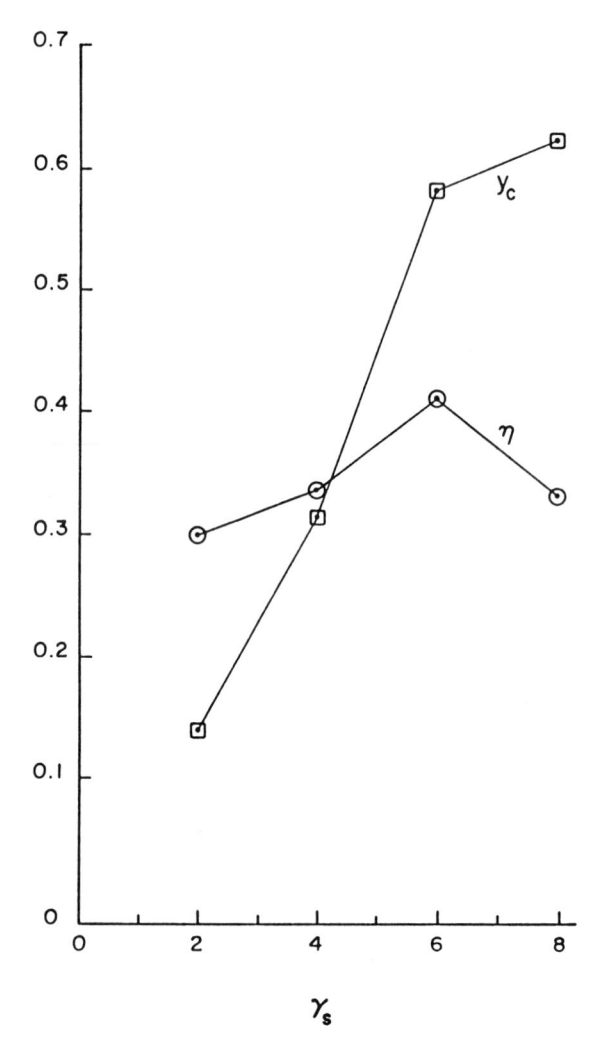

FIGURE 7-2 Variation of biomass carbon yield, y_c, and available electron yield, η, with the reductance degree of the substrate for formate, formaldehyde, methanol, and methane. The data are based on Tsuchiya et al. (1982) and Harwood and Pirt (1972).

retical maximum values calculated based on the work of Anthony (1978) is 0.53 for formaldehyde and 0.54 for methanol and formate. The somewhat-lower true growth yield based on free energy, when the substrate is formate, is not surprising; while formate has a greater free energy per equivalent of available electrons, not all of that energy is effectively used in the conversion process (Lee et al. 1984).

TABLE 7-6 Estimates of True Growth Yield, η_{max}, and Maintenance Parameter, m_e, Based on Parameter Estimates of van Verseveld and Stouthamer (1978) for *Paracoccus denitrificans* Growing on Methanol

	η_{max}		m_e, h^{-1}	
Data Used	Point Estimate	95% Confidence Interval	Point Estimate	95% Confidence Interval
$1/\eta$	0.339	0.306, 0.379	0.0673	0.0277, 0.107
$(\eta + \epsilon)/\eta$	0.370	0.338, 0.408	0.0712	0.0396, 0.100
$(y_c + d)/\eta$	0.326	0.308, 0.347	0.0119	−0.0079, 0.0317

Results reported by Lee et al. (1984). Results are based on Form II and the cell composition of $C_6H_{10.8}N_{1.5}O_{2.9}$ as reported by van Verseveld (1979).

Values of the true growth yield and maintenance coefficient have been estimated for *Paracoccus denitrificans* by van Verseveld and Stouthamer (1978). Their results are presented in available electron units in Table 7-6. Values of true growth yield based on methanol as substrate are similar to several of the values reported in the other tables. However, for the ribulose bisphosphate pathway used by *P. denitrificans*, the theoretical yield estimated by Anthony (1978) is smaller than it is for the ribulose monophosphate and serine pathways. The results of van Verseveld (1979) are about 70% of the theoretical available electron yield value of 0.485, which is obtained from Anthony's value of 0.55 g/g. If one compares the results of van Verseveld with the results of Dostálek and Molin (1975), the ratio of the estimated true growth yield to the respective maximum value predicted by Anthony is about 0.7 for both cases (Lee et al. 1984).

Amano et al. (1983) have reported results with *Methylomonas methanolovorans* growing on methanol. The consistency of their data is relatively good. Conversion of their estimates of true growth yield and maintenance coefficient to available electron units gives $\eta_{max} = 0.44$ and $m_e = 0.029$ h^{-1} (Lee et al. 1984). This value of the true growth yield is similar to that reported for *M. methanolica* for the data of Dostálek and Molin (1975).

7.3 ANAEROBIC GROWTH AND PRODUCT FORMATION

In this section, yield data on the anaerobic production of microbial biomass and methane from several substrates is presented and reviewed. Erickson (1988b) has reviewed biomass yields associated with anaerobic digestion for several other substrates such as acetate, hydrogen, and carbon dioxide. In that work and in this work, the stoichiometry of Equation 7.13 is assumed in order to estimate the biomass-available electron yield, η, from the biomass

TABLE 7-7 Anaerobic Growth Yields of Methylotrophs on Various Substrates

Substrate	γ_s	Organism	$Y_{x/s}$ (g cells per g mol substrate)	$Y_{x/p}$ (g cells per g mol product)	η	Reference and Remarks
Methanol	6	*Methanosarcina* strain 227		5.10	0.095	Smith and Mah (1980); complex media
		Methanosarcina TM-1		4.50	0.085	Zinder and Mah (1979); complex media
		Methanosarcina barkeri	3.4–4.5		0.094–0.124	Scherer and Sahm (1981); complex media
				3.3	0.0637	Stadtman (1967)
				7.2	0.129	Weimer and Zeikus (1978); complex media
				4.36	0.083	Mountfort and Asher (1979); complex media, 1.25 mm sulfide
				1.90	0.038	12.5 mm sulfide
				5.5	0.102	Walther et al. (1981)
Formate	2	*Methanobacterium formicicum*	1.2	4.8	0.098	Schauer and Ferry (1980); complex media
			1.4		0.116	Chua and Robinson (1983); complex media, true growth yield
		Methanobrevibacter arboriphilus	0.81	3.52	0.067	Morii et al. (1983); complex media
Methylamine	6	*Methanosarcina barkeri*		5.8	0.105	Walther et al. (1981)
Dimethylamine	6			6.2	0.113	Walther et al. (1981)
Trimethylamine	6			6.0	0.110	Walther et al. (1981)
Carbon monoxide	2	*Peptostreptococcus productus*	0.95		0.078	Vega et al. (1989a); complex media
				6.40	0.117	Vega et al. (1989b); complex media

Results from Erickson (1988b) and other sources.

and methane production data. The available electron balance allows one to write

$$\eta = \frac{\sigma_b \gamma_b Y_{x/p}/12}{8 + \sigma_b \gamma_b Y_{x/p}/12} \quad (7.35)$$

The average values $\sigma_b = 0.462$ and $\gamma_b = 4.291$ are used in calculating η.

Table 7-7 lists the growth yields for several substrates which support the anaerobic growth of methylotrophs. The available electron yields vary from 0.038 to 0.129 for methanol; however, there is little variation from substrate to substrate. All of the yields for the other substrates fall within this range. Under anaerobic conditions, the product-available electron yield

$$\xi_p = 1 - \eta$$

is often between 0.9 and 1.0. The largest biomass-available electron yield reported in Table 7-7, $\eta = 0.129$, corresponds to $\eta_{th} = 0.55$; thus, little additional biomass yield improvement can be expected.

Sufficient data to estimate the true growth yield and maintenance coefficient in Equation 7.19 have been obtained by Chua and Robinson (1983). For *Methanobacterium formicicum* growing on formate, they found a maintenance coefficient of 6.8 mmol per g dry weight h, which is equal to $m_e = 0.082$ h^{-1}; the true growth yield is 0.116 in available electron units (Erickson 1988b).

Erickson (1988b) reports somewhat higher biomass growth yields for mixed substrates such as methanol and acetate, and free energy yields as high as 0.76.

Erickson and Fan (1988) have reviewed the anaerobic biodegradation of toxic and hazardous substances. Many chlorinated aliphatic hydrocarbons can be biodegraded under anaerobic conditions; however, growth yield data have not been collected in most of these studies.

7.4 CONCLUSIONS

Growth yield data can be examined and compared using available electron concepts. This allows data with different substrates to be compared. For γ_s (reductance degree) of six or less, growth yields appear to be energy limited. Under anaerobic conditions, reported growth yields are relatively good and little yield improvement can be expected. Under aerobic conditions, the better growth yields for methanol are approximately 70% of the values predicted by Anthony (1978).

REFERENCES

Allais, J.J., and Baratti, J. (1983) *J. Ferment. Technol.* 61 (4), 339–345.
Amano, Y., Takada, N., Sawada, H., Sakuma, H., and Terui, G. (1983) *Biotechnol. Bioeng.* 25, 2735–2755.

Anthony, C. (1978) *J. Gen. Microbiol.* 104, 91-104.
Anthony, C. (1982) *The Biochemistry of Methylotrophs.* Academic Press, London.
Anthony, C., and Jones, C.W. (1987) in *Microbial Growth on C_1 Compounds* Proceedings of 5th International Symposium (van Verseveld, H.W., and Duine, J.A., eds.), pp. 195-202, Martinus Nijhoff, Dordrecht, The Netherlands.
Barrett, E.L., and Kwan, H.S. (1985) *Annu. Rev. Microbiol.* 39, 131-149.
Bouwer, E.J., and McCarty, P.L. (1983a) *Appl. Environ. Microbiol.* 45, 1286-1294.
Bouwer, E.J., and McCarty, P.L. (1983b) *Appl. Environ. Microbiol.* 45, 1295-1299.
Bouwer, E.J., and McCarty, P.L. (1985) *Biotechnol. Bioeng.* 27, 1564-1571.
Brunner, W., Staub, D., and Leisinger, T. (1980) *Appl. Environ. Microbiol.* 40, 950-958.
Chua, H.B., and Robinson, J.P. (1983) *Arch. Microbiol.* 135, 158-160.
Cooney, C.L., Makiguchi, N. (1977) *Biotechnol. Bioeng. Symp.* 7, 65-76.
de Bont, J.A.M., van Dijken, J.P., and Harder, W. (1981) *J. Gen. Microbiol.* 127, 315-323.
Dostálek, M., and Molin, N. (1975) in *Single-Cell Protein II* (Tannenbaum, S.R., and Wang, D.I.C., eds.), pp. 385-401, MIT Press, Cambridge, MA.
Ehhalt, D.H. (1976) in *Symposium on Microbial Production and Utilization of Gases (H_2, CH_4, CO)* (Schlegel, H.G., Gottschalk, G., and Pfennig, N., eds.), pp. 13-22, Akademie der Wissenschaften, Göttingen, FRG.
Erickson, L.E. (1980) *Biotechnol. Bioeng.* 22, 451-456.
Erickson, L.E. (1987) in *Thermal and Energetic Studies of Cellular Biological Systems* (James, A.M., ed.), pp. 14-33, Wright, Bristol, England.
Erickson, L.E (1988a) in *Handbook on Anaerobic Fermentations* (Erickson, L.E., and Fung, D.Y.C., eds.), pp. 119-146, Marcel Dekker, New York.
Erickson, L.E. (1988b) in *Handbook on Anaerobic Fermentations* (Erickson, L.E., and Fung, D.Y.C., eds.), pp. 325-344, Marcel Dekker, New York.
Erickson, L.E., and Fan, L.T. (1988) in *Handbook on Anaerobic Fermentations* (Erickson, L.E., and Fung, D.Y.C., eds.), pp. 695-732, Marcel Dekker, New York.
Erickson, L.E., and Fung, D.Y.C. (eds.) (1988) *Handbook on Anaerobic Fermentations*, Marcel Dekker, New York.
Erickson, L.E., and Oner, M.D. (1983) *Ann. N.Y. Acad. Sci.* 413, 99-113.
Erickson, L.E., Minkevich, I.G., and Eroshin, V.K. (1978a) *Biotechnol. Bioeng.* 20, 1595-1621.
Erickson, L.E., Kuvshinnikov, V.D., Minkevich, I.G., and Eroshin, V.K. (1978b) *J. Ferment. Technol.* 56, 524-531.
Erickson, L.E., Yang, S.S., and Oner, M.D. (1988) in *Handbook on Anaerobic Fermentations* (Erickson, L.E., and Fung, D.Y.C., eds.), pp. 463-497, Marcel Dekker, New York.
Fogel, M.M., Taddeo, A.R., and Fogel, S. (1986) *Appl. Environ. Microbiol.* 51, 720-724.
Ghisalba, O., and Küenzi, M. (1983a) *Experientia* 39, 1257-1263.
Ghisalba, O., and Küenzi, M. (1983b) *Experientia* 39, 1264-1271.
Goldberg, I., Rock, J.S., Ben-Bassat, A., and Mateles, R.I. (1976) *Biotechnol. Bioeng.* 18, 1657-1668.
Hamer, G., Pal, H.S., and Hamdan, I.Y. (1979) *Biotechnol. Lett.* 1, 9-14.
Hartmans, S., Schmuckle, A., Cook, A.M., and Leisinger, T. (1986) *J. Gen. Microbiol.* 132, 1139-1142.
Harwood, J.H., and Pirt, S.J. (1972) *J. Appl. Bacteriol.* 35, 597-607.

References

Held, W., Schlanderer, G., Reimann, J., Dellweg, H. (1978) *Eur. J. Appl. Microbiol. Biotechnol.* 6, 127–132.
Henson, J.M., Yares, M.V., and Cochran, J.W. (1989) *J. Ind. Microbiol.* 4, 29–35.
Jara, P., Allais, J.J., and Baratti, J. (1983) *Eur. J. Appl. Microbiol. Biotechnol.* 17, 19–23.
Kanagawa, T., and Mikami, E. (1989) *Appl. Environ. Microbiol.* 55 (3), 555–558.
Knowles, C.J. (1976) *Bacteriol. Rev.* 40, 652–680.
LaPat-Polasko, L.T., McCarty, P.L., and Zehnder, A.J.B. (1984) *Appl. Environ. Microbiol.* 47, 825–830.
Large, P.J., and Bamforth, C.W. (1988) *Methylotrophy and Biotechnology*, Longman Scientific and Technical, Harlow, Essex, England.
Leak, D.J., and Dalton, H. (1986) *Appl. Environ. Microbiol.* 23, 470–476.
Lee, H.Y., Erickson, L.E., and Yang, S.S. (1984) *J. Ferment. Technol.* 62 (4), 341–351.
Luedeking, R., and Piret, E.L. (1959a) *Biotechnol. Bioeng.* 1, 393–412.
Luedeking, R., and Piret, E.L. (1959b) *Biotechnol. Bioeng.* 1, 431–459.
Lin, D., Nishio, N., and Nagai, S. (1989) *J. Ferment. Bioeng.* 68 (2), 88–91.
Linton, J.D., and Vokes, J. (1978) *FEMS Microbiol. Lett.* 4, 125–128.
Mell, H., Bronder, M., and Kröger, A. (1982) *Arch. Microbiol.* 131, 224–228.
Meyer, O. (1985) in *Microbial Gas Metabolism* (Poole, R.K., and Dow, C.S., eds.), pp. 131–151, Academic Press, London.
Minkevich, I.G., and Eroshin, V.K. (1973) *Folia Microbiol. (Praha)* 18, 376–385.
Minkevich, I.G., Eroshin, V.K., Alekseena, T.A., and Tereshchenko, A.P. (1977) *Mikrobiol. Promst.* (in Russian) 2 (144), 1–3.
Morii, H., Nishihara, M., and Koga, Y. (1983) *Agric. Biol. Chem.* 47, 2781–2789.
Mountfort, D.O., and Asher, R.A. (1979) *Appl. Environ. Microbiol.* 37, 670–675.
Oldenhuis, R., Vink, R.L.J.M., Janssen, D.B., and Witholt, B. (1989) *Appl. Environ. Microbiol.* 55 (11), 2819–2826.
Oner, M.D., Erickson, L.E., and Yang, S.S. (1984) *Biotechnol. Bioeng.* 26, 1436–1444.
Oner, M.D., Erickson, L.E., and Yang, S.S. (1988) in *Handbook on Anaerobic Fermentations* (Erickson, L.E., and Fung, D.Y.C., eds.), pp. 293–323, Marcel Dekker, New York.
Parsons, F.Z., and Lage, G.B. (1985) *J. Am. Water Works Assoc.* 77 (5), 52–59.
Patel, S.A., and Erickson, L.E. (1981) *Biotechnol. Bioeng.* 23, 2051–2067.
Pirt, S.J. (1975) *Principles of Microbe and Cell Cultivation*, Wiley, New York.
Roels, J.A. (1983) *Energetics and Kinetics in Biotechnology*, Elsevier, New York.
Schär, H.P., Holtzmann, W., Ramos-Tombo, G.M., and Ghisalba, O. (1986) *Eur. J. Biochem.* 158, 469–475.
Schauer, N.L., and Ferry, J.G. (1980). *J. Bacteriol.* 142, 800–807.
Scherer, P., and Sahm, H. (1981) *Appl. Microbiol. Biotechnol.* 12, 28–35.
Sheehan, B.T., and Johnson, M.T. (1971) *Appl. Microbiol.* 21 (3), 511–515.
Smith, M.R., and Mah, R.A. (1980) *Appl. Environ. Microbiol.* 39, 993–999.
Stadtman, T.C. (1967) *Annu. Rev. Microbiol.* 21, 121–142.
Strand, S.E., and Shippert, L. (1986) *Appl. Environ. Microbiol.* 52, 203–205.
Stucki, G., Brunner, W., Staub, D., and Leisinger, T. (1981) in *Microbial Degradation of Xenobiotics and Recalcitrant Compounds* (Leisinger, T., Cook, A.M., Hutter, R., and Nuesch, J., eds.), pp. 131–137, Academic Press, London.
Suylen, G.M.H., and Kuenen, J.G. (1986) *Antonie van Leeuwenhoek (J. Microbiol.)* 52, 281–293.

Tsuchiya, Y., Nishio, N., Roldán, H.M., and Nagai, S. (1982) *J. Ferment. Technol.* 60, 333–341.
van Dijken, J.P., and Harder, W. (1975) *Biotechnol. Bioeng.* 27, 15–30.
Vary, P.S., and Johnson, M.J. (1967) *Appl. Microbiol.* 15 (6), 1473–1478.
Vega, J.L., Clausen, E.C., and Gaddy, J.L. (1989a). *Biotechnol. Bioeng.* 34 (6), 774–784.
Vega, J.L., Antorrena, G.M., Clausen, E.C., and Gaddy, J.L. (1989b) *Biotechnol. Bioeng.* 34 (6), 785–793.
van Verseveld, H.W. (1979) Ph.D. dissertation, Free University, Amsterdam, The Netherlands.
van Verseveld, H.W., and Stouthamer, A.H. (1978) *Arch. Microbiol.* 118, 21–26.
Walther, R., Fiebig, K., Fahlbusch, K., et al. (1981) in *Microbial Growth on C_1 Compounds* (Dalton, H., ed.), pp. 146–151, Heyden, London.
Weast, R.C. (ed.) (1975) *CRC Handbook of Chemistry and Physics*, pp. D274–D279, CRC Press, Boca Raton, FL.
Weimer, P.J., and Zeikus, J.G. (1978) *Arch. Microbiol.* 119, 49–57.
Wilson, J.T., and Wilson, B.H. (1985) *Appl. Environ. Microbiol.* 49, 242–243.
Yang, S.S., Solomon, B.O., Oner, M.D., and Erickson, L.E. (1984) *Technometrics* 26, 355–361.
Zinder, S.H., and Mah, R.A. (1979) *Appl. Environ. Microbiol.* 38, 996–1008.

CHAPTER 8

Mixed Substrates and Mixed Cultures

T. Egli
C.A. Mason

8.1 Introduction

In most natural environments, microorganisms are rarely present as pure cultures but are usually part of a mixed culture existing in an environment containing an enormous spectrum of different nutrients potentially capable of satisfying identical physiological requirements. This is especially true in the case of heterotrophic microorganisms for which a wide range of carbon and energy compounds are simultaneously present at low concentrations. In contrast to the natural environment, laboratory batch-type growth represents an extreme case of growth conditions where microorganisms are exposed to abnormally high nutrient levels. It has been demonstrated in several cases over the last decade that the phenomena of enzyme regulatory induction/repression, which have been so nicely worked out for batch cultivation, are rarely relevant under natural environmental growth conditions (reviewed by Harder and Dijkhuizen 1976 and 1982; Wanner and Egli 1990). Indeed, it has been shown for many microorganisms that simultaneous utilization of substrates occurs during cultivation in continuous culture and that this behavior can be taken as the rule rather than the exception under natural growth conditions, thereby conferring selective advantages. In this respect, methylotrophs are no exception, and a number of studies have been carried out using methylotrophs to study the phenomena of mixed substrate growth.

In this context it is necessary to define what is understood by the term "mixed substrate utilization" as it will be used in this chapter: it means the simultaneous utilization by a microorganism of two or more substrates which fulfill identical physiological functions, e.g., as a source of carbon, energy, or nitrogen (Table 8–1). With respect to methylotrophs it should be pointed out that this definition also includes the case of a non-methylotroph utilizing methanol as an energy source. The term mixed substrate utilization is difficult to define when a carbon substrate, e.g., as described for choline by Levering et al. (1981), is internally broken down into carbon skeletons, because this can lead to the concerted induction of catabolic pathways normally, during classical single substrate-limited growth, not active simultaneously in such cells. This latter aspect will not be considered here.

Similarly the presence of "pure" cultures is the exception under natural conditions, and only a few extreme environments have been described that contain mono-bacterial species. More common is the coexistence of a range of different physiological types of microorganisms fulfilling different ecological roles and existing either in harmony or antagonistically in constant dynamic variation. Understanding of this competition lags significantly behind understanding of pure cultures because of the complexity of such systems and due to the ease of carrying out and interpreting experiments from pure cultures. Nevertheless a good working knowledge of mixed culture

TABLE 8–1 Some Examples of Mixtures of Substrates Able to Potentially Fulfill the Same Physiological Function in the Growth of C_1-utilizing Microorganisms

Mixed Substrate	Physiological Function	Type of Organism
Glucose + methanol	Mixed carbon/energy sources	Facultative methylotroph
Glucose + methane	Mixed carbon/energy sources	Facultative methanotroph
Glucose + methanol	Mixed energy sources	C_1-dissimilating non-methylotroph
Methanol + formaldehyde	Mixed carbon/energy sources	Obligate or facultative methylotroph
Methanol + formate	Mixed energy sources	Obligate methylotroph or C_1-dissimilating non-methylotroph
Methylamine + ammonia	Mixed nitrogen sources	Methazotroph

interactions is required, for example, in industrial applications, where satellite organisms are often present, and it is of fundamental importance to study the competitiveness of these contaminants under process conditions. On the other hand, efficient and effective application of mixed cultures are desirable objectives in processes such as wastewater treatment.

Defined culture conditions (both batch and continuous) in combination with selected mixtures of substrates have been shown to be beneficial for improving culture performance and that these can be further used to manipulate metabolism so that microorganisms can be exploited for product formation (Harder et al. 1987; Harder and Brooke 1990). In this review, aspects of mixed cultures and mixed substrate growth are summarized for the case of the methylotrophic bacteria and yeasts. Especially in the case of the methylotrophic yeasts a considerable amount of information has been built up over the last 20 years, which has now found application in the industrial sector (Tani et al. 1987) (see Chapter 11).

8.2 METHYLOTROPHIC BACTERIA

8.2.1 Mixed Substrate Utilization by Pure Cultures

8.2.1.1 Obligate Methylotrophs. Many obligate methylotrophic bacteria (including methanotrophs) are known to simultaneously assimilate small amounts of carbon from multi-carbon compounds, which do not support growth on their own, when growing on a one-carbon (C_1) substrate. However, when non-C_1 compounds were assimilated, their fraction rarely exceeded 5–10% of the total carbon assimilated into new biomass. Typical examples of this include the assimilation of ethanol, acetate, or amino acids by *Methylococcus capsulatus* growing exponentially on methane (Ecclestone and Kelly 1972 and 1973) or the two obligate methylotrophic strains 4B6 and C2A1 growing in batch culture on trimethylamine (Colby and Zatman 1975). In both cases, label from ethanol or acetate was assimilated primarily into the lipid fraction of the organisms. Although carbon balances are incomplete in all experiments reported, and the labelled CO_2 produced was not measured, some of the data suggests that the metabolized carbon originating from these non-growth substrates was primarily assimilated and little was oxidized to CO_2. In the case of methanotrophs, it has been clearly demonstrated that assimilation of carbon from non-growth substrates was dependent on the presence of an energy source (Patel et al. 1975 and 1979). Utilization of mixtures of C_1 compounds by obligate methylotrophs in continuous culture has been described for *Methylomonas* L3 growing on methanol/formaldehyde or formate (Hirt et al. 1978; Papoutsakis et al. 1981). Not unexpectedly the growth yield on the mixture was less than that on methanol alone, and this is consistent with the fact that formaldehyde and formate yield less biochemical energy than methanol (Anthony 1987). Carbon from formate could not be fixed, since formaldehyde is the assimilatory entry point for carbon in all of the methylotrophic bacteria which assimilate C_1 substrates via the ribulose monophosphate pathway. The results from such growth experiments suggested that *Methylomonas* L3 was unable to use the reducing equivalents derived from the oxidation of formate as a supplementary source of energy. This is in contrast to the obligate methylotroph *Pseudomonas* C (which also uses the ribulose monophosphate cycle for formaldehyde fixation) which was reported to grow on mixtures of methanol and formate. At a dilution rate of 0.3 h^{-1} the methanol-based yield coefficient increased by up to 12% with increasing fraction of formate supplied, the latter being oxidized via a NADH-dependent formate dehydrogenase (Ben-Bassat et al. 1980). When the molar ratio of formate to methanol exceeded 0.4, no further enhancement of the yield coefficient occurred. Virtually identical assimilation patterns for ^{14}C-labelled formate and $^{14}CO_2$ suggested that formate was not assimilated at the level of formaldehyde but served purely as an additional source of energy. In the same way, glucose could be partially utilized as an energy source in the presence of methanol with only a minor fraction (2%) of the metabolized glucose being assimilated into biomass (Samuelov and Goldberg 1982a and 1982b). The metabolic significance of co-metabolism of glucose by *Pseudomonas* C in this case remains undefined.

8.2.1.2 Restricted Methylotrophs.

The restricted methylotrophs are a group of organisms capable of utilizing both C_1 compounds and a limited range of multi-carbon substrates. Small amounts of non-growth substrates, when supplied together with a growth supporting C_1 carbon source, can be metabolized by restricted methylotrophs and partially incorporated into biomass in the same manner as described above for the case of the obligate methylotrophs (Colby and Zatman 1975; Egli and Harder 1984). The best studied examples of mixed substrate growth by a restricted methylotroph are probably those of various strains of *Hyphomicrobium* species. For example, significant amounts of ^{14}C-labelled acetate, succinate, pyruvate, or malate were assimilated during growth on methanol (or ethanol) in batch culture by *Hyphomicrobium* X (Harder et al. 1975; Harder and Attwood 1978). However, a more detailed study of the growth of *Hyphomicrobium* sp. strains X and ZV622 in both methanol- or ethanol-limited chemostat culture revealed that addition of succinate, malate, aspartate, or a mixture of amino acids resulted in no significant increase in biomass (Meiberg 1979). Details concerning the metabolic fate of these compounds is lacking. Enzymatic analysis indicated that the absence of pyruvate dehydrogenase in these organisms, as in obligate methylotrophs (Anthony 1987), was the reason for the restricted range of growth substrates which could be utilized. This was confirmed later by cloning pyruvate dehydrogenase from *Escherichia coli* into *Hyphomicrobium* X, after which growth on succinate and pyruvate was possible (Dijkhuizen et al. 1984). A detailed study on the utilization of mixtures of methylamine and ethanol by *Hyphomicrobium* X demonstrated that in batch culture, this organism was able to simultaneously assimilate both substrates under conditions where the initial concentration of methylamine was lower than approximately 9 mM. At higher methylamine concentrations, synthesis of ethanol dehydrogenase was repressed which resulted in the preferential utilization of methylamine (Brooke and Attwood 1983). When the methylamine concentration dropped below this level, ethanol utilization occurred. The data indicates that it was not the dependence on the C:N ratio which caused repression of ethanol dehydrogenase. In chemostat culture, both substrates were utilized simultaneously and to completion, independently of both mixture composition and dilution rate up to $D = 0.06$ h^{-1}. During mixed substrate growth, the biomass yield was additive. The specific activities of enzymes involved in the metabolism of the two substrates correlated well with the specific consumption rates for methylamine and ethanol (Brooke and Attwood 1985). It might be of interest to point out that in this case, methylamine did not serve simply as a carbon source but also as a nitrogen source (despite the presence of excess ammonia in the growth medium).

Recently, both the growth of the thermotolerant *Bacillus* sp. NCIB 12522 on mixtures of methanol and formaldehyde and its susceptibility to pulses of formaldehyde were investigated by Al-Awadhi and co-workers (Al-Awadhi 1989; Al-Awadhi et al. 1990a, 1990b, and 1990c). Although ^{14}C-labelled formaldehyde was not used, the data indicated that during growth in carbon-limited chemostat cultures ($D = 0.2$ h^{-1}) this thermotolerant methylotroph was capable of assimilating methanol and formaldehyde simultaneously up to a formaldehyde fraction in the mixture of 35% w/w. The calculated yields for methanol and formaldehyde were 0.47 and 0.38 respectively, which are consistent with the fact that assimilation of formaldehyde proceeds via the ribulose monophosphate pathway. The biomass yield on the mixture was additive. When formaldehyde or methanol were added as pulses to steady state cultures of *Bacillus* sp. NCIB 12522 growing on methanol, the bacteria were incapable of assimilating the excess added substrate, which was dissimilated entirely to CO_2. The size of the pulse affected the survival of the culture, e.g., when methanol pulses resulting in a final concentration of approximately 300 mg/l were given to a methanol-limited chemostat culture operating at $D = 0.2$ h^{-1}, the complete conversion of methanol to formaldehyde and the subsequent death and washout of the culture was observed (Al-Awadhi et al. 1990c). Cultures grown on mixtures of methanol/formaldehyde were more resistant to pulses of methanol.

8.2.1.3 Facultative Methylotrophs.
Colby and Zatman (1975) demonstrated that various strains of facultative methylotrophs were able to assimilate a range of heterotrophic carbon sources during batch growth with meth-

ylamine with no indication of sequential utilization. Here methylamine probably served as a combined carbon/nitrogen source or primarily as a nitrogen source, and the surplus carbon was assimilated to adjust the C:N ratio as necessary. Attention should be drawn to the fact that even in the presence of NH_4^+, the use of a combined C/N compound as the nitrogen source can permit faster growth (Wanner and Egli 1990). Simultaneous utilization of methanol and succinate has also been reported for *Pseudomonas* sp. strain AM1 and *Methylobacterium organophilum* XX growing in batch culture (O'Connor and Hanson 1977; O'Connor 1981). Recently, growth of the facultative methylotroph *Nocardia* sp. 239 in chemostat cultures on mixtures of methanol with glucose, ethanol, or acetate has been reported (de Boer et al. 1990). Simultaneous utilization of both substrates irrespective of the mixture composition or enforced dilution rate occurred, although no details were given. Anaerobic acidogenic bacteria capable of growth on reduced C_1 compounds have also been described, in which CO_2 serves as a co-substrate during methanol fermentation. The rate of growth of *Eubacterium limosum* growing on methanol/CO_2 mixtures could be influenced by addition of formate, and there are some indications that the type of fermentation end-products excreted by such anaerobic methylotrophs can be manipulated by using different substrate mixtures for fermentative growth (Loubière et al. 1987; Lindley et al. 1987).

By far the best-studied example of mixed substrate growth of a facultative methylotroph is that of the fast-growing, versatile methylotroph, *Arthrobacter* sp. strain P1. Its growth on mixtures of acetate and methylamine in both batch and continuous cultures and under transient state growth conditions has been investigated (Levering and Dijkhuizen 1985; Levering et al. 1986a). When it is supplied in batch culture with both acetate and methylamine in the absence of any other source of nitrogen, methylamine served both as a nitrogen and carbon/energy source while acetate was simultaneously utilized as a carbon/energy substrate. The uptake of both substrates was carefully controlled in a way such that the intracellular carbon:nitrogen ratio was regulated to provide a balanced supply to meet internal biosynthetic requirements without excretion of excess ammonia from the oxidation of methylamine (Levering and Dijkhuizen 1985). Additional experiments showed that acetate strongly influenced both methylamine utilization at the level of uptake (Dijkhuizen et al. 1982) and the activity of the key enzyme involved in amine oxidation (Levering and Dijkhuizen 1985). In batch culture, concentrations of more than 5 mM acetate resulted in almost complete repression of methylamine metabolism, thereby inhibiting nitrogen assimilation via the simultaneous utilization of methylamine. Relief from this repression resulted in acceleration of growth and excretion of ammonium into the culture medium (Figure 8–1). In continuous culture, *Arthrobacter* P1 simultaneously utilized the two substrates over a wide range of dilution rates under carbon-limited growth conditions irrespective of the ratio of substrates in the medium reservoir and the presence or absence of

8.2 Methylotrophic Bacteria

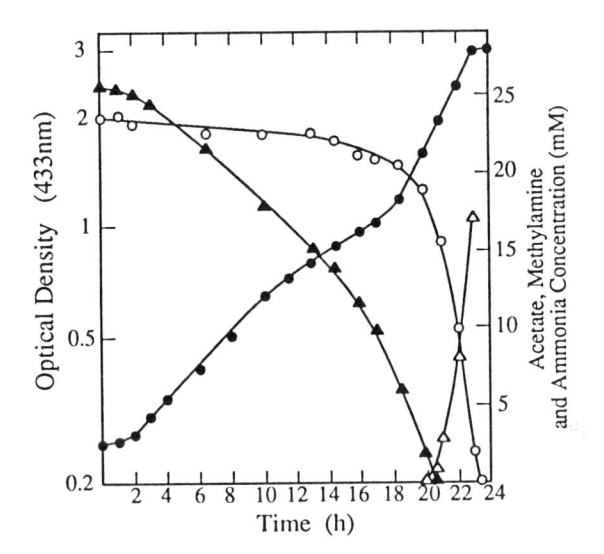

FIGURE 8-1 Batch growth of *Arthrobacter* P1 in a synthetic medium on a mixture of acetate and methylamine in the absence of ammonia. Growth given as OD$_{433}$ (●). Concentration of: acetate (▲), methylamine (○), and ammonia (△). Reproduced with permission from Levering and Dijkhuizen (1985).

ammonium (Levering and Dijkhuizen 1985). When increasing concentrations of methylamine were added to the medium reservoir of an acetate-limited continuous culture of *Arthrobacter* P1 growing at a constant dilution rate of 0.1 h^{-1}, induction of the methylamine-related metabolic enzymes occurred. Approximately half-maximum enzyme-specific activities resulted with acetate (25 mM)/methylamine (10 mM) mixtures. In contrast, when increasing concentrations of acetate were added to the medium reservoir of a methylamine-limited (25 mM) chemostat culture, there was no induction of the two glyoxylate shunt enzymes isocitrate lyase and malate synthase until the concentration of acetate in the feed exceeded 10 mM. These results imply that during growth with mixtures containing low fractions of acetate, this substrate was used solely as an additional source of energy and was not assimilated. Nevertheless, at all mixture compositions, the bacterial biomass produced was approximately the sum of that expected from yield data obtained during growth on each substrate independently. Pulse-wise addition of acetate or glucose to methylamine- or formaldehyde-limited cells of *Arthrobacter* P1 growing in the chemostat ($D = 0.1$ h^{-1}) resulted in their rapid assimilation and accelerated growth (Levering et al. 1986a). However, their addition also caused a transient repression of the key enzymes of methylamine or formaldehyde metabolism (amine oxidase, hexulose phosphate synthase). Transient accumulation of the C$_1$ substrate occurred only in the case of an acetate pulse to a methylamine-limited culture, while in all other cases residual concentrations of methylamine or formaldehyde remained undetectable during the pulse. When formaldehyde was continuously added to cultures of *Arthrobacter* P1 growing exponentially in batch culture on either glucose or acetate, enzymes of formaldehyde metabolism were rapidly induced and the C_1 substrate was simultaneously utilized. On the other hand, when methylamine was added in place of formaldehyde, simultaneous utilization only resulted with glucose, while with acetate the methylamine accumulated in the culture medium with no apparent effect on the synthesis of C_1 metabolic enzymes (Levering et al. 1986b).

8.2.2 Single and Mixed Substrate Utilization by Mixed Cultures of Methylotrophic Bacteria

Interest in the use of mixed bacterial cultures consisting either exclusively of methylotrophs or of a mixture of both methylotrophs and nonmethylotrophs has concentrated on two industrial applications of such systems. The first is the production of single cell protein (SCP), which is concerned predominantly with the growth of a mixed culture on methanol as the sole carbon source (see Chapter 9). The second application is in biological wastewater treatment systems, and is therefore concerned with the use of mixed bacterial cultures on a mixed carbon energy substrate. Since both applications are designed with precise objectives in mind, research has focused on quite different aspects of growth and physiology under the different conditions considered.

8.2.2.1 Mixed Bacterial Cultures Growing on Methanol.
Interest in mixed cultures of methylotrophs with respect to SCP production was initiated because they frequently exhibited higher growth yields than pure cultures of methylotrophs (Harrison 1978). During cultivation of obligate methylotrophs, by-products of metabolism are frequently released in considerable quantities (Wilkinson et al. 1974; Harrison 1976; Papoutsakis et al. 1978) such that either accumulation of inhibitory substances or the contamination of the process culture by potentially pathogenic opportunistic microorganisms is possible. Intentional inclusion of nonpathogenic bacteria scavenging the excretion products and/or competitively excluding the pathogens in a defined mixed culture has therefore attracted special interest.

Characterization of both mesophilic (Crémieux et al. 1977) and thermotolerant (Snedcor and Cooney 1974) methylotrophic mixed cultures have been reported. The composition of methylotrophic mixed cultures can vary considerably and typically is composed of three to four species, which can either all be methylotrophs (Goldberg and Mateles 1975; Rokem et al. 1980) or a mixture where methylotrophs and non-methylotrophs co-exist (Crémieux et al. 1977; Harrison 1976). The data indicate the predominant presence of Gram-negative species in mesophilic mixtures (but see Bitzi 1986;

Bitzi et al. 1991), while thermotolerant consortia have been shown to consist of sporulating strains (Snedcor and Cooney 1974; Dijkhuizen et al. 1988; Al-Awadhi et al. 1989). However, it is necessary to be cautious since classical isolation methods were used in all of these studies to isolate and differentiate between the various members of the mixed culture communities, and one should bear in mind that analysis by this procedure does not necessarily reflect true culture composition. The use of agar cultivation procedures usually results in the preferential enrichment (additional selection step) of the most rapidly and easily growing members of the mixture.

Goldberg and Mateles (1975) described a mixed culture of three methylotrophic strains of *Pseudomonas* (C, 1, and 135). The strains differed from one another in the nature of the assimilation pathways for formaldehyde fixation, *Pseudomonas* C using the ribulose monophosphate (RuMP) pathway, whereas strains 1 and 135 fixed C_1 units via the serine pathway. The study revealed that the composition of this mixed culture of methylotrophs, when growing on methanol as the only carbon energy source in continuous culture, changed with growth rate with one strain or another completely dominating the mixed culture at different dilution rates. In subsequent experiments with defined binary cultures it was demonstrated that at low dilution rates ($D = 0.1$ h^{-1}), the strains using the serine pathway and exhibiting a low K_s value and low μ_{max}, namely *Pseudomonas* sp. strain 1 or strain 135, dominated the culture, whereas at a dilution rate of 0.3 h^{-1}, the RuMP strain, for which the K_s value and μ_{max} were higher, was able to outcompete both of the serine pathway strains (Goldberg et al. 1976; Rokem et al. 1980). These binary mixtures of *Pseudomonas* C and strains 1 or 135 always resulted in dominance (>99% of total population) of one strain over the other. In contrast, Crémieux et al. (1977) reported results for a stable mixed culture composed of four methylotrophs growing in a chemostat where three of the strains together always comprised 20% of the total.

Rokem et al. (1980) investigated the interaction of a culture of the obligately methylotrophic bacterium *Pseudomonas* C in methanol-limited chemostat culture with heterotrophic pathogenic bacteria. Concocting mixed cultures composed of *Pseudomonas* C and either of the pathogenic bacteria *Staphylococcus aureus* or *Salmonella typhimurium*, both of which are unable to utilize methanol as a sole carbon energy source, resulted in mixed cultures in a chemostat where between only 1‰ and 1% of the pathogens persisted irrespective of the dilution rate applied. A strain of *Escherichia coli* was more persistent in a similar experiment and comprised 2-5% of the total population. Interestingly, addition of an undefined mixture of soil bacteria in such experiments resulted in almost complete elimination of *Staphylococcus aureus* or of *Salmonella typhimurium*. The results indicated that the two pathogenic bacteria stuck preferentially to the fermentor wall and that addition of soil bacteria caused competitive exclusion of the pathogenic contaminants and facilitated their loss from the system.

8.2.2.2 Mixed Bacterial Cultures Growing on Mixed Substrates. The second major application of methylotrophic mixed cultures is in the area of wastewater treatment. Industrial effluents typically contain only a restricted range of compounds. For example, wastewater from the petrochemical industry has been shown to contain less than 100 different organic pollutants of which 10 key compounds represented more than 80% of the pollutant load (Hamer 1983). Wilkinson and Hamer (1979) showed that a triple-moiety, mesophilic mixed culture with methanol, phenol, and acetone as growth substrates degraded the individual specific substrates in a strictly neutralistic manner. This is probably a consequence of the very specific nature of these substrates, each of which are metabolized via completely independent pathways.

More complex interactions were observed in a mixed culture enriched in the chemostat for the aerobic biodegradation of a mixture of methanol, methylene chloride, acetone, and isopropanol (Bitzi et al. 1991). The culture was composed of two effective substrate-degrading strains and two satellite organisms unable to grow on any of the supplied substrates. One of the primary substrate-degrading strains was characterized as an acetate/isopropanol-degrading non-methylotroph which constituted 60% of the total population. This strain did not grow on methanol or methylene chloride. The second strain was a facultative methylotroph which was unable to grow on any substrate in this mixture other than methanol. Although neither strain was capable of growing on methylene chloride, this compound was eliminated by the mixed culture. The growth characteristics of the two main strains were investigated in both pure and defined binary mixed cultures using single and multiple substrate combinations. The mixed culture containing the satellite organisms performed significantly better with respect to both growth and solvent degradation than the artificially reconstructed binary culture. More detailed investigations of the growth of the non-methylotrophic strain showed that when acetone or isopropanol were supplied, accumulation of methanol occurred in the culture supernatant under stress conditions, e.g., during batch culture (Figure 8-2), at high dilution rates in the chemostat or under oxygen limitation (Bitzi 1986; Bitzi et al. 1991). This indicated that under stress, the breakdown of isopropanol and acetone proceeded partially via C_2 and C_1 intermediates. Although this organism was unable to grow with methanol as the only source of carbon and energy, experiments with ^{14}C-labelled methanol demonstrated, that in the presence of a growth-supporting substrate, it was indeed capable of utilizing methanol as an additional source of energy. When this bacterium was grown on mixtures of methanol/isopropanol up to a methanol-weight fraction of 50%, all of the methanol was oxidized to CO_2 with a concomitant increase in biomass. Supplying higher fractions of methanol in the feed resulted in drastic accumulation of methanol and led to culture washout.

The competition between two thermotolerant solvent-utilizing strains, a methylotrophic Gram-positive rod (*Bacillus* sp. NCIB 12522) and a Gram-

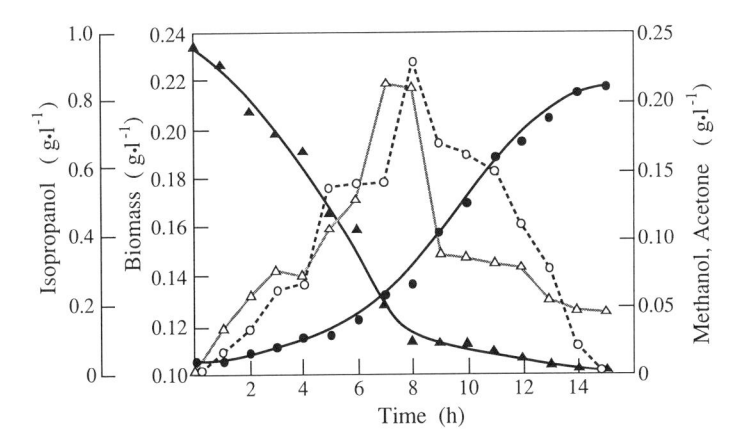

FIGURE 8-2 Batch growth of strain IP in a synthetic medium with isopropanol as the sole source of carbon and energy. Dry biomass produced (●). Concentration of: isopropanol (▲), acetone (○), and methanol (△). Reproduced with permission from Bitzi (1986).

negative non-methylotrophic coccus, was investigated in chemostat culture at a fixed dilution rate (Al-Awadhi et al. 1990a). The strains were grown at $D = 0.3$ h^{-1} with a mixture of methanol (substrate for the methylotroph only), ethanol (substrate for the non-methylotrophic coccus only), and isopropanol (substrate for both strains). The culture composition was analyzed directly by microscopic observation to determine the relative fraction of the two bacteria under different growth conditions. As expected, by increasing the proportion of methanol in the feed, the relative fraction of the methylotroph increased, whereas an increase in the proportion of ethanol resulted in a decrease in the fraction of the methylotroph. When the concentration of isopropanol was increased, the composition of the culture remained constant indicating that both strains were utilizing isopropanol at a fixed ratio as a growth substrate under these conditions.

8.3 METHYLOTROPHIC YEASTS

While methylotrophic yeasts can grow on methanol, their preferred substrates are simple sugars. For SCP production, one major disadvantage of using the methylotrophic yeasts was always that they exhibited both low growth rates and low yields during growth with methanol, and in this respect they were never able to compete with methylotrophic bacteria. This aspect was investigated during the latter part of the 1970s with the hope of improving the competitiveness of the methylotrophic yeasts for use in SCP production processes (see Chapter 9). One advantage of using methylotrophic yeasts stems from their unique biochemical pathways for methanol assimilation and from both the regulation and compartmentalization of the enzymes produced, which can be exploited for the production of various chemicals and as a source of specific enzymes related to C_1 metabolism (Dijkhuizen et al. 1985; also see Chapter 11). Studies of the regulation of the C_1 enzymes have indicated that the methylotrophic yeasts are particularly versatile in that increased rates of production of enzymes and the diversion of fluxes of carbon into product-forming routes were possible as a result of the simultaneous utilization of methanol and sugars.

With few exceptions, the bulk of knowledge concerning mixed substrate utilization by methylotrophs stems from studies of the methylotrophic yeasts belonging to the genera *Hansenula, Candida,* and *Pichia.* Some workers have suggested that *Hansenula* be reclassified within *Pichia* (Kurzman 1984; also see Chapter 2), nevertheless, to avoid confusion the original nomenclature will be used here.

8.3.1 Utilization of Methanol/Sugar Mixtures During Batch Cultivation

In all methylotrophic yeasts, batch growth on mixtures of glucose and methanol results in a classical diauxic growth pattern, with glucose being preferentially utilized and with concomitant repression of enzymes associated with methanol metabolism (Sahm and Wagner 1973; Sakai et al. 1987). Eggeling and Sahm (1978) were the first to report derepression (i.e., synthesis of a specific protein in the absence of the inducing agent) of methanol-dissimilatory enzymes during exponential growth of *Hansenula polymorpha* on sugars such as sorbitol, glycerol, ribose, and xylose. Subsequently, this derepression in the exponential growth phase has been confirmed for both C_1-dissimilatory and -assimilatory enzymes and was found to be consistent amongst all genera of methylotrophic yeasts (Egli et al. 1982a; de Koning et al. 1987; Volfová et al. 1988). The extent of (de)repression by different carbon substrates varied considerably, with ethanol showing the highest repression potential followed by glucose. During growth with these two carbon sources no synthesis of C_1 enzymes occurred. The repression potential of the other sugars as indicated by expression of alcohol oxidase, the key enzyme in methanol metabolism, was generally lower than that with glucose or ethanol and follows in a sequence with ethanol $>$ glucose $\gg$ ribose $>$ sorbitol $>$ glycerol $>$ xylose $>$ xylitol. Noteworthy is the observation that the level of alcohol oxidase in both *Hansenula* and *Candida* species during growth with xylose or xylitol, can reach levels of 60–100% of those found in methanol-grown cells, despite the fact that this enzyme has no apparent function in the absence of methanol (de Koning et al. 1987; Volfová et al. 1988).

Judging from the regulation of the synthesis of C_1-specific enzymes, simultaneous utilization of mixtures involving methanol plus sugars other

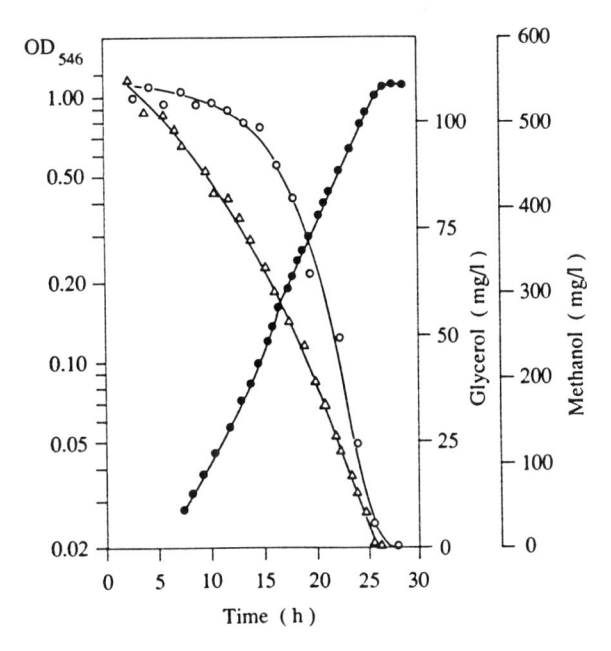

[Figure showing plot with OD546, Glycerol (mg/l), and Methanol (mg/l) versus Time (h)]

FIGURE 8-3 Batch growth of *Hansenula polymorpha* CBS 4732 in a synthetic medium with glycerol (○) and methanol (△) as the sole source of carbon and energy. Cells were precultured on glycerol only. Growth given as OD$_{546}$ (●). Reproduced with permission from Wanner and Egli (1990).

than glucose should be possible. Recently, this has been confirmed experimentally for mixtures of xylose/methanol for *Candida boidinii* and *Hansenula polymorpha* (de Koning et al. 1987; Volfová et al. 1988), glycerol/methanol for *H. polymorpha* (Figure 8-3), and for a glucose catabolite-repression-insensitive mutant of *C. boidinii* growing on mixtures of glucose and methanol (Sakai et al. 1987). For both *H. polymorpha* and *C. boidinii* growing with sugar/methanol mixtures, the presence of methanol resulted in a further increase in C$_1$-related enzyme activities to levels equivalent to those found in methanol-grown cells (de Koning et al. 1987; Sakai et al. 1987; Volfová et al. 1988). In none of these examples was a stimulatory effect on growth rate observed. This is in contrast to an interesting observation made by Babel (1982) who reported increased growth rates for cultures of *H. polymorpha* growing with mixtures of xylose/methanol ($\mu_{max(xylose)}$ = 0.12 h^{-1} compared to $\mu_{max(mixture)}$ = 0.21 h^{-1}) and of glycerol/methanol ($\mu_{max(glycerol)}$ = 0.16 h^{-1} compared to $\mu_{max(mixture)}$ = 0.22 h^{-1}).

8.3.2 Utilization of Methanol/Sugar Mixtures During Continuous Cultivation

In continuous cultures of *H. polymorpha* and *Kloeckera* sp. 2201 growing under either glucose- or sorbose-limited conditions the extent of derepression of C$_1$-dissimilatory enzymes increased with decreasing growth rates, probably as a result of decreasing residual sugar concentrations (Egli and Fiechter 1978; Egli et al. 1980; Eggeling and Sahm 1981). As would have been expected from the batch data, derepression also occurred during growth in the chemostat with a variety of other sugars, such as sorbitol, glycerol, and xylose (Eggeling and Sahm 1980; Tani et al. 1988; de Koning et al. 1990b). Differences in the extent of derepression observed in *C. boidinii* and *H. polymorpha* strains were mainly limited to the synthesis of alcohol oxidase where stronger control was exerted in *C. boidinii* (Egli et al. 1980). No synthesis of alcohol oxidase nor of formate dehydrogenase was observed under conditions where ethanol was the growth-limiting carbon source (Eggeling and Sahm 1980; Tani et al. 1988). In contrast, formaldehyde dehydrogenase seems to be least affected by repression in all the yeasts, and even during growth with ethanol small amounts of this enzyme have been detected in both continuously and batch-grown cells (Eggeling and Sahm 1980; de Koning et al. 1987). Although it is not known whether the derepression pattern of C_1-assimilatory enzymes is similar to that of the dissimilatory enzymes, data from fixed dilution rate experiments (Egli et al. 1983; de Koning et al. 1990b) suggest that the assimilatory enzymes should behave in an analogous fashion.

Derepression of methanol-dissimilatory enzymes at low growth rates during growth of methylotrophs on a variety of different sugars implies that simultaneous utilization of sugar/methanol mixtures should be possible in continuous culture. This has been tested for various sugar/methanol combinations and yeast species, and both growth kinetics and regulation of the synthesis of enzymes involved in methanol metabolism have been studied in considerable detail. Generally, it was found that at low dilution rates (ca. 0.05–0.20 h^{-1}) both sugar and methanol were utilized to completion. Such utilization of mixtures of carbon sources was only possible under strict carbon limitation, and simultaneous utilization ceased when nitrogen became the growth-limiting nutrient (Egli 1982; Egli and Quayle 1986).

Addition of methanol in increasing amounts to the feed of continuous cultures of *H. polymorpha* and *Kloeckera* sp. 2201 growing at a dilution rate of approximately 0.14 h^{-1} on glucose as the sole carbon source, resulted in increasing induction of methanol metabolic enzymes involved in the xylulose monophosphate pathway for formaldehyde fixation (exclusively found in the methylotrophic yeasts), until a plateau was reached (Figure 8–4). Depending on the enzyme and the yeast, a level of 10–50% of methanol in the mixture was sufficient to achieve maximum expression of enzyme specific activity, corresponding to that found in methanol-grown cells at the same growth rate. This pattern was found for enzymes involved in both assimilation and dissimilation of methanol, whereas enzymes involved in glycolysis and in the tricarboxylic acid (TCA) cycle responded in the opposite manner (Egli et al. 1982b and 1983; Egli and Lindley 1984). Virtually identical patterns have been observed for the synthesis of C_1-related enzymes for *H. polymorpha* growing at a constant dilution rate in the chemostat with

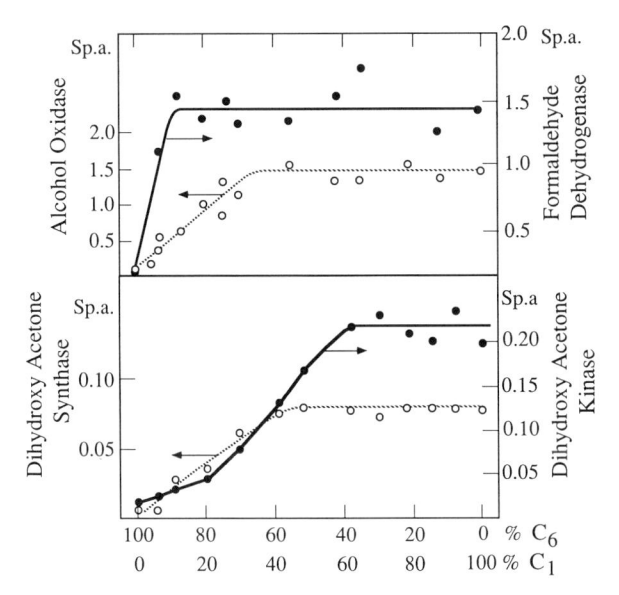

FIGURE 8-4 Specific activities of four enzymes involved in methanol assimilation and dissimilation in cell-free extracts of *Kloeckera* sp. 2201 grown carbon-limited in chemostat culture at a fixed dilution rate of 0.14 h^{-1}, as a function of the composition (given in %, w/w) of the methanol/glucose (C_1/C_6) mixture supplied as carbon source in the inflowing medium. Data with permission from Egli et al. (1982b and 1983).

xylose/methanol mixtures (de Koning et al. 1990b). At the phenomenological level, the coordinated regulation of all of the enzymes responsible for the rearrangement and regeneration of the acceptor molecule for assimilation of C_1 skeletons, such as fructose 1,6-bisphosphatase, ribose 5-phosphate isomerase, and ribulose 5-phosphate epimerase, corresponds strikingly with the increased flux of carbon handled by these enzymes (Egli et al. 1983). Moreover, the simultaneous presence of 6-phosphofructokinase and fructose 1,6-bisphosphatase, which should theoretically result in futile cycling of ATP, has been observed. Presently it is not known how the cells cope with the simultaneous presence of those two enzymes, regulate the flux of carbon up and down the glycolytic pathway, and at the same time ensure constancy of the efficiency of assimilation of carbon from both carbon sources (see below).

Simultaneous utilization of glucose/methanol and sorbitol/methanol mixtures have been observed to occur over a wide range of dilution rates with *H. polymorpha* and *Kloeckera* sp. 2201 and, depending on the mixture composition, complete utilization of both substrates up to a dilution rate of approximately 0.35 h^{-1} has been reported (Egli 1980; Eggeling and Sahm 1981; Egli et al. 1982c and 1986). The general pattern observed during growth with mixtures of sugar/methanol was complete utilization of both carbon sources at low growth rates, whereas at enhanced growth rates (defined as $D_{transition} = D_t$) the cells switched to preferential utilization of the sugar as shown in Figure 8–5 for a culture of *H. polymorpha* growing in the chemostat with a series of mixtures of glucose/methanol. Particularly noteworthy in this example was the strict dependence of D_t on the mixture composition in the feed, where D_t increased with decreasing proportions of methanol in the mixture. This resulted in the utilization of methanol at dilution rates considerably higher than the μ_{max} observed for this yeast during growth on methanol as the sole carbon source (0.2 h^{-1}). The data indicate that the maximum specific growth rate for *H. polymorpha* growing on glu-

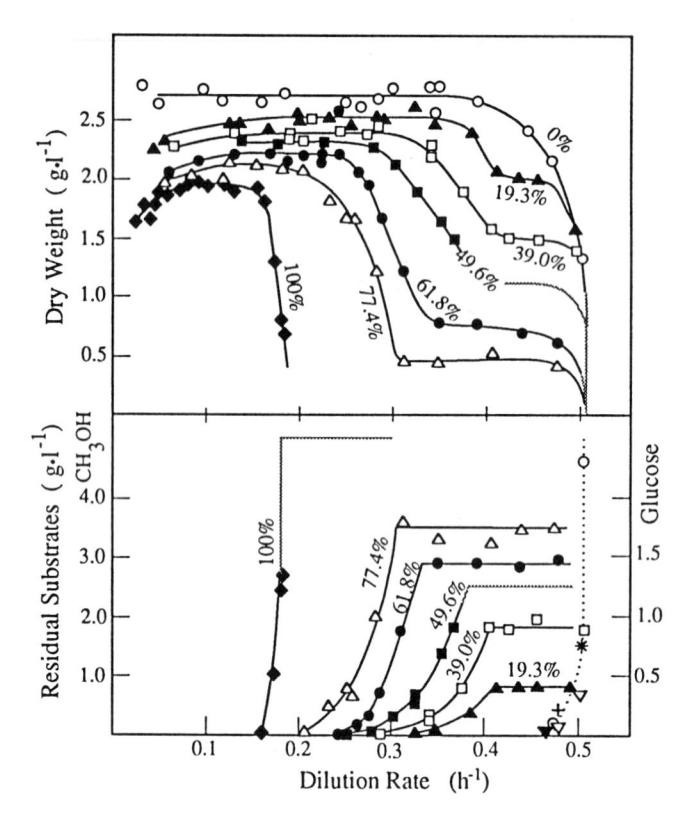

FIGURE 8–5 Cell dry weight formed and residual concentrations of carbon substrates during growth of *Hansenula polymorpha* CBS 4732 in a chemostat as a function of the composition (given in %, w/w) of the methanol/glucose (C$_1$/C$_6$) mixture supplied in the inflowing medium. For simplicity, only the methanol fraction is indicated to characterize the mixture composition. Total concentration of substrate (C$_1$ + C$_6$) in the feed was always 5 g l^{-1}. Symbols for residual glucose are: 0% C$_1$ (○); 19.3% C$_1$ (▽); 39.0% C$_1$ (✶); 61.8% C$_1$ (+); 77.4% C$_1$ (▼). Reproduced with permission of John Wiley and Sons, Inc., from Egli et al. (1986).

8.3 Methylotrophic Yeasts

cose was not influenced by the presence of methanol (Egli et al. 1986). Recently, these data have been confirmed (Giuseppin et al. 1988a), and a kinetic model describing both biomass and residual substrate concentrations during continuous cultivation of *H. polymorpha* with mixtures of glucose and methanol has been proposed (Giuseppin 1989).

The ability to utilize mixtures of glucose and methanol appears to be restrained within two metabolic limits. This becomes apparent when the specific methanol consumption rate, $q(C_1)$, shown in Figure 8-6 is examined for *H. polymorpha* growing in a chemostat culture with various different

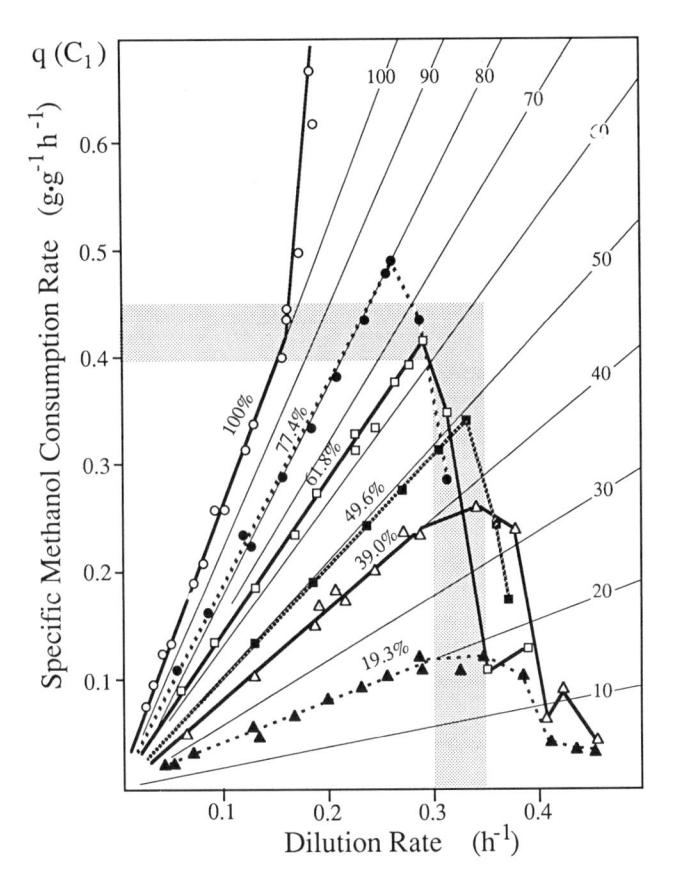

FIGURE 8-6 Specific methanol consumption rate, $q(C_1)$, during growth of *Hansenula polymorpha* CBS 4732 in the chemostat as a function of the methanol/glucose (C_1/C_6) mixture composition supplied in the inflowing medium. Individual mixtures are characterized by their weight percent of methanol only. The fine lines represent theoretical q values for the different C_1/C_6 mixtures assuming complete utilization of both carbon sources and additive formation of biomass from methanol and glucose with yield coefficients of 0.38 and 0.55, respectively. Reproduced with permission of John Wiley and Sons, Inc., from Egli et al. (1986).

mixture compositions (Egli et al. 1986). The data indicate that there is a maximum flux of methanol carbon that cells can optimally metabolize, which is in the range of 0.40–0.45 g-methanol (g-dry weight)$^{-1}$ h^{-1}. This can be seen from the results from growth with methanol as the sole carbon source, where $q(C_1)$ increased linearly with dilution rate until $q(C_1)_{max}$, was reached. At higher dilution rates, the fraction of methanol dissimilated increased dramatically, possibly because detoxification was necessary. Theoretical calculations (Figure 8-6) indicate that during growth with mixtures of methanol/glucose, the $q(C_1)_{max}$ threshold will be reached at higher growth rates as the fraction of methanol in the mixture decreases. It is evident from the results shown for the mixtures containing either 61.8% or 77.4% methanol, that an attempt to push the specific methanol consumption rate above this limit by increasing the dilution rate, causes the cells to repress methanol utilization and to switch over to growth on glucose only. With mixtures containing less than approximately 50% methanol, $q(C_1)$ increased linearly up to a dilution rate of 0.30–0.35 h^{-1} until transition from mixed to single substrate utilization occurred. The precise physiological nature of this apparent second limit is less clear but it might be affected by concentrations of metabolites resulting from glucose metabolism.

Under conditions where both substrates were utilized to completion, the growth yield for both *C. boidinii* and *H. polymorpha* has been shown to be additive with mixtures of xylose/methanol, glucose/methanol, glycerol/methanol and sorbitol/methanol (Eggeling and Sahm 1980 and 1981; Egli et al. 1982b; Giuseppin et al. 1988a; Tani et al. 1988; de Koning et al. 1990b). Using mixtures of ^{14}C-methanol and glucose, growth yields of *H. polymorpha* and *Kloeckera* sp. 2201 were essentially constant at 0.36–0.37 (g-dry weight g-methanol^{-1}) and 0.55–0.56 (g-dry weight g-glucose^{-1}) for mixtures which contained more than 20% (w/w) methanol. Supplying lower fractions of methanol resulted in a reduction in the yield coefficient for methanol to approximately 0.30 while that for glucose was unaffected (Egli 1980; Egli et al. 1982b). Similar results have also been obtained with ^{14}C-methanol/sorbitol mixtures (Eggeling and Sahm 1981). These findings are not in agreement with the results reported by Müller and co-workers (1983, 1985a, and 1988) who found a stimulation of overall biomass yield, up to 25% of that theoretically expected assuming additive growth yields, during growth of *H. polymorpha* strains of MH20 and MH26 with mixtures of glucose/methanol. This effect was most pronounced at a dilution rate of 0.1 h^{-1} and a mixture composition of methanol to glucose of 7:1 (w/w) (Müller et al. 1983). It was suggested that these results are due to enhanced efficiency in energy production from reducing equivalents produced from methanol metabolism, thus enabling more efficient growth on glucose, but this effect seems to be restricted to the particular strains of *H. polymorpha* studied.

A further point of interest concerns the fact that both methylotrophic and non-methylotrophic yeasts can improve the rate of carbon assimilation from a primary substrate when they are provided with an additional source of energy. This has been demonstrated for *Pichia pinus* (growing on methanol), *H. polymorpha* strains MH20 and MH26, and *C. utilis* (all growing on glucose) using formate, which is oxidized to CO_2, as an additional energy source. When glucose was the growth substrate, the maximum increase in the growth yield was in the range of 23–51% (Babel et al. 1983; Bruinenberg et al. 1985; Müller et al. 1985b), and the maximum yield increase during growth with methanol was reported to be 49% (Müller et al. 1986).

8.3.2.1 Kinetic Aspects. Growth on mixtures of two substrates such as methanol and glucose poses an interesting question with respect to growth kinetics. Whereas single substrate-(methanol)-limited growth at a constant dilution rate is characterized by a constant specific rate of substrate consumption, $q(C_1)$, the presence of a second substrate causes $q(C_1)$ to become a function of the substrate mixture supplied in the inflowing medium (Figure 8–7). Assuming that $q(C_1)$ is a function of the extracellular substrate concentration and can be described according to the Monod relationship:

$$q(C_1) = q(C_1)_{max} \cdot s/(K_s + s) \tag{8.1}$$

it can be deduced from Equation 8.1, that in order to change $q(C_1)$ at a constant dilution rate, either changes in the residual substrate concentration, s, or in $q(C_1)_{max}$ have to be considered (where K_s is assumed to be constant). The results for the residual concentration of methanol, for $q(C_1)$ and for the generalized pattern of specific activities of enzymes involved in methanol metabolism shown in Figure 8–7 demonstrate that the cell makes use of both strategies in order to adjust $q(C_1)$ to the substrate mixture (Egli et al. 1983). When the proportion of methanol in the inflowing medium was less than approximately 50%, the cells responded by changing their content of methanol-assimilatory and -dissimilatory enzymes while maintaining a constant residual methanol concentration in the chemostat. In contrast, during growth with mixtures containing more than 50% methanol, the cell maintains a constant enzyme capacity such that a further increase in $q(C_1)$ was brought about by increasing concentrations of methanol. Noteworthy is that the residual methanol concentration was always much lower during growth with the glucose/methanol mixed substrate than during growth with methanol alone. Since multiple substrates are usually present in most natural environmental systems, the fact that reduced residual substrate concentrations were found in this example during growth with mixed substrates, might be of ecological relevance.

8.3.3 Applied Aspects of Mixed Substrate Utilization by Methylotrophic Yeasts

Four different ways have been described by which methanol in combination with other substrates can be used for product formation with methylotrophic yeasts: (1) Methanol can be used as a low-cost feedstock of the production

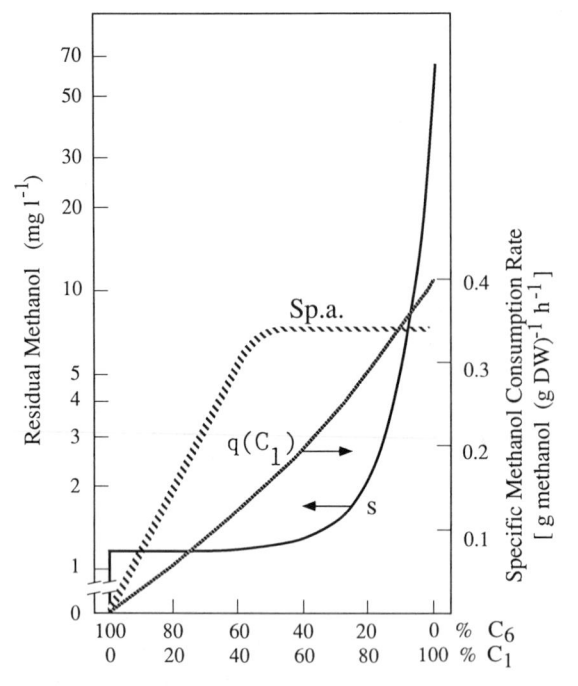

FIGURE 8-7 Residual methanol concentration (s), specific methanol consumption rate, $q(C_1)$, and idealized pattern of specific activities of enzymes involved in methanol metabolism (Sp. a.) during growth of *Kloeckera* sp. 2201 in chemostat culture at a dilution rate of 0.14 h^{-1}, as a function of the composition (given in %, w/w) of the methanol/glucose (C_1/C_6) mixture supplied as carbon source in the inflowing medium. Idealized curves with permission based on Egli et al. (1982b and 1983).

of cells as a supplementary proteinaceous source for animal feed. (2) Mixtures of methanol and other substrates can be used for the production of metabolites or enzymes related (or unrelated) to methanol metabolism. (3) Use can be made of the promoter of alcohol oxidase for high-level expression of recombinant genes where a second substrate is used for growth of such cells and methanol serves merely as an inducer for the expression of the introduced gene. (4) It is also possible to use methanol metabolism to generate reducing equivalents and ensure reducing conditions within the cell to drive certain reactions. In the following sections each of these will be discussed.

8.3.3.1 Single Cell Protein. Soon after their discovery, methylotrophic yeasts were considered for use as an alternative to bacteria for the production of SCP from methanol. The widespread interest in the use of methylotrophic yeasts for this purpose led to a rapid advance in knowledge concerning high-cell-density cultivation of these microorganisms. For example, Phillips Petroleum Corp. has developed a process where cell densities of 120–150 g dry cell weight l^{-1} of a selected strain of *Pichia pastoris* can be relatively easily attained (Wegner and Harder 1987; Harder and Brooke 1990). Use of other substrates such as molasses or whey as alternative feedstocks in the various production systems provides the potential for using mixed substrates in continuous culture processes (Linton and Niekus 1987).

8.3.3.2 Enzyme and Metabolite Production

Alcohol Oxidase. Oxidases such as alcohol oxidase, have many applications, including generation of hydrogen peroxide as a bleaching agent in detergents, oxygen scavenging in food and other products, synthesis of organic compounds, analytical detection of various alcohols, and use in enzyme mixtures for treatment of wastewaters (Woodward 1986; Unilever 1987a and 1987b; Henkel & Cie GmbH 1977; Unichika Co. Ltd. 1986; Verduyn et al. 1984; Herzberg and Rogerson 1985; Philips Petroleum Co. 1980 and 1984; Sakai and Tani 1987a and 1987b; Behringwerke 1974; Yonehara and Tani 1987). Especially in the case of alcohol oxidase from methylotrophic yeasts, the advantage of continuous culture operation compared to batch cultivation has been well documented, and this is a typical example where optimization of product (enzyme) formation is achieved mainly at the physiological level by selecting the appropriate mixture composition and growth conditions. At low dilution rates in a methanol-limited chemostat this enzyme can amount to 30–40% of the total protein (van Dijken et al. 1976; Giuseppin 1988). A further increase in alcohol oxidase productivity was reported by Egli (1980), who grew *Hansenula polymorpha* in continuous culture on mixtures of glucose and methanol (Figure 8–8). Under such conditions, productivities in the region of 400–450 units of alcohol oxidase (g-dry weight h)$^{-1}$ can be achieved, thus improving the productivity of the process by a factor of approximately 60 compared to the 7.7 U (g-dry weight h)$^{-1}$ maximally possible under batch culture conditions using methanol as the only growth substrate, assuming a fermentation cycle time of 48 h (Giuseppin et al. 1988b).

Enhanced specific activities (oversynthesis) of alcohol oxidase during growth on mixtures of glucose/methanol, sorbose/methanol, or glycerol/methanol were reported by Eggeling and Sahm (1980) during carbon-limited growth of *H. polymorpha* in a chemostat ($D = 0.1$ h^{-1}). However, these results have not been confirmed (Giuseppin et al. 1988b; Tani et al. 1988), and a possible explanation is that the so-called oversynthesis was due to higher residual methanol concentrations in methanol-limited continuous cultures than normally observed, leading to reduced levels of alcohol oxidase activity (van Dijken et al. 1976).

While the traditional approach has relied on wild-type strains, applications using mutants with deletions and/or specific metabolic blocks to

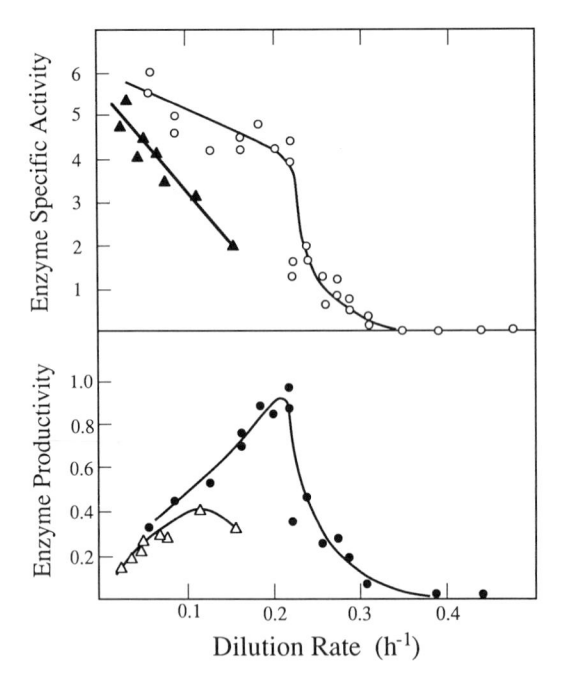

FIGURE 8-8 Specific activity (U mg protein^{-1}) and enzyme productivity (U mg-protein^{-1} h^{-1}) for alcohol oxidase during growth of *Hansenula polymorpha* CBS 4732 with methanol (▲, △) or with a mixture of glucose/methanol (38.8%/61.2%, w/w; ●, ○) as the growth-limiting carbon sources. Data from Egli (1980). Alcohol oxidase specific activities of methanol-grown cells were taken from van Dijken et al. (1976).

direct metabolism into product-forming routes can improve the overall efficiency of product formation, saving costs at the level of downstream processing. Many examples of this approach are known, including the production of catalase-free alcohol oxidase in *H. polymorpha* (Giuseppin et al. 1987 and 1988a; Giuseppin 1989). Catalase is a contaminant in alcohol oxidase production resulting in the need for further purification. The approach adopted was to isolate a catalase-deficient mutant, which theoretically should be incapable of growing on methanol. By using glucose as the growth substrate in continuous culture and by including formaldehyde or formate, which are inducing agents that do not generate the toxic compound hydrogen peroxide, induction of alcohol oxidase synthesis and of other methanol-dissimilating enzymes was possible. Different ratios of formaldehyde (or formate) to glucose were examined for their effectiveness with respect to induction of alcohol oxidase. As was found with the wild-type strain growing on glucose/methanol mixtures (Figure 8-8), increasing the proportion of the C_1 substrate resulted in higher steady state alcohol oxidase levels. When the molar ratio of formaldehyde to glucose exceeded the critical value of 2.2, instability in the system caused by formaldehyde toxicity resulted in reduced alcohol oxidase production and washout of *H. polymorpha* from the bioreactor. Addition of formate caused an induction of alcohol oxidase, however, much higher levels of alcohol oxidase were attained when formaldehyde was used as the inducer, and at a formaldehyde to glucose molar ratio of 1.8, the alcohol oxidase level was 58% of that found in wild-type strains growing under the same conditions (Giuseppin et al. 1988b). Surprisingly, utilization of methanol by the catalase-deficient strain of *H. polymorpha* was also found to be possible in the presence of glucose. Subsequent studies revealed that under these conditions the problem of decomposing the hydrogen peroxide formed during methanol oxidation by alcohol oxidase could be resolved by induction of high levels of the mitochondrial-cytochrome-*c* peroxidase activity (Giuseppin et al. 1988c; Verduyn et al. 1988).

Formate Dehydrogenase. A second enzyme of the dissimilatory pathway for methanol in methylotrophic yeasts that is of commercial interest is formate dehydrogenase. The interest in formate dehydrogenase is mainly in its use as a cofactor (NADH) regenerating system where enzymatic reductions catalyzed by NADH-dependent dehydrogenases can be continuously supplied with NADH with concomitant production of CO_2 as the sole by-product from formate. The advantage of having CO_2 as an end product of the reaction supplying reducing equivalents is that it can easily be removed from the reaction mixture, thereby ensuring continued production of reducing equivalents by preventing the NADH-generating reaction from reaching equilibrium. Increased productivity of formate dehydrogenase can be achieved in an analogous manner by growth of methylotrophic yeasts on mixtures of methanol and glucose in continuous cultures (Egli and Harder 1984).

Production of Cells with Specific Catalytic Activity. Metabolite production by methylotrophic yeasts can be enhanced by separating the growth and product formation phases. During the first phase, cells are grown in the presence of mixed substrates such that they contain optimum catalytic activity. During the second stage, the high catalytic potential of the resting cells, or parts of these cells, is used for product formation. A good example of this is the production of aldehydes from various alcohols by using cells with high levels of alcohol oxidase (Sakai and Tani 1987a and 1987b). Several cases have been described where whole cells have been used for the biocatalytic synthesis of metabolites, such as in the production of various aldehydes (Tani et al. 1985a and 1985b; Sakai and Tani 1986, 1987a, and 1987b) or sorbitol (Tani and Vongsuvanlert 1987; Vongsuvanlert and Tani 1988). For efficient product formation, both high catalyst concentration and continuous synthesis of fresh catalyst are required to counteract the loss of catalyst due to inactivation by the reaction products. In the case of formaldehyde production, the isolation of a glucose-catabolite-repression-insensitive mutant enabled first-phase chemostat cultivation with mixtures of glucose/methanol so that a high alcohol oxidase productivity of the cells resulted in maximum productivities of aldehyde formation, which were 3.8 times higher than when the parent strain was grown on methanol alone (Tani et al. 1988; Sakai and Tani 1988). Optimum alcohol oxidase activity/formaldehyde productivity was obtained under these conditions when a glucose to methanol ratio of 1–1.6:0.4 was used.

Exploitation of an enzyme activity not directly involved in methanol metabolism, where use is made of the methanol system as a source of producing reducing equivalents, has been described for the production of sorbitol. Sorbitol production is achieved by reduction of fructose by sorbitol dehydrogenase, however the reaction is reversible so that production can only occur when the cellular NADH:NAD ratio is high. In this process, growth on mixtures of xylose and methanol in batch-culture produced cells with high levels of activity of both methanol-dissimilating enzymes and of sorbitol dehydrogenase. Optimum sorbitol production from fructose or glucose was then obtained when these cells were incubated with methanol and/or formate as a source of reducing equivalents (Tani and Vongsuvanlert 1987). Fructose could be converted to sorbitol at an efficiency of 95% in this manner.

Dihydroxyacetone and Glycerol. In addition to those metabolites already mentioned, another interesting system has been proposed for the production of dihydroxyacetone and glycerol by Kato et al. (1986) and de Koning (1989). The basic idea is to prevent the flux of carbon through the cyclic assimilation pathway by introducing metabolic blocks at key points and to use the action of dihydroxyacetone synthase to convert xylulose-5-phosphate and formaldehyde to dihydroxyacetone and glyceraldehyde-3-phosphate. Prevention of further metabolism of dihydroxyacetone and excretion was then achieved by the use of mutants. Mutants of *H. polymorpha* blocked in dihydroxyacetone kinase (Kato et al. 1986; de Koning et al. 1987, 1990a and 1990b). For optimum production in this system it was not sufficient to block only the phosphorylation of dihydroxyacetone by deletion of dihydroxyacetone kinase but, in addition, a second metabolic block in glycerol kinase was required to inhibit an unexpected bypass converting dihydroxyacetone via glycerol and glycerol-3-phosphate to dihydroxyacetone 3-phosphate. Using such a glycerol-kinase/dihydroxyacetone-kinase-negative mutant, it was possible to achieve significant accumulation of trioses when the cells were supplied with methanol plus a source for the replenishment of xylulose-5-phosphate, the acceptor molecule for the fixation of formaldehyde. Highest productivity of trioses could be achieved when using resting cells fed intermittently with methanol and xylose (de Koning et al. 1990a). The results indicate that the ratio of dihydroxyacetone to glycerol produced by these mutants was dependent on the internal ratio of NADH:NAD. Conditions which ensured a high degree of reduction of the cellular NADH/NAD pool resulted in the predominant excretion of glycerol.

Expression of Heterologous Proteins. It has been shown that at low dilution rates in continuous culture, the enzyme alcohol oxidase can account for 30–40% of the total cellular protein (van Dijken et al. 1976). This indicates that the alcohol oxidase promoter is very powerful, and this property has been used for high-level expression of heterologous proteins (see Chapter 13). Since the alcohol oxidase structural gene is usually deleted in such systems, the use of mixed substrates permits the independent switching on of the alcohol oxidase promoter during growth on a second substrate. Using this system, methylotrophic yeasts are being used as hosts for the expression of various heterologous proteins. Production of hepatitis B surface antigen, human tumor necrosis factor (TNF), invertase, and bovine lysozyme have been investigated in *Pichia pastoris*. Details on the construction of these strains have been described by Thill et al. (1987), Cregg et al. (1987), Tschopp et al. (1987), Digan et al. (1988), and Sreekrishna et al. (1988 and 1989). The general principle on which they operate is the same for all constructs, namely that the structural gene for alcohol oxidase (*AOX1*) is deleted and replaced with that for the heterologous protein such that the latter comes under the control of a methanol-inducible promoter. The processes of growth and product formation are separated by the fact that these strains only grow weakly on methanol as their sole substrate, and they therefore require a growth-supporting carbon-energy source. Poor growth may occur due to the presence of another weakly expressed alcohol oxidase gene (*AOX2*). Besides the traditional approach, i.e., pre-growing the cells on glucose or glycerol and subsequently exposing them to methanol for up to 200 h to achieve induction of expression of the heterologous protein (Tschopp et al. 1987; Sreekrishna et al. 1988; Cregg et al. 1987), a different approach involving growth on mixed substrates has been used for the production of TNF using continuous culture techniques, where, following an initial growth phase on glycerol alone until steady state conditions were reached, the culture was switched to a medium consisting of 5% (w/v) glycerol plus 1% (v/v) methanol. The resulting expression of TNF reached a steady state level within 18 h and was maintained over an extended period of time (73 h) at a dilution rate of 0.05 h^{-1} (Sreekrishna et al. 1988). Comparison of productivities attainable under either batch or continuous culture conditions can be extrapolated from the data given by Sreekrishna et al. (1988) and suggests that, under continuous culture/mixed substrate conditions, enhanced productivity can be obtained due to the prolongation of the production phase although the concentration of the TNF protein produced was only ca. 75% of the total soluble protein possible under batch conditions using pre-growth on glucose followed by induction with methanol.

REFERENCES

Al-Awadhi, N. (1989) Ph.D. thesis, no. 8810, Swiss Federal Institutes of Technology (ETH), Zürich, Switzerland.
Al-Awadhi, N., Egli, T., Hamer, G., and Wehrli, E. (1989) *System. Appl. Microbiol.* 11, 207–216.
Al-Awadhi, N., Hamer, G., and Egli, T. (1990a) *Bioprocess Eng.* 5, 39–45.
Al-Awadhi, N., Egli, T., Hamer, G., and Mason, C.A. (1990b) *Biotechnol. Bioeng.* 36, 816–820.
Al-Awadhi, N., Egli, T., Hamer, G., and Mason, C.A. (1990c) *Biotechnol. Bioeng.* 36, 821–825.
Anthony, C. (1987) in *Carbon Substrates in Biotechnology* (Stowell, J.D., Beardsmore, A.J., Keevil, C.W., and Woodward, J.R., eds.), pp. 93–118, IRL Press, Oxford.
Babel, W. (1982) in *Abhandl. Akad. Wiss. DDR (N2) "Biotechnologie"* (Ringpfeil, M., ed.), pp. 183–188, Akademie-Verlag, Berlin.
Babel, W., Müller, R.H., and Markuske, K.D. (1983) *Arch. Microbiol.* 136, 203–208.
Behringwerke, A.G. (1974) French patent FP 2218346.
Ben-Bassat, A., Goldberg, I., and Mateles, R.I. (1980) *J. Gen. Microbiol.* 116, 213–223.
Bitzi, U. (1986) Ph.D. thesis, no. 8118, Swiss Federal Institutes of Technology (ETH), Zürich, Switzerland.
Bitzi, U., Egli, T., and Hamer, G. (1991) *Biotechnol. Bioeng.* 37 (in press).
Brooke, A.G., and Attwood, M.M. (1983) *J. Gen. Microbiol.* 129, 2399–2404.
Brooke, A.G., and Attwood, M.M. (1985) *FEMS Microbiol. Lett.* 29, 251–256.
Bruinenberg, P.M., Jonker, R., van Dijken, J.P., and Scheffers, W.A. (1985) *Arch. Microbiol.* 142, 302–306.
Colby, J., and Zatman, L.J. (1975) *J. Gen. Microbiol.* 90, 169–177.
Cregg, J.M., Tschopp, J.F., Stillman, J., et al. (1987) *Bio/technology* 5, 479–485.
Crémieux, A., Chevalier, J., Combet, M., et al. (1977) *Eur. J. Appl. Microbiol.* 4, 1–9.
de Boer, L., Euverink, G.J., van der Vlag, J., and Dijkhuizen, L. (1990) *Arch. Microbiol.* 153, 337–343.
de Koning, W. (1989) Ph.D. thesis, University of Groningen, Groningen, The Netherlands.
de Koning, W., Gleeson, M.A.G., Harder, W., and Dijkhuizen, L. (1987) *Arch. Microbiol.* 147, 375–382.
de Koning, W., Weusthuis, R.A., Harder, W., and Dijkhuizen, L. (1990a) *Appl. Microbiol. Biotechnol.* 32, 693–698.
de Koning, W., Weusthuis, R.A., Harder, W., and Dijkhuizen, L. (1990b) *Yeast* 6, 107–115.
Digan, M.E., Tschopp, G., Grinna, L., et al. (1988) *Dev. Ind. Microbiol.* 29, 59–65.
Dijkhuizen, L., de Boer, L., Boers, R.H., Harder, W., and Konings, W.N. (1982) *Arch. Microbiol.* 133, 261–266.
Dijkhuizen, L., Harder, W., de Boer, L., et al. (1984) *Arch. Microbiol.* 139, 311–318.
Dijkhuizen, L., Hansen, T.A., and Harder, W. (1985) *Trends Biotechnol.* 3, 262–267.
Dijkhuizen, L., Arfman, N., Attwood, M.M., et al. (1988) *FEMS Microbiol. Lett.* 52, 209–214.

References

Ecclestone, M., and Kelly, D.P. (1972) *J. Gen. Microbiol.* 71, 541–554.
Ecclestone, M., and Kelly, D.P. (1973) *J. Gen. Microbiol.* 75, 211–221.
Eggeling, L., and Sahm, H. (1978) *Eur. J. Appl. Microbiol. Biotechnol.* 5, 197–202.
Eggeling, L., and Sahm, H. (1980) *Arch. Microbiol.* 127, 119–124.
Eggeling, L., and Sahm, H. (1981) *Arch. Microbiol.* 130, 362–365.
Egli, T. (1980) Ph.D. thesis, no. 6538, Swiss Federal Institutes of Technology, Zürich, Switzerland.
Egli, T. (1982) *Arch. Microbiol.* 131, 95–101.
Egli, T., and Fiechter, A. (1978) in *Abstract of the 12th Int. Symp. Microbiol.*, p. 125, International Association of Microbiology Societies, Munich.
Egli, T., and Harder, W. (1984) in *Microbial Growth on C_1 Compounds* (Crawford, R.L., and Hanson, R.S., eds.), pp. 330–337, American Society for Microbiology, Washington DC.
Egli, T., and Lindley, N.D. (1984) *J. Gen. Microbiol.* 130, 3239–3249.
Egli, T., and Quayle, J.R. (1986) *J. Gen. Microbiol.* 123, 1779–1788.
Egli, T., van Dijken, J.P., Veenhuis, M., Harder, W., and Fiechter, A. (1980) *Arch. Microbiol.* 124, 115–121.
Egli, T., Haltmeier, T., and Fiechter, A. (1982a) *Arch. Microbiol.* 131, 174–175.
Egli, T., Käppeli, O., and Fiechter, A. (1982b) *Arch. Microbiol.* 131, 1–7.
Egli, T., Käppeli, O., and Fiechter, A. (1982c) *Arch. Microbiol.* 131, 8–13.
Egli, T., Lindley, N.D., and Quayle, J.R. (1983) *J. Gen. Microbiol.* 129, 1269–1281.
Egli, T., Bosshard, C., and Hamer, G. (1986) *Biotechnol. Bioeng.* 28, 735–741.
Giuseppin, M.L.F. (1988) Ph.D. thesis, Technical University, Delft, The Netherlands.
Giuseppin, M.L.F. (1989) *Biotechnol. Bioeng.* 33, 524–535.
Giuseppin, M.L.F., van Eijk, H.M.J., Hellendoorn, M., and van Almkerk, J.W. (1987) *Eur. J. Appl. Microbiol. Biotechnol.* 27, 31–36.
Giuseppin, M.L.F., van Eijk, H.M.J., and Bes, B.C.M. (1988a) *Biotechnol. Bioeng.* 32, 577–583.
Giuseppin, M.L.F., van Eijk, H.M.J., Verduyn, C., and van Dijken, J.P. (1988b) *Eur. J. Appl. Microbiol. Biotechnol.* 28, 14–19.
Giuseppin, M.L.F., van Eijk, H.M.J., Bos, A., Verduyn, C., and van Dijken, J.P. (1988c) *Eur. J. Appl. Microbiol. Biotechnol.* 28, 286–292.
Goldberg, I., and Mateles, R.I. (1975) *J. Bacteriol.* 124, 1028–1029.
Goldberg, I., Rock, J.S., Ben-Bassat, A., and Mateles, R.I. (1976) *Biotechnol. Bioeng.* 18, 1657–1668.
Hamer, G. (1983) *Instn. Chem. Engrs. Symp. Ser.* 77, 87–101.
Harder, W., Matin, A., and Attwood, M.M. (1975) *J. Gen. Microbiol.* 86, 319–326.
Harder, W., and Attwood, M.M. (1978) *Adv. Microb. Physiol.* 17, 303–359.
Harder, W., and Brooke, A.G. (1990) in *Yeast Biotechnology and Biocatalysis* (Verachtert, H., and De Mot, R., eds.), pp. 395–428, Marcel Dekker, New York and Basel, Switzerland.
Harder, W., and Dijkhuizen, L. (1976) in *Continuous Culture, Applications and New Fields*, vol. 6, (Ellwood, A.C.R., Evans, C.G.T., and Melling, J., eds.) pp. 297–314, Ellis Harwood, Chichester, UK.
Harder, W., and Dijkhuizen, L. (1982) *Philos. Trans. R. Soc. London Ser. B.* 297, 459–479.
Harder, W., Trotsenko, Y.A., Bystrykh, L.V., and Egli, T. (1987) in *Microbial Growth on C_1 Compounds* (Van Verseveld, H.W., and Duine, J.A., eds.), pp. 139–149, Martinus Nijhoff Publishers, Dordrecht, The Netherlands.

Harrison, D.E.F. (1976) *Chem. Technol.* 6, 570–574.
Harrison, D.E.F. (1978) *Adv. Appl. Microbiol.* 24, 129–164.
Henkel & Cie GmbH (1977) German Patent GP 557623.
Herzberg, G.R., and Rogerson, M. (1985) *Anal. Biochem.* 149, 354–357.
Hirt, W., Papoutsakis, E., Krug, E., Lim, H.C., and Tsao, G.T. (1978) *Appl. Environ. Microbiol.* 36, 56–62.
Kato, N., Kobayashi, H., Shimao, M., and Sakasawa, C. (1986) *Appl. Microbiol. Biotechnol.* 23, 180–186.
Kurzman, C.P. (1984) *Antonie van Leeuwenhoek* 50, 209–217.
Levering, P.R., and Dijkhuizen, L. (1985) *Arch. Microbiol.* 142, 113–120.
Levering, P.R., Binnema, D.J., van Dijken, J.P., and Harder, W. (1981) *FEMS Microbiol. Lett.* 12, 19–25.
Levering, P.R., Croes, L.M., Tiesma, L., and Dijkhuizen, L. (1986a) *Arch. Microbiol.* 144, 272–278.
Levering, P.R., Croes, L.M., and Dijkhuizen, L. (1986b) *Arch. Microbiol.* 144, 279–285.
Lindley, N.D., Loubière, P., Pacaud, S., Mariotto, C., and Goma, G. (1987) *J. Gen. Microbiol.* 133, 3557–3563.
Linton, J.D., and Niekus, H.G.D. (1987) in *Microbial Growth on C_1 Compounds* (Van Verseveld, H.W., and Duine, J.A., eds.), pp. 263–271, Martinus Nijhoff Publishers, Dordrecht, The Netherlands.
Loubière, P., Pacaud, S., Goma, G., and Lindley, N.D. (1987) *J. Gen. Appl. Microbiol.* 33, 463–470.
Meiberg, J.B.M. (1979) Ph.D. thesis, University of Groningen, Groningen, The Netherlands.
Müller, R., Markuske, K.D., and Babel, W. (1983) *Z. Allg. Mikrobiol.* 23, 375–384.
Müller, R., Uhlenhut, G.J., and Babel, W. (1985a) *Arch. Microbiol.* 143, 77–81.
Müller, R., Markuske, K.D., and Babel, W. (1985b) *Biotechnol. Bioeng.* 27, 1599–1602.
Müller, R.H., Sysoev, O.V., and Babel, W. (1986) *Appl. Microbiol. Biotechnol.* 25, 238–244.
Müller, R., Uhlenhut, G.J., and Babel, W. (1988) *Acta Biotechnol.* 4, 319–326.
O'Connor, M.L. (1981) in *Microbial Growth on C_1 Compounds* (H. Dalton, ed.), pp. 294–300, Heyden and Son, London.
O'Connor, M.L., and Hanson, R.S. (1977) *J. Gen. Microbiol.* 101, 327–332.
Papoutsakis, E., Lim, H.C., and Tsao, G.T. (1978) *AIChE. J.* 24, 406–417.
Papoutsakis, E., Hirt, W., and Lim, H.C. (1981) *Biotech. Bioeng.* 23, 235–242.
Patel, R.N., Hoare, S.L., and Taylor, B.F. (1975) *J. Bacteriol.* 123, 382–384.
Patel, R.N., Hoare, S.L., and Hoare, D.S. (1979) *Antonie van Leeuwenhoek J. Microbiol. Serol.* 45, 499–511.
Philips Petroleum Co. (1980) U.K. patent EP 0019937.
Philips Petroleum Co. (1984) U.S. patent US 4485016 and EP 0071990.
Rokem, J.S., Goldberg, I., and Mateles, R.I. (1980) *J. Gen. Microbiol.* 116, 225–232.
Sahm, H., and Wagner, F. (1973) *Arch. Mikrobiol.* 90, 263–268.
Sakai, Y., and Tani, Y. (1986) *Agric. Biol. Chem.* 50, 2615–2620.
Sakai, Y., and Tani, Y. (1987a) *Agric. Biol. Chem.* 51, 2617–2620.
Sakai, Y., and Tani, Y. (1987b) *J. Ferment. Technol.* 65, 489–491.
Sakai, Y., and Tani, Y. (1988) *Appl. Environ. Microbiol.* 54, 485–489.
Sakai, Y., Sawai, T., and Tani, Y. (1987) *Appl. Environ. Microbiol.* 53, 1812–1818.

Samuelov, I., and Goldberg, I. (1982a) *Biotechnol. Bioeng.* 24, 731–736.
Samuelov, I., and Goldberg, I. (1982b) *Biotechnol. Bioeng.* 24, 2605–2608.
Snedcor, B., and Cooney, C.L. (1974) *Appl. Microbiol.* 27, 1112–1117.
Sreekrishna, K., Potenz, R.H.B., Cruze, J.A., et al. (1988) *J. Basic Microbiol.* 28, 265–278.
Sreekrishna, K., Nelles, L., Potenz, R., et al. (1989) *Biochem.* 28, 4117–4125.
Tani, Y., Sakai, Y., and Yamada, H. (1985a) *J. Ferment. Technol.* 63, 443–449.
Tani, Y., Sakai, Y., and Yamada, H. (1985b) *Agric. Biol. Chem.* 49, 2699–2706.
Tani, Y., and Vongsuvanlert, V. (1987) *J. Ferment. Technol.* 65, 405–411.
Tani, Y., Yonehara, T., Sakai, Y., and Yoon, B.D. (1987) in *Microbial Growth on C_1 Compounds* (Van Verseveld, H.W., and Duine, J.A., eds.), pp. 263–271, Martinus Nijhoff Publishers, Dordrecht, The Netherlands.
Tani, Y., Sawai, T., and Sakai, Y. (1988) *Biotechnol. Bioeng.* 32, 1165–1169.
Thill, G., Davis, G., Stillmann, C., et al. (1987) in *Microbial Growth on C_1 Compounds* (Van Verseveld, H.W., and Duine, J.A., eds.), pp. 289–296, Martinus Nijhoff Publishers, Dordrecht, The Netherlands.
Tschopp, J.F., Sverlow, G., Kosson, R., Craig, W., and Grinna, L. (1987) *Bio/technology* 5, 1305–1308.
Unichika Co. Ltd. (1986) Japanese patent JP 217522.
Unilever (1987a) U.K. patent EP 87/00295 (WO 87/07639).
Unilever (1987b) U.K. patent EP 87/242007.
van Dijken, J.P., Otto, R., and Harder, W. (1976) *Arch. Microbiol.* 111, 137–144.
Verduyn, C., van Dijken, J.P., and Schefers, W.A. (1984) *Int. Microbiol. Meth.* 42, 15–25.
Verduyn, C., Giuseppin, M.L.F., Scheffers, W.A., and van Dijken, J.P. (1988) *Appl. Environ. Microbiol.* 54, 2086–2890.
Volfová, O., Korínek, V., and Kyslíková, E. (1988) *Biotechnol. Lett.* 10, 643–648.
Vongsuvanlert, V., and Tani, Y. (1988) *Agric. Biol. Chem.* 52, 1817–1824.
Wanner, U., and Egli, T. (1990) *FEMS Microbiol. Rev.* 75, 19–44.
Wegner, G.H., and Harder, W. (1987) *Ant. v. Leeuwenhoek* 53, 29–36.
Wilkinson, T.G., and Hamer, G. (1979) *J. Chem. Tech. Biotechnol.* 29, 56–67.
Wilkinson, T.G., Topiwala, H.H., and Hamer, G. (1974) *Biotechnol. Bioeng.* 16, 41–59.
Woodward, J.R. (1986) *Microbiol. Sciences* 3, 9–11.
Yonehara, T., and Tani, Y. (1987) *J. Ferment. Technol.* 65, 255–260.

PART III

Uses of Methylotrophs

CHAPTER 9

Single Cell Protein Production from C_1 Compounds

L. Ekeroth

J. Villadsen

9.1 OVERVIEW OF SINGLE CELL PROTEIN USE

9.1.1 Use of Single Cell Protein—Past, Present, and Future

9.1.1.1 Malnutrition, Protein Deficiency, and Ecosystem Misuse. Protein deficiency in human diets is a problem in certain regions of the world, as are food deficiencies in general. An inadequate distribution system (political or otherwise) for the abundant food resources of the industrial countries has often been blamed, but the problem is not that simple. A rapidly progressing devastation of the ecosystems of developing countries is occurring—in part as a response to a demand for foreign currency derived from export of hardwood, but primarily to meet urgent demands for additional farmland and fuel.

Large-scale malnutrition problems in general and protein deficiency in particular seems to be connected to overexploitation (and other misuse) of local natural resources. The cumulative effect of all habitat misuse (including humus layer thinning all over the world caused by "modern" crop cultivation) will be a progressive decline in the potential level of food production of the world. Thus, we can expect a declining upper limit to the world-wide

206 Single Cell Protein Production from C₁ Compounds

agricultural production potential at the same time that the global population is increasing rapidly.

9.1.1.2 What Can Be Done? In the developing countries it is well recognized by social workers, health officers, and district engineers as well as by foreign experts that the problems, especially at the village level, often defy solution by otherwise logical and reasonable methods proposed by international bodies.

Events seem to unfold, following their own logical course, whether it is within the single family that destroys a nearby habitat to feed a large number of children, or whether it is within the government that condones destructive processes on a large scale in an attempt to contain an endemic drain of foreign currency.

At the present, redistribution of assets such as food may be an adequate short-term solution if it is combined with preservation of ecosystems in the form of national forests or parks. In the longer perspective, the world's food production should, to whatever extent possible, be directed away from ecosystem-destructive procedures like overfishing and forest destruction. Production of single cell protein, SCP, (protein-rich, microbial biomass produced for feed or food purposes) based on an abundant but often sorely misused natural resource—the vast natural gas fields which have been and continuously are being discovered around the world—could become an important nondestructive factor in closing the food and especially the protein gap that afflicts large numbers of people in developing countries. At least one may hope to slow down the most irreversible ecosystem destruction and create a situation in which the necessary social changes towards an ultimate social and ecological improvement can develop. Social security is important, because it takes away one incentive for families to have too many children: fear of old-age poverty.

9.1.1.3 SCP in Human Diets. The direct use of SCP as an ingredient in human diets was the *leit-motif* of much research in the 1950s and 1960s. One may also today consider an admixture of some industrially produced SCP in bread to be an improvement on the recipes used in some cities of the developing world (fish meal of doubtful origin is sometimes an accepted substitute for flour), but in general the idea that SCP should be used on a large scale for human consumption has been abandoned, largely due to the fear that it may contain toxic compounds. There is no really logical reason for this, since lactic acid bacteria are ingested by consumers of fermented milk products all over the world, and since every loaf of bread contains natural SCP from the yeast used in the baking process. The only known constituent of SCP as such, which due to its high concentration (about 9% by weight in untreated bacterial SCP), may cause problems is the group of nucleic acids named purines. In humans, excessive intake of purines can cause gout, especially if liquid intake is low. But with sufficient liquid intake there is no ill effect. Also, since liver, meat, eggs, roe, and legumes are also rich in purines, neither baker's yeast and fermented milk products, nor bacterial SCP has a particularly exposed position.

9.1.1.4 SCP as a Feedstuff for the Future. There are plenty of indications that large-scale SCP production may indirectly, as a feedstuff for animals, become a blessing for the poorer countries of the world. In addition, there are several good reasons why SCP production may also be advantageous for more-developed countries, which are fully equipped to give their population a rich nutrient diet, to feed livestock, and even to export agricultural products on a large scale.

One obvious advantage of the use of SCP is that it is possible to produce SCP of a specific nutritional quality that supplements the nutritional composition of other feedstuffs, or which may give a feedstuff combination of superior quality or price. The invariability of an industrially produced protein: free from price fluctuations caused by climatic events, and always of the same composition and free of mycotoxins—these are powerful incentives for companies in industrialized countries to invest the considerable capital that goes into an SCP plant.

Furthermore, in the long perspective it is doubtful that even industrial countries can claim to be self-sufficient in feedstuffs. Industrial fisheries are for good reasons being hemmed in from all sides by environmental protection regulations, and quarrels over quota arrangements will become progressively more virulent. This has an immediate effect on poultry farming, mink farming, the pet food industry, and on the rapidly growing aquaculture industry.

The export of salmon and sea trout from Norway was until recently supported by an abundant supply of an available feed fish meal produced and exported from Norway. Now Norway is a net importer of fish meal and is eager to find substitutes for its declining production of fish meal. The production of SCP using natural gas as a carbon source is an obvious solution, considering Norway's vast North Sea gas fields and knowing that SCP can replace fish meal on a one-to-one basis. The effort of Asian countries to upgrade the nutritional quality of the human diet for their burgeoning populations by an explosive development of aquafarming enterprises is in danger of collapsing, if substitutes for fish meal—the most important protein source for most farmed fish—are not found. Here SCP could be extremely useful.

Even the endless open country devoted to agriculture in North America, in southern South America, and in Europe may not be permanent in a world which is becoming increasingly conscious of the dangers of irreversible and catastrophic climatic changes. Reforestation programs have been suggested for both ecological and long-term economical reasons. Scientific farming of trees in large areas of the industrialized world may also incidentally put a brake on the wasteful deforestation of developing countries, and the attempts of ecologically fragile tropical nations to produce paper pulp for developed countries by replacing natural habitats for pine forests. But reforestation means decreasing the feedstuff production, and therefore there is a need for development of nonagricultural feedstuff production.

There are many countries, especially in gas-rich regions like the Middle East, who should be eager to produce SCP. Their natural gas comes in quantities and at a price which would enable them to export SCP worldwide. The infrastructure is already available at many industrial sites, such as Al Jubail in Saudi Arabia, which have been planned and built to answer the needs of the 21st century. These countries could also find immediate outlets for SCP to support their growing livestock industry, and they could realize their long-cherished goal of diversifying an originally petrochemically based export industry.

9.1.2 Various Routes to SCP Production

SCP can be produced in a variety of ways, and some "SCP" is not even produced from single cell organisms. The various routes to produce SCP are discussed below.

One way to discuss SCP production is to distinguish between SCP made on renewable substrates and SCP made on exhaustible substrates. This distinction makes sense for planners trying to optimize the use of resources, but it may also be confusing because some SCP processes may be carried out using both renewable and exhaustible substrates.

Therefore, we will view SCP production the other way—that is, we will focus on the processes, and for each process, we will discuss potential raw materials, be they renewable or exhaustible.

A first distinction is between autotrophic and heterotrophic SCP—between processes having a net gain of organic carbon and processes having a net loss of organic carbon.

9.1.2.1 Autotrophic SCP Production.

Autotrophic SCP production makes use of the fact that most other crops use up only a small percentage of the total energy that can be assimilated from the sunlight. SCP crops may use the available solar energy almost quantitatively, yielding a much larger productivity per area cultivated. Usually filamentous (i.e., not *single* cell) algae or cyanobacteria are grown because they are easier to harvest than nonfilamentous cells. Cyanobacteria like *Spirulina* can assimilate atmospheric nitrogen, but the major limitation of the process is the need for adequate amounts of carbon dioxide for photosynthetic growth. Bicarbonate or carbonate may be used, depending on the pH buffer capacity of the culture water and on the addition of acids if economically feasible.

Autotrophic SCP contains much pigment that may be seen either as a desired raw material or as a nuisance reducing the usefulness of the SCP. *Spirulina* is used and has been used for hundreds of years as a supplement in human diets. However, in order to get a reasonably pure culture under nonsterile conditions, *Spirulina* has to be cultivated at high pH, which hampers productivity somewhat. Rates of 50 metric tons ha^{-1} $year^{-1}$ can be achieved (Olsen and Allermann 1987).

This process requires only cheap substrates, and the process may be run at even higher productivity if mixed algal populations are acceptable.

9.1.2.2 Non-C_1 Heterotrophic SCP Production. Heterotrophic SCP can be grown both on carbon sources that could have been used for other purposes and on organic waste. This distinction is not clear, however, because certain carbon sources of some inherent value may be considered unsuitable for technical reasons because they are not available in pure form.

One such substrate is the n-alkanes ("paraffins" C_{14} to C_{18}) found in crude mineral oil. Though valuable in themselves, they increase the viscosity of the oil and make pumping more expensive and refining difficult (Sharp 1989). Attempts have been made to extract the alkanes directly from the oil through the action of yeasts like *Candida lipolytica* grown directly in emulsified oil. This is feasible, but the yeasts produce emulsifiers that are difficult to remove from the purified mineral oil. One alternative is to extract the alkanes using molecular sieves, but this raises the question of profitability of the SCP production, because production of SCP is then not directly linked to purification of the mineral oil.

Water that is heavily contaminated with organic waste products has also been used as substrate for SCP production, e.g., sulfite liquor from paper mills and wastewater from food- or feed-processing factories. The quality of such SCP depends heavily on the quality of the wastewater, and also there is the obvious problem of coupling two processes serially: changing the first affects the second.

On the other hand, very high quality substrates such as starch hydrolysate can be used to produce SCP that can be ingested in bulk quantities by humans. An example is the "Mycoprotein" developed by Ranks Hovis McDougall (RHM) and Imperial Chemical Industries (ICI)—the organism cultivated being the filamentous fungus *Fusarium graminearum* (Sharp 1989). The purine content of the SCP is reduced by heating and washing.

9.1.2.3 Methylotrophic SCP Production. All the other useful SCP processes are based on the use of methylotrophs. The most successful Western one has been the ICI process for making "Pruteen" using the bacterium *Methylophilus methylotrophus* grown on methanol.

Methanol is manufactured from a number of organic carbon sources, both renewable ones like wood and exhaustible ones like natural gas, giving potential robustness to the use of methanol as an SCP substrate. Unfortunately, the prices of these carbon sources have all tended to increase compared to the prices of alternative feedstuffs, making the use of methanol problematic unless the SCP is of high value, justifying the price of the methanol. Also, methanol necessitates the use of sterile culture conditions. This adds significantly to production costs.

Sterility is far less important when methane is used as the carbon substrate, and cheap methane sources such as biogas from waste can be used. With biogas, two or more processes are coupled serially as was the case for use of wastewater for production of SCP (see Section 9.1.2.2). Separating the SCP culture liquid from the wastewater, which is the main idea of using biogas as a carbon source, does, however, give greater robustness to this procedure compared to production of SCP directly in wastewater. Considerable research efforts are being devoted to this subject—not primarily in order to produce SCP, but rather to improve the treatment of the waste material.

9.2 INDUSTRIAL SCP PRODUCTION USING C_1 COMPOUNDS

9.2.1 Production Requirements

Industrial scale production of methylotrophic bacteria for SCP is primarily an oxidation process. Methane or methanol is the usual C_1 substrate, and oxygen must be supplied.

9.2.1.1 Culture Requirements.
Some methylotrophs require only oxygen + methane (or methanol), water, nitrogen-free minerals, and atmospheric nitrogen for growth. However, ammonia, or some mineral nitrogen source (NH_4^+, NO_3^-), or both, is usually added, since it gives better growth and yield on the most expensive substrate, the C_1 source. The nutrients must be well mixed, because uneven distribution in the fermentor, especially of the C_1 substrate, oxygen and/or the nitrogen source, can cause reduced yield and even process instability or termination.

The process also requires the following: the appropriate temperature, salinity, composition of mineral and organic substances in the culture medium (this is a static requirement in addition to the requirement of substrate feed), oxygen tension, carbon dioxide level, and an appropriate degree of mixing of the culture broth.

The use of methylotrophic bacteria also requires protection against the following: introduced and produced inhibitory substances, introduced microbial competitors, introduced predatory organisms, and introduced pathogens—either to the methylotrophs themselves, to "helping" microorganisms cocultivated with the methylotrophs, or the consumers of the SCP.

Sterile conditions must be used if methanol is the C_1 substrate, because methanol is readily used as a substrate by many organisms, such as some microbial competitors and pathogens. In contrast, if methane is used as the C_1 substrate, sterility is needed only as protection against predators and pathogens to the SCP culture. In fact, the admixture of harmless heterotrophs to the methanotrophic culture is useful for keeping the available level of other carbon sources below the requirements of microorganisms that are potentially pathogenic to the consumers of the SCP. An additional advantage of adding heterotrophs to the culture is that "contaminated" methane, e.g., natural gas, can be used as a carbon source, since inhibitory contaminants or substances produced from contaminants by the methanotrophs can be removed by the admixed heterotrophs. An example is ethane in natural gas; it is oxidized by the methanotrophs to inhibitory acetate, which is readily consumed by the admixed heterotrophs.

9.2.1.2 Minimizing Costs. Cost considerations are also extremely important. It has already been mentioned that ammonia or a mineral nitrogen source is preferred for reasons of cost, although some methylotrophs are capable of using atmospheric nitrogen.

Also, since methylotrophs produce a lot of heat (about 2×10^4 kJ/kg dry matter during growth on methane), cheap cooling is essential. Therefore, the use of methylotrophs that can grow at high temperatures will be advantageous, since a high growth temperature drastically reduces the cost of cooling, e.g., by seawater. Hence, a culture capable of growing at high temperatures should usually be chosen, and the plant should be located where there is easy access to cooling water, e.g., close to the sea.

If natural gas is used as the C_1 substrate, very efficient mass-transfer equipment—with respect to both fixed costs and operation costs—should be selected. Both the gas and the air should be dispersed into the culture liquid as fine bubbles; dispersing the liquid as droplets in a continuous gas phase or using liquid films would demand highly expensive fermentors.

Also, down-stream processes, especially removal of water, should be as economical as possible, calling for the highest possible culture density.

9.2.2 Fermentor Design Principles

The main objectives in the design of an aerated fermentor are to achieve good mixing and good mass transfer between bubbles and liquid.

9.2.2.1 Mixing in the Fermentor. In an aerated fermentor, the bubbles themselves will cause effective mixing of the liquid at the millimeter scale.

The fermentor design should therefore focus on mixing at larger scale and on creating and maintaining evenly distributed bubbles of the desired size. The cost of keeping a certain degree of mixedness in the liquid increases with the volume. Hence, in terms of the mixing cost, a fully mixed large fermentor is considerably more expensive to run per unit volume than a laboratory fermentor.

If one makes the somewhat unrealistic assumption that mixing is achieved through isotrophic turbulence in the fermentor liquid, a rough estimate of how the energy consumption increases with volume is as follows:

$$\left(\frac{P}{V}\right) \propto \frac{(\text{volume})^{2/3}}{l} \tag{9.1}$$

where P/V is the power input (W/m³) and l is the wave length of the turbulence created by the power input.

For a linear scale-up of the fermentor, whereby the size of the turbulence-creating device and hence presumably the wave length of primary turbulence are both proportional to (volume)$^{1/3}$, one obtains:

$$\left(\frac{P}{V}\right) \propto (\text{volume})^{1/3} \tag{9.2}$$

implying that the specific energy consumption required for a given degree of total mixing increases linearly with the scale of the fermentor.

Industrial SCP production is performed in large fermentors—of about 20–2,000 m³. This large size would result in extremely high energy consumption costs if mixing was done on the full scale of the fermentor.

However, costs can be greatly reduced by the use of only local mixing, if the fermentor processes are also controlled locally, so that mixing on the full scale of the fermentor is not necessary. However, establishing effective local control of all process variables by installing a large number of measurement devices throughout the entire volume of the large fermentor would be prohibitively expensive. Fortunately, many process variables only change very slowly from their set point values, and these variables can be controlled by rather infrequent measurements. One way of obtaining this is by circulation of the total fermentor volume past a single measurement and control unit. One fermentor design based on this concept is called the "Loop Reactor" (LR, Figure 9–1). Loop reactors can be designed for very large working volumes without excessive expenditure for measurement and control equipment, and this translates into a considerable saving on both capital costs and operational costs.

9.2.2.2 Liquid Circulation Time and Control Problems.
The circulation time of a large loop reactor may, however, be large compared to the time constants for some of the important reactions in the culture, i.e., it may

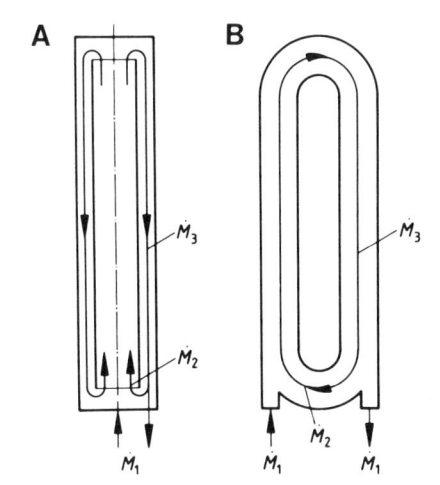

FIGURE 9-1 Liquid circulation patterns in two Loop Reactors (LR). (A) With internal circulation; (B) with external circulation. $\dot{M}_1$ is liquid flow rate in and out of the loop, $\dot{M}_3$ is liquid flow rate through the main part of the loop, and $\dot{M}_2 = \dot{M}_3 - \dot{M}_1$. Reproduced with permission from Blenke (1985).

exceed a critical maximum period between successive control actions for some of the variables. The most noteworthy example of this phenomenon in a large loop is the "pressure cycle fermentor" erected by ICI at Billingham, England. This fermentor was built for the production of SCP on methanol using the bacterium *Methylophilus methylotrophus*. The fermentor is very large (about 60 m high), and the organism will grow only in the presence of a small surplus of methanol, above which the biomass yield on methanol, Y_{sx}, decreases—so methanol had to be deployed at many places along the flow path of the loop. Actually, more than one thousand methanol inlets were required for optimal performance (Schügerl 1985). It was a major technological feat that ICI was able to successfully commission the production unit in the early 1980s. Today the unit still produces SCP, however, not from C_1 substrate, but using hydrolyzed starch to make mycoprotein for food additives, as was briefly mentioned previously.

Similarly, excess NH_3 or NH_4^+ can be oxidized into inhibitory nitrite by some methylotrophs, indicating that the addition of these N sources must be carefully controlled and that good mixing is required in the inlet regions.

As a final problem in loop reactors, it is known that facultative anaerobes such as baker's yeast, *Saccharomyces cerevisiae*, can change from anaerobic to aerobic metabolism and back again quite rapidly. This may cause trouble if, due to the circulation in the loop, the dissolved oxygen tension in the fermentor liquid varies at a frequency near the maximum turnover frequency between aerobic and anaerobic metabolism by the yeast, forcing the culture into a synchronous growth pattern that is always out of phase with the oxygen tension, and hence resulting in low production rate and poor biomass yield.

One cannot a priori exclude the possibility of anaerobic metabolism in a microorganism capable of utilizing compounds like methane or methanol, because the microorganism may be able to use other compounds, including substances present inside the cells at the onset of a sudden drop in the dissolved oxygen level.

If a transient anaerobic metabolism is actually taking control of the culture with detrimental effects on yield and productivity, one must either choose a long circulation time in the loop, giving the microorganisms sufficient time to adjust themselves to the changing oxygen levels, or a short circulation time, which will keep the microorganisms from even attempting to switch their metabolism.

With the exception of certain critical control jobs, such as the fine tuning of the N level in the fermentor, most other controls can be carried out satisfactorily by letting the fermentor liquid pass measurement and control equipment in a bulk flow manner.

9.2.2.3 Bubble Behavior in Loops. The loop design has the further advantage of permitting plug flow right from an inlet of a substrate, e.g., methane, to the outlet of products and substrate remnants and under those fermentation conditions there is no mixing in the direction of the flow. In this way one obtains both a high concentration of the substrate near the inlet, giving a high production rate there, without reducing the utilization of the substrate. If switching of metabolism either does not take place or has no detrimental effect, then the plug flow characteristic of the loop design will, at a given energy consumption, give a very favorable combination of productivity and substrate utilization—better than what could possibly be achieved in a fully mixed fermentor.

This is partly due to energy saving in plug flow with only local mixing (as approximated by Equation 9.1) and partly to the fact that bubbles of different composition, which will continuously merge in the full-scale mixed fermentor, will be physically separated in the loop fermentor. The extent to which the coalescence and break-up of bubbles take place can for practical purposes only be assessed empirically, because these phenomena depend heavily on fermentor design (evenness or unevenness of turbulence), on the gas hold up (total volume fraction of bubbles in bubble/liquid mixture), and on the physical and chemical properties of the liquid—increasing viscosity promotes bubble coalescence, and salts and surfactants (the latter often produced by the culture) inhibit bubble coalescence.

Bubbles rising in a liquid create a turbulence in the liquid, and the turbulence level will, at any given set of liquid parameters (viscosity, surface tension, liquid homogeneity/heterogeniety), determine an equilibrium size of the bubbles. This is true for any turbulence level, regardless of how the turbulence is created.

The influence of salts on bubble coalescence in otherwise pure water can be seen in Figure 9-2. This figure shows the results of an experiment in which small bubbles were produced and afterwards allowed to merge freely. The equilibrium diameter of the bubbles after merging was deter-

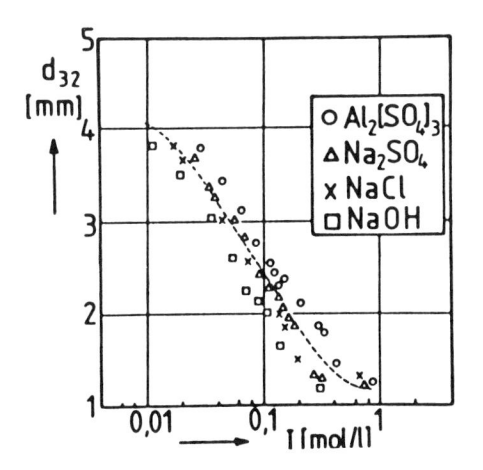

FIGURE 9-2 Diameter of merged bubbles in water containing various electrolytes versus ionic strength. Reproduced with permission from Keitel and Onken (1982).

mined as a function of ionic strength of the water. The measurements deviate for diameters near 1 mm, indicating that perhaps the bubbles were originally produced at this size.

9.2.2.4 Loop Fermentor Designs. The main difference between various loop fermentor designs is related to the methods used to obtain bulk flow, bubble formation, and local mixing.

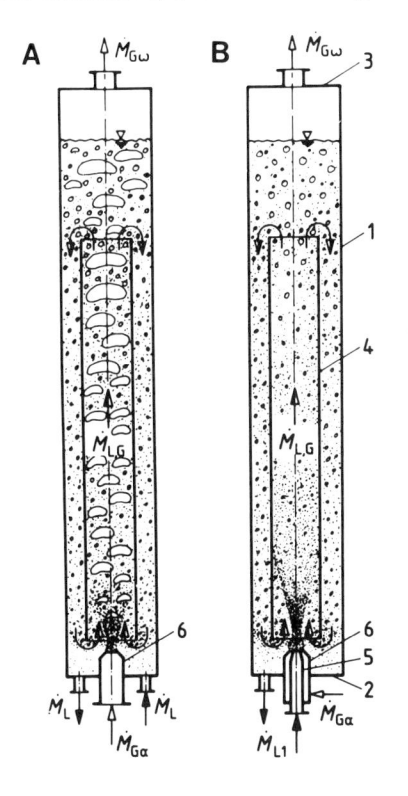

FIGURE 9-3 Diagram of (A) an Air-lift Loop Reactor (ALR) and (B) a Jet Loop Reactor (JLR). The numbers correspond to the following components: 1, jacket; 2, bottom; 3, lid; 4, draft tube; 5, liquid nozzle; 6, gas sparger. $\dot{M}_L$ and $\dot{M}_{L1}$ correspond to $\dot{M}_1$ in Figure 9-1, and $\dot{M}_{G\alpha}$ is flow rate of gas phase (air) into and $\dot{M}_{Gw}$ is flow rate of gas phase out of the loop. Reproduced with permission from Blenke (1985).

Through aeration, bulk flow can be obtained by a density difference between the ascending and the descending part of the loop. A fermentor based on this principle is called an "Air-lift Loop Reactor" (ALR, Figure 9–3A).

Another possibility is to use a mechanical liquid pump located in a separate pumping loop. This loop ends in ports through which the liquid enters into the reaction loop at a considerable speed producing "jets"—hence the name "Jet Loop Reactor" (JLR, Figure 9–3B).

A third way of obtaining bulk flow is by means of a mechanical pump installed inside the loop. If the pump is in the form of a propeller, this design is called the "Propeller Loop Reactor" (PLR, Figure 9–4).

The advantage of the ALR is that no mechanical liquid pump is employed. So, there are no capital costs for this pump, no maintenance, and no problems of keeping a shaft seal tight. The disadvantage of the ALR is that air has to be compressed in order to drive the liquid flow in the fermentor. The work of compression adds an extra energy expenditure compared to a simple mechanical pumping of fermentor liquid. The cost of compression is, however, small, if the air is introduced in the descending liquid as in ICI's deep shaft design (Figure 9–5) for wastewater purification. At the start-up of liquid circulation, air must be introduced into the ascending liquid in order to obtain the correct direction of flow.

It takes some time, depending on gas hold up and on the physical properties of the liquid, before the bubbles have joined each other or fragmented to such a degree that a stable distribution of bubble sizes around

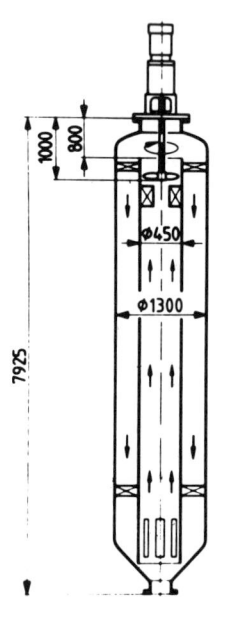

FIGURE 9–4 Diagram of a Propeller Loop Reactor (PLR). Reproduced with permission from Blenke (1985).

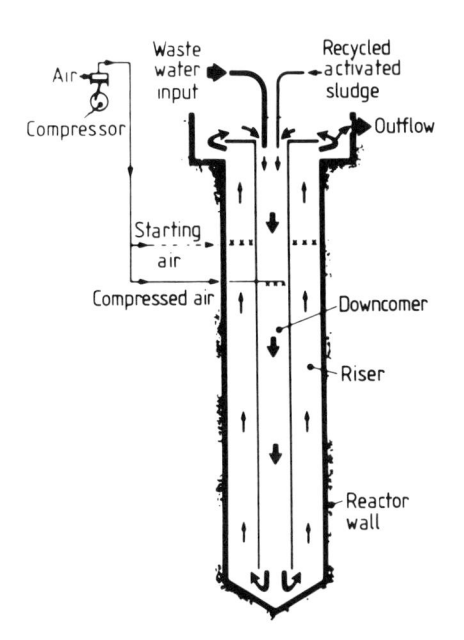

FIGURE 9-5 ICI deep shaft fermentor for biological wastewater purification. Reproduced with permission from Blenke (1985).

the equilibrium size has been reached. Loop fermentors for SCP production will typically be several meters high, and this will typically be sufficient for the bubbles to reach equilibrium size in less than half of their residence time in the liquid. Some medium-scale mixing of the liquid can be obtained at the air inlet ports, and an ALR can thus be constructed completely without any other mechanical mixing structures than the above. An example of this type of reactor is the Uhde Hoechst pilot fermentor. Figure 9-6 gives a flow sheet of a pilot plant for this fermentor.

ICI decided to use the other mixing method in their pressure cycle fermentor (an ALR design) at Billingham (Figure 9-7): a row of perforated (or sieve) plates was inserted into the ascending tube of the fermentor. The plates create additional turbulence for better mixing and renewed fragmentation of bubbles.

JLRs and PLRs can be constructed to give roughly the same performance as ALRs, because the liquid pumps in these designs do not necessarily play an important role in mixing and bubble fragmentation.

9.2.2.5 Using JLRs for High Productivity. Especially in the case of a JLR, one does, however, have a powerful tool for mixing and bubble fragmentation. The liquid jets can be associated directly with inlet ports for gas or air, so that the gas or air is fragmented in a turbulence field created by collision of a jet of perhaps 5–50 m/s with relatively stagnant liquid. Bubble

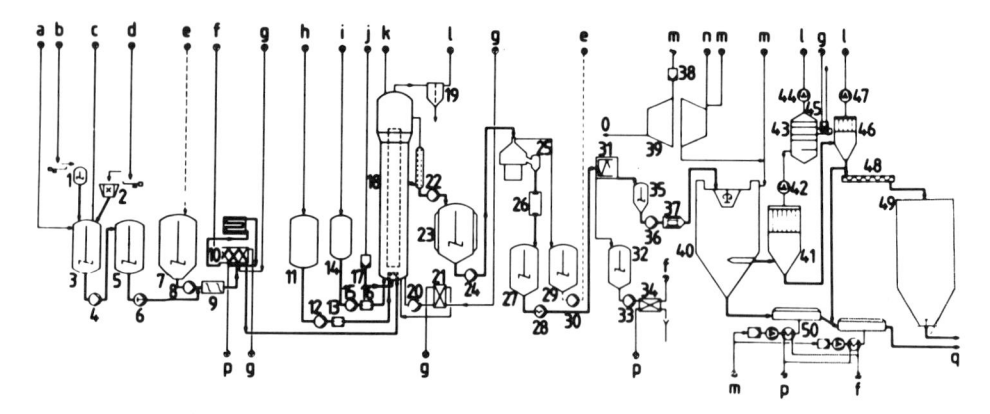

FIGURE 9-6 A flow sheet for the Uhde Hoechst process for SCP production: a pilot plant using a loop reactor (JLR or ALR) designed to produce 1,000 tons of SCP per year. The numbers indicate the following parts of the plant: 1, mixing vessel for trace elements; 2, crusher; 3, mixing vessel for nutrient solution; 4, pump; 5, nutrient solution reservoir; 6, pump; 7, recycle water reservoir; 8, pump; 9, mixing device; 10, thermal sterilization section; 11, methanol reservoir; 12, pump; 13, sterilizing filter; 14, NH$_3$ reservoir; 15, pump; 16, sterilizing filter; 17, sterilizing filter; 18, fermentor; 19, separator; 20, pump; 21, cooler; 22, pump; 23, harvest tank; 24, pump; 25, separator; 26, conditioning section; 27, concentrating vessel; 28, pump; 29, clear liquid vessel; 30, pump; 31, decanter; 32, wastewater tank; 33, pump; 34, wastewater conditioning section; 35, concentrating vessel; 36, pump; 37, thermolysis facility; 38, filter; 39, air compressor; 40, drier; 41, filter; 42, ventilator; 43, scrubber/cooler; 44, ventilator; 45, cooler; 46, solids separator; 47, ventilator; 48, screw conveyor; 49, silo; 50, granulation facilities. The letters correspond to the following components: a, H$_3$PO$_4$; b, trace elements; c, H$_2$O; d, nutrient salts; e, recycle water; f, steam; g, cooling water; h, methanol; i, NH$_3$; j, process air; k, inoculant; l, waste air; m, air; n, gas; o, process air to the fermentor; p, condensate; q, bagging, storage and dispatch. Reproduced with permission from Blenke (1985).

diameters can be chosen almost at will—but there is, of course, a price to be paid.

The purpose of reducing the bubble sizes is to increase the production, since smaller bubbles have a larger specific area, a (m^2/m^{-3}), and thus higher rate of mass transfer into and out of the liquid. Increased fragmentation requires increased power, and this is where the operation costs enter the picture. Unfortunately, the two factors have different weights. The mass transfer, N_g (kg m^{-3} sec^{-1}), is related to the specific power input, P/V (W m^{-3}), according to the formula:

$$N_g = k_L a \cdot (c_g^* - c_L) \propto \left(\frac{P}{V}\right)^x \cdot (c_g^* - c_L) \qquad (9.3)$$

where k_L (m · sec^{-1}) is a mass transfer coefficient that varies with the prop-

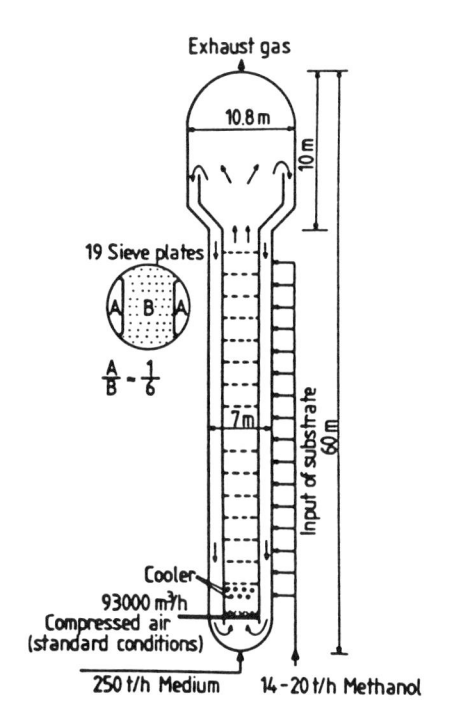

9.2 Industrial SCP Production Using C₁ Compounds 219

FIGURE 9-7 Diagram of a pressure cycle fermentor. Reproduced with permission from Blenke (1985).

erties of the liquid and with the specific power input. c_g^* is the liquid phase concentration of the transported gas just inside the gas-liquid interface. c_g^* is related to the bulk gas phase concentration c_g of the solute via an absorption equilibrium which is only insignificantly modified by gas phase transport resistance. c_L is the bulk liquid phase solute concentration.

Unfortunately, the value of exponent x in Equation 9.3 is less than 1. Values from 0.4 to 0.8 are given in the literature; 0.65 is probably the best estimate for an SCP culture. Thus the energy expenditure increases more rapidly with P/V than does $k_L a$.

Still, if capital costs, costs of substrate, and costs of down-stream processing are considered together with energy costs, the indication is that the optimal $k_L a$ is very large. Mass transfer is often the slowest kinetic process of the total SCP system, and when $k_L a$ is high, one can sustain a dense culture in a small fermentor. This saves both operational costs in the downstream processes and investment costs.

9.2.2.6 Energy Consumption and Gas Utilization. Table 9-1 compares the energy efficiency in kg O_2 transferred/kWh for several different fermentor designs. The figures cannot be assumed to represent major differences between various fermentor principles, because some of the measurements were

TABLE 9-1 Energy Efficiencies of Oxygen Transfer in Various Fermentor Types

Fermentor Type	E_{O2} (kg O_2/kWh)	Reference
Stirred tank with:		
Tubular impeller	1.4	Zlokarnik (1980)
Turbine stirrer	2.0–2.5	Zlokarnik (1980)
Propeller	0.8–1.1	Zlokarnik (1980)
Nozzle tower loop	2.1	Adler and Fiechter (1983)
Plunging jet loop	0.88	Zlokarnik (1980)
	3.0	Bin and Smith (1982)
	2.0	Meijel Sewage Treatment Plant
	3.3	Böhnke (1970)
Plunging jet channel	4.0	Müllner (1973)
Multistage tower loop	1.6	Viesturs et al. (1981)
Single-stage air lift tower loop (ICI pressure cycle fermentor)	1.5 at 6.6 kW m^{-3}	Walker and Wilkinson (1979) (SCP production)
	2.0 at 1.5 kW m^{-3}	Hines (1978)
Single stage air lift tower loop	2.0	Uhde Brochure (SCP production)
	2.0	Euzen et al. (1978)
	3.3–3.5	Uhde Hoechst brochure (wastewater treatment)
Deep shaft (ICI)	3 at 1 kW m^{-3}	Hines (1978)
Tower loop with perforated plate	3.39	Zlokarnik (1980)
Static mixer (kenics)	3.3	Zlokarnik (1980)
Tower loop with slit jet	3.6	Zlokarnik (1980)
Tower loop with injector nozzle	3.6	Zlokarnik (1980)
Tower loop with porous plate	4.0	Zlokarnik (1980)

Source: With permission from Schügerl (1985).

made at noncomparable experimental conditions. It is, however, noteworthy, that ICI's pressure cycle fermentor with an air inlet at the bottom is less efficient than ICI's deep shaft fermentor, with an air inlet in the descending part of the loop. One should also note that Zlokarnik (1980) obtained a high energy efficiency in various loop fermentor designs compared to stirred tanks. The stirred tanks are typical examples of total-volume mixed fermentors.

If bubbles are of equal size and move with the same speed from air inlet (solute concentration $c_{g,in}$) to air outlet (solute concentration $c_{g,out}$) without merging with other bubbles of different gas composition, and if the gas volume from the fermentor equals the inlet gas volume, then the term c_g in Equation 9.3 can be calculated as a logarithmic mean:

$$c_g = \frac{c_{g,in} - c_{g,out}}{\ln c_{g,in} - \ln c_{g,out}} \tag{9.4}$$

where $c_{g,in}$ is the gas concentration at the gas inlet and $c_{g,out}$ is the gas concentration at the fermentor's gas outlet.

In a fully mixed fermentor:

$$c_g = c_{g,out} \tag{9.5}$$

9.2.2.7 Operation of Aerated Loops. The speed with which bubbles rise in the fermentor is crucial to the design of the fermentor loop and the operation of the fermentors. Bubbles should have an adequate residence time to give a high utilization of solute, and bubbles should also be allowed enough time to separate from the liquid at the top of the fermentor. Figure 9-8 shows the terminal speed of bubbles rising in dilute fermentation media as a function of bubble size. Interesting enough, it appears that bubbles in the diameter range of 2-25 mm all have a terminal speed of 20-35 cm/sec. This greatly simplifies the design and operation of a loop for bubbles in this commonly encountered size range.

As discussed earlier, in a JLR it is possible to produce much smaller bubbles. In this case, a very productive fermentor is obtained, but one has to allow bubbles to either merge with each other or leave the liquid at a surface with only minor liquid speeds. Above 30% gas hold up, the bubble merging process accelerates rapidly, leading to an inhomogeneous swarm of larger and smaller bubbles.

Experiments have been made with JLR models containing water with high ionic strength (Blenke 1979). The linear water velocity in the loop probably exceeded 60 cm/sec, and the liquid jets approached 60 m/sec. Thus, small bubbles were produced, and since the loop head was not widened to enhance bubble escape, bubbles must have been recycled in the loop whenever the liquid speed was high. The high ionic strength must have partially inhibited bubble merging, and the system may have behaved in a highly complex manner with qualitatively changing characteristics at changing liquid speeds and gas flows. It is therefore difficult to generalize from Blenke's data, but they may serve as a good example of an extreme situation with gas hold up values of up to 70% (Figure 9-9) and specific interface areas approaching 3000 m^2/m^3 (Figure 9-10). In the figures, V_G is the aeration rate, D_1 is the diameter of the liquid jet nozzle, P_L/V is the power of the liquid jet divided by the reactor volume, and Re_1 is the jet speed $\times$ D_1 $\times$ (liquid kinematic viscosity)$^{-1}$.

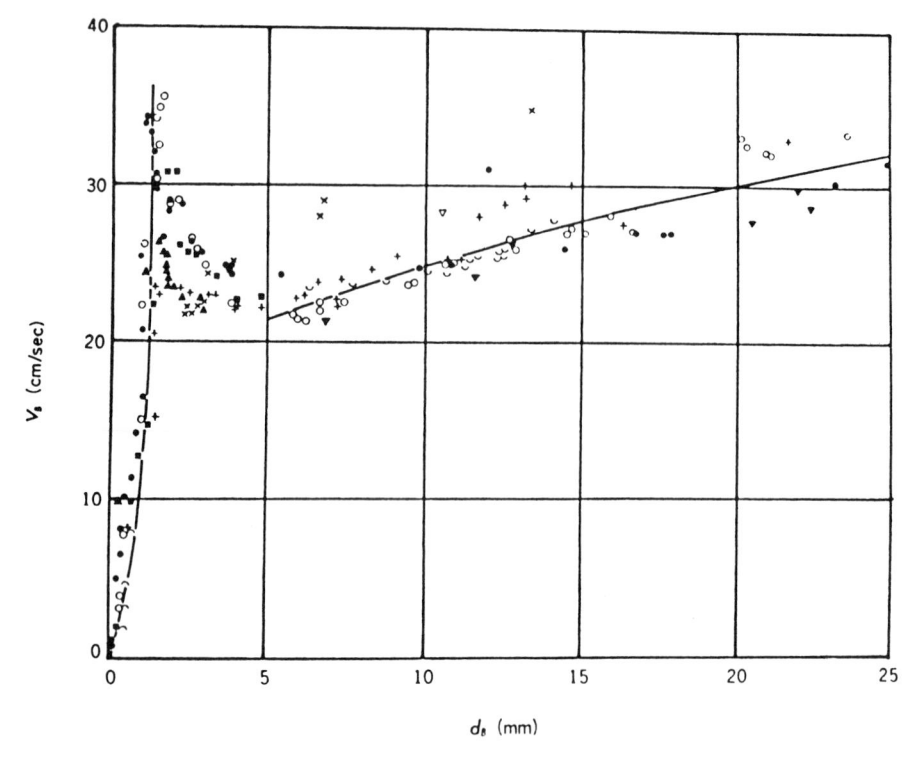

FIGURE 9-8 Ascending terminal velocity of single bubbles in water; the ordinate is the ascending terminal velocity, v_B, of single bubbles, while the abscissa represents bubble diameter, d_B. The various symbols used in the figure indicate different sources of original data (Wilson 1953). Reproduced with permission from Aiba et al. (1965).

9.2.3 Down-Stream Processing

Currently, down-stream processing of methylotrophic SCP is geared primarily to its use as a feedstuff supplement. Its texture is relevant only to considerations of binding properties in compounded feedstuffs, dust hazards, etc. The more important factors are its nutritional quality (both chemical composition and digestibility), and the level of pathogens toxic substances, allergens, and gustatory repellents. Down-stream processing can improve digestibility and eliminate pathogens, and—at a high cost—toxic substances, allergens, and repellents.

Down-stream processing can also deteriorate the biomass in terms of the content of important nutrients and its digestibility. Sometimes toxic substances, allergens, and repellents may even be produced, and mishandling may increase the rate of decay of the biomass.

9.2.3.1 Water Removal.
The usual first step in SCP processing is to eliminate most of the culture water in order to produce a concentrated paste.

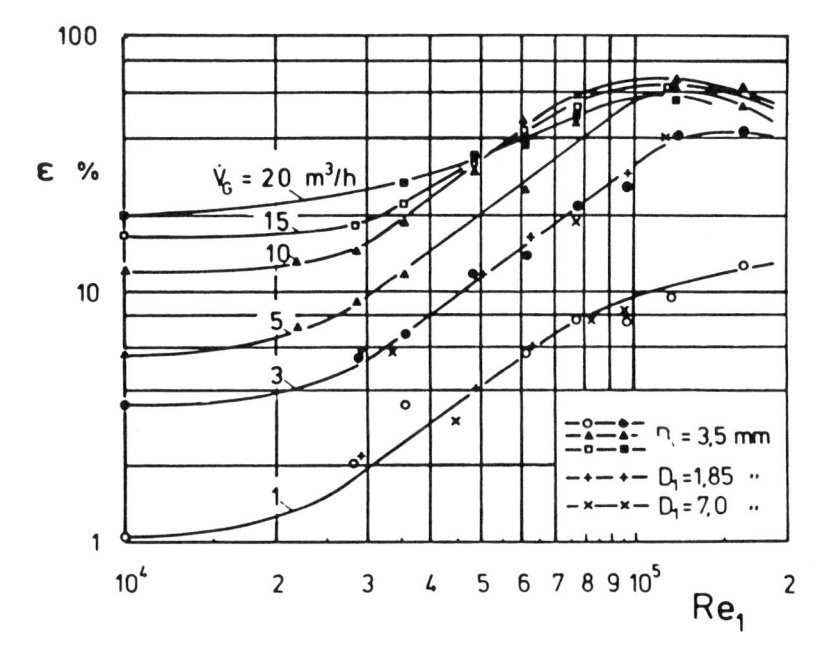

FIGURE 9-9 Gas hold up, ϵ, in a JLR as a function of the jet Reynolds number, Re_1, for different gas flow rates, V_G. Reproduced with permission from Blenke (1979).

The choice is between centrifugation or filtration or perhaps some combination of the two. Both microfilters and decanter centrifuges can be used as the final step in the concentration process, both giving a paste of about 20% dry matter.

Usually one would not want anoxic conditions to prevail for long in the culture liquid, since toxic substances and repellents may be formed under these conditions. Consequently water removal should be carried out fairly rapidly, and the raw culture broth to be concentrated should be as biomass-rich as possible, reducing the amount of water to be removed.

It is possible to inhibit the culture prior to concentration, also without any heating or cooling, but if the culture water is to be recycled to the fermentor, then there is only a limited choice of inhibition methods.

One may enhance water removal and minimize energy expenditure for this by agglomeration of the biomass prior to any further treatment. Again the choice of methods is limited if the water is to be recycled.

9.2.3.2 Heat Treatment. Sterility of the final product can be achieved by a heat treatment, which will also increase its digestibility by lysing the cells (thermolysis). Heat treatment will, however, also denature some proteins and possibly will increase their allergenic properties. A low level of aller-

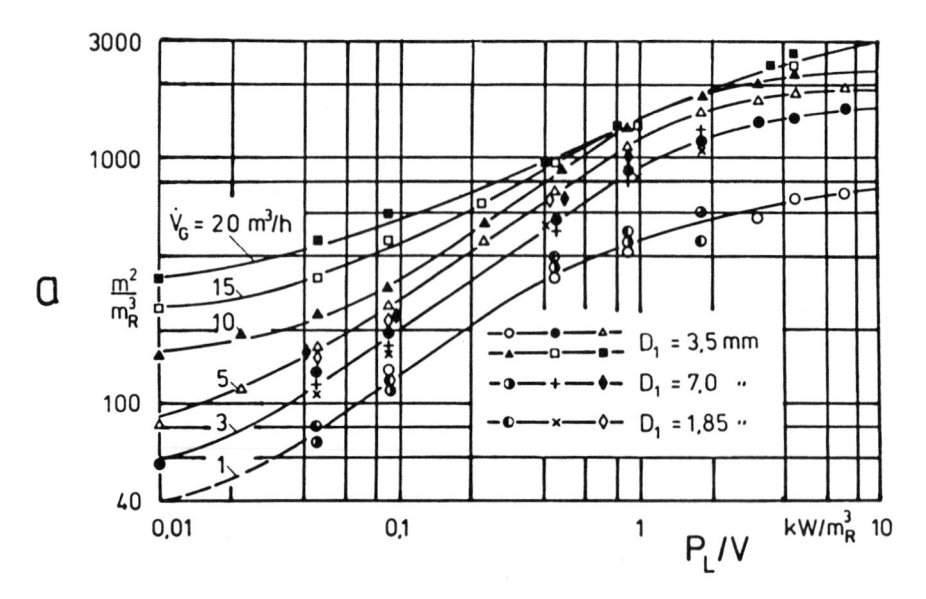

FIGURE 9-10 Specific interfacial area, a, in a JLR as a function of specific liquid jet power input, P_L/V. Reproduced with permission from Blenke (1979).

genicity might not be detrimental (most feeds and foods are slightly allergenic), but in any case the product should be tested for any noticeable allergic reactions in animals. Overheating the biomass leads to reduced levels of some vitamins and also to reduced digestibility of the protein, but a carefully executed heat treatment may be beneficial through activation of RNAses, enzymes that will reduce the nucleotide content of the SCP. SCP fully suitable for human consumption in bulk quantities may be produced in this way (Sharp 1989).

9.2.3.3 Drying. Further modifications of the biomass can be made after the heat treatment, or the product may be dried (for example, in a spray drier, or an extruder) and packaged. The storage stability is excellent. No deterioration of the product can be detected even after one year of storage in plastic sacks.

9.3 IS THERE A FUTURE FOR SCP PRODUCTION?

9.3.1 Availability of C_1 Compounds

9.3.1.1 Natural Gas. At present the known reserves of natural gas are increasing at a rate of about 5 trillion cubic meters per year. Since the total production is around 2 trillion cubic meters per year, this means that each year about 7 trillion cubic meters of recoverable natural gas are found (British Petroleum 1989). The global reserves of recoverable, natural gas are at present about 115 trillion cubic meters. How this gas should be used is, of course, a matter of priorities. Relevant questions are

1. How quickly should we exploit exhaustible resources like natural gas?
2. What is the best use of exhaustible resources?
3. How much oxidation of natural gas to CO_2 (as a result of combustion, of chemical processes leading to degradable and combustible waste, and of the action of methylotrophs producing SCP used to feed animals) can be tolerated by our global ecology?

These questions and a host of other questions have to be asked and answered, both by local authorities and world-wide. None of the questions can be answered independently, and in particular the third question cannot be tackled without considering all contributions of CO_2 to the ecosystems. Some kind of quota arrangement is needed, calling for global cooperation.

9.3.1.2 Biogas. If the question of the availability and use of methane-containing natural gas gives rise to concern, one must also consider that large amounts of renewable methane are becoming available now. It is well known that anaerobic breakdown of organic matter leads to the formation of a mixture of methane and carbon dioxide known as biogas. The proportions of the two gases is variable, especially in the presence of sunlight where the carbon dioxide may be assimilated by anaerobic photosynthesizing bacteria.

Much research and development work has been carried out on the anaerobic breakdown of organic matter in anaerobic digesters. The technology for design and operation of these digesters is rapidly maturing, and in the long run, biogas will become an obvious alternative to natural gas for production of SCP. The capital cost for establishing large-scale biogas units is formidable, however, and this implies that the extent to which biogas will replace natural gas will be largely determined by political decisions. The use of biogas as a substitute for oil or natural gas even in small, decentralized heating stations is still controversial, but its use in the production of feed for animals seems almost an ideal solution to a major environmental problem.

9.3.1.3 Gas Quality and SCP. From the viewpoint of SCP production there is no big difference between natural gas and biogas as substrate. Both methane sources can be converted to methanol, if one prefers this substrate for SCP production, and both methane sources can readily be utilized by methanotrophic bacteria, because the contaminants of both natural gas and biogas can be removed by other bacteria that can be added to the culture. The only potential obstacle in using biogas is that the hydrogen sulfide it contains might be used by a microorganism having inferior or detrimental SCP qualities. The large CO_2 content of biogas from digesters (about 30%) does not have any detrimental effect on the methanotrophic bacterium *Methylococcus capsulatus*, which is commonly used as an SCP producer due to its high temperature optimum (45°C).

9.3.1.4 Production Potential. The degree to which the world's reserves of natural gas can contribute to the world's production of feedstuffs can be estimated from the fact that just 1% of the world's present production of natural gas equals a production of about 10 million tons of SCP per year, which is about 10% of the world's present production of soya meal.

9.3.2 Marketable Products and Major Markets

9.3.2.1 Competing Protein Sources. The major protein sources in the feed market today are plant seed products: soybean meal, peanut (groundnut) meal, cottonseed meal, rapeseed meal, and sunflowerseed meal, and animal products: fish meal, meat and bone meal, shrimp meal, skimmed milk powder, and whey. In weight, the plant seed products far exceed the animal protein production.

These sources vary greatly in quality. Fish meal may be made either on fresh or on somewhat degraded fish, and cooking and drying procedures may be carried out at different temperatures. Plant seed meal may be from whole or dehulled seeds, and heat treatment may be used, e.g., to denature protease inhibitors in soybeans.

9.3.2.2 Prices. Prices vary greatly depending on quality, and hence the perspective for use of SCP is a cost-benefit perspective that must be viewed in the light of a wide array of feedstuff niches, raw material prices, and treatment prices. Considering the present price of natural gas in a number of producing nations, it seems obvious that C_1-SCP can outcompete almost any other available protein raw material on a cost-per-organic-N basis.

Neither protein prices nor gas prices are, however, set by market forces alone. This is especially true for gas prices which for political reasons (energy saving programs, taxation schemes, self-sufficiency in raw materials) often bear no relation to an abstract price based on its availability in the gas field or to production costs. Also, plant seed protein prices are regulated according to the agricultural policies of certain important nations, in addition to depending on plant oil prices. Hence, it is very difficult to predict the future price of any competing source of protein. Also, gas prices are seldom directly coupled to protein prices, so that changes in the relative price levels may appear to be almost random.

9.3.2.3 Quality of C$_1$-SCP. It is also necessary to compare the benefits and drawbacks of the various protein sources—both in terms of raw material quality and in terms of suitability for any upgrading process.

Comparisons are not easily made at present, because Western production of CH$_4$-SCP has not yet been put to the harsh economic test of large scale (>100,000 tons/year) production, whereas Western CH$_3$OH-SCP (Pruteen) production has come to a standstill for what can only be interpreted as economic reasons: the rising price of methanol and the falling price of feedstuffs.

Some experience is, however, available both from Western and from Eastern European productions of bacterial CH$_4$-SCP and alkane-yeast SCP.

First of all, alkane-yeast SCP has proved to be very valuable as a feed component in probably the most demanding area of all: nutrition of fish fry (Appelbaum and Uland 1979). This implies good digestibility and high nutritional quality. Similar results have been obtained in Eastern Europe when feeding carp fry on a feed product containing 35–50% CH$_4$-SCP.

The successful application of such high percentages of SCP to fish fry feed should probably be seen in conjunction with the high growth rate of the fry, which can almost double their weight every day. In this case, the large amounts of nucleotides and other highly growth-related substances will be a clear advantage, and considering the short retention time of feed in the digestive tract of fish fry, the high digestibility of both alkane-yeast SCP and bacterial CH$_4$-SCP is also a very important advantage.

The market for fish fry feed is, however, small. Individual fish fry weigh only 1 mg at the commencement of feeding and their digestive tracts are fully developed at just 100–1,000 mg body weight.

Fish fry feed may consequently be seen mainly as a highly revealing indicator for testing the digestibility and nutritional value of SCP and other feedstuffs. For comparison, most kinds of carp fry weigh at least 20 mg before they can get much nutrition from boiled soybean crush or from other mainly plant-based compound feeds (Ekeroth, unpublished observations).

Investigations have also been carried out on the utilization of SCP by major feedstuff consumers like poultry, pigs, calves, and carnivorous fish. ICI's Pruteen was tested thoroughly on broilers, pigs, calves, and fish and generally found to be fully acceptable in limited doses, typically even increasing growth performance. Purine nucleotides did not accumulate, and their breakdown and excretion was apparently no problem for the animals in question.

A high percentage of Pruteen in chicken feed did show a coincidence (up to 1%) of liver enlargement, a problem that could be circumvented by a peroxide treatment of the Pruteen before drying (Sharp 1989).

A Soviet form of CH_4-SCP called "Gaprin" has also been thoroughly tested in large cooperatives in the Soviet Union and elsewhere. The results for pigs, calves, and poultry were very satisfactory according to the information published by the government export organization "Medexport."

As in the case of Pruteen, the addition of Gaprin to feed formulas seems favorable. Poultry feed, normally having 40% of its total protein from fish meal, could without negative effects be substituted by feed with 20–30% protein from fish meal and 10–20% protein from Gaprin. In pig feed, Gaprin could fully replace high-protein components like skimmed milk powder, fish meal, and various plant seed meals. For calves, Gaprin could be used as a partial skimmed milk replacer in composite milk substitutes, and in later feeds, Gaprin could partially replace whey (Medexport 1985).

Of major relevance is the test for carcinogenicity using white rats and mice. According to the available information, no carcinogenicity or cocarcinogenecity has been observed for any C_1-based SCP.

On the basis of all the available evidence it thus seems that C_1-SCP is highly nutritive and safe as a limited feed supplement.

9.3.2.4 Future Market Considerations. As pointed out in Section 9.1.1, the objectives of SCP production may vary according to the prevailing conditions in any given country. The main product will probably be bulk SCP, since this can readily be utilized by many consumer animals. A main point to be noted is that some animal productions call for high-quality protein sources: skimmed milk powder for calves and fish meal for poultry, shrimp, and fish. Skimmed milk is, however, better used as a valuable human food, and the production of fish meal requires fish catches that could threaten ocean ecosystems, in which case a good substitute for fish meal could be beneficial for marine ecology in general and for the catchable stocks of high-quality sea fish for human consumption in particular, in addition to being beneficial to the users of high-quality protein feeds.

Even optimistic assessments of future catches of sea fish (Pike 1989) indicate that these will soon become stagnant or actually decrease. It is well known that overfishing is taking place, with industrial catches contributing to the decline of all catchable fish species, not only the industrially relevant ones.

So even though optimists foresee only a minor decline in the world's production of fish meal during the 1990s, surely ecologists (including sea bird experts), politicians worried about feeding the world's increasing population, and fishermen catching fish for human consumption would all agree that substituting fish meal by other protein sources would increase the amount of valuable, proteinaceous human food given to us by the sea, without decreasing our production of other human foods.

The Food and Agricultural Organization (FAO) in Rome has expressed a need for the world's production of fish through fish farming to double during the next decade (to about 10 million tons per year) in order to supplement the stagnating or declining catches of wild fish. It would be a tragic mistake to feed farmed fish major amounts of caught ones (to the level of up to 5 kg total feed per kg farmed fish), thus reducing wild catches of edible fish as the price for increasing production of farmed fish. This is pointless. The increasing need for fish feed should to whatever extent possible be met by nonfish proteins like SCP.

9.3.3 SCP and Public Opinion

Most people feel a natural concern about the quality of the food they eat— concern about its palatability, nutritional content, and possible content of pathogens and toxic substances. As animal feed highly influences the quality of animal-based foodstuffs, similar concern is directed towards feedstuffs. Thus, SCP should be competitive, not only with regard to its actual quality but also in the view of the people who purchase SCP or who consume animal-based foods from animals fed SCP.

9.3.3.1 Feedstuff Toxicity. The presence in food of dangerous manmade poisons like dioxin and some natural poisons like aflatoxin and botulin is a very serious matter that receives much attention, but we should remember that chemical "warfare" or at least defense is a very widespread phenomenon in nature. To a large extent immobile organisms like plants base their survival on production of more or less toxic substances. Thus almost all food contains some level of toxins. It is not a coincidence, therefore, that both man and animals have very large livers capable of detoxifying an immense variety of substances.

One should distinguish between substances posing threats to health even in "naturally" occurring doses—heavy metals, dioxin, aflatoxin, botulin— and minor poisons normally fully neutralizable in the liver. If one does not make this distinction, then almost nothing in this world can be considered "safe" feed or food; almost everything constitutes a "health hazard," but the difference lies in the severity of the hazard.

9.3.3.2 A Common Basis for Feedstuff Evaluation. Realizing that traditional feed- or foodstuffs may be hazardous to the health, one should not automatically see new products as possible threats. Rather, they should be seen as alternatives, perhaps reducing the risks posed by traditional feed- or foodstuffs. Consequently the testing of new products should always be carried out in conjunction with similar tests of the traditional products that are to be supplemented or replaced by the new product. In this way the two *really* interesting questions can be answered: is the new product better than, comparable to, or worse than the old product, and is a combination of the two better than one alone?

Such clarity of procedures has, unfortunately, not been used in the case of SCP. There have been stringent and understandable demands for extensive testing of SCP, but similar tests of traditional feedstuffs have not been required.

In addition, the earlier mentioned distinction between real hazards and minor problems has not been made. Alkane-yeast SCP has been accused both of carcinogenicity—which is serious if it is true—and of having a relatively high content of fatty acids with an odd number of carbon atoms—which is an irrelevant criticism, because such fatty acids are readily oxidized by the consumer just like other fatty acids. The only difference is that, in addition to a number of molecules of acetyl-CoA produced during the oxidation of fatty acids with an odd number of carbon atoms, there is also the production of one molecule of propionyl-CoA, which joins the propionyl-CoA produced by oxidation of protein. These molecules enter the citric acid cycle of metabolism as succinyl-CoA (Stryer 1975).

Naturally, the confusion regarding SCP properties that seems to prevail in the authorizing bodies has had a negative effect on the acceptance of SCP, making farmers and farming organizations reluctant to use SCP as a feed supplement.

If all feedstuff components were subjected to the same tests, preferably on both minimal, average, and optimal qualities of all, then a true and informative picture could be obtained. It might turn out that SCP—produced in a carefully controlled industrial process—is even in its minimal qualities much safer than the minimal qualities of other feedstuff raw materials, which may contain toxins, e.g., fungal toxins (Foley 1962; Badiali et al. 1968; Kommedahl and Windels 1981; Marasas et al. 1981; Marasas 1982; Abbas et al. 1984; Zhen 1984; Wilson et al. 1985; Nelson and Wilson 1986; Jaskiewicz et al. 1987; Abbas et al. 1988), and perhaps SCP would rank highly even when competing with the average or best qualities of other feedstuff components.

REFERENCES

Abbas, H.K., Mirocha, C.J., and Shier, W.T. (1984) *Appl. Environ. Microbiol.* 48, 654–661.
Abbas, H.K., Mirocha, C.J., Kommedahl, T., et al. (1988) *Phytopathology* 78, 1258–1260.
Adler, I., and Fiechter, A. (1983) *Chem.-Ing.-Tech.* 55, 322–323.
Aiba, S., Humphrey, A.E., and Millis, N.F. (1965) *Biochemical Engineering*, University of Tokyo Press, Tokyo.
Appelbaum, S., and Uland, B. (1979) *Aquaculture* 17, 175–179.
Badiali, L., Abou-Youssef, M.H., Radwin, A.L., Hamdy, F.M., and Hildebrandt, P.K. (1968) *Am. J. Vet. Res.* 29, 2029–2035.

References

Bin, K., and Smith, J.M. (1982) *Chem. Eng. Commun.* 15, 367–383.
Blenke, H. (1979) in *Advances in Biochemical Engineering* (Ghose, T.K., Fiechter, A., and Blakebrough, N., eds.), pp. 121–214, Springer–Verlag, Berlin.
Blenke, H. (1985) *Biotechnology—A Comprehensive Treatise in Eight Volumes*, vol. 2 (Rehm, H.-J., and Reed, G., eds), VCH Verlagsgesellschaft, Weinheim, Germany.
British Petroleum (1989) *Statistical Review of World Energy*, Government and Public Affairs Department, London.
Böhnke, B. (1970) in *Adv. Water Poll. Res., Proc. Int. Conf., 5th*, (Jenkins, S.H., ed.), paper II-9, TD 420.A27. Pergamon, Oxford.
Euzen, J.P., Trambouze, P., and van Landeghem, H. (1978) in *Chemical Reaction Engineering, Houston, 5th Int. Symp. (ACS Symp. Ser., no. 65)*, pp. 153–162.
Foley, D.C. (1962) *Phytopathology* 52, 870–872.
Hines, D.A. (1978) *Prep. Eur. Congr. Biotechnol., 1st*, 1978, 55–64.
Jaskiewicz, K., van Rensburg, S.J., Marasas, W.F., and Gelderblom, W.C. (1987) *JNCI* 78, 321–325.
Keitel, G., and Onken, V. (1982) *Chem. Eng. Commun.* 17, 85–98.
Kommedahl, T., and Windels, C.E. (1981) in *Fusarium: Diseases, Biology and Taxonomy* (Nelson, P.E., Toussoun, T.A., and Cook R.J., eds.), pp. 94–103, Pennsylvania State University Press, University Park, PA.
Marasas, W.F.O. (1982) in *Cancer of the Esophagus*, vol. 1 (Pheiffer, C.J., ed.), pp. 29–40, CRC Press Inc., Boca Raton, FL.
Marasas, W.F., Wehner, F.C., van Rensenburg, S.J., and van Schalkwyck, D.J. (1981) *Phytopathology* 71, 792–796.
Medexport (1985) *Gaprin—A High-Value Protein and Vitamin Supplement for Feeds*, Ministry for Medicine and Bioproduction, Moscow (in Russian).
Meijel Sewage Treatment. *Dutch State Mines Brochure.*
Müllner, J. (Wagner Biro GmbH) (1973) Austrian Patent 319 864.
Nelson, P.E., and Wilson, T.M. (1986) in *Mycotoxins and Phytotoxins* (Styen, P.S., and Weggaar, R., eds.), pp. 535–544, Elsevier Science Publishing, Amsterdam.
Olsen, J., and Allermann, K. (1987) in *Basic Biotechnology* (Bu'Lock, J.D., and Kristiansen, B., eds.), pp. 285–308, Academic Press, London.
Pike, I.A. (1989) *The Availability of Fish Meal and Fish Oil for Fish Farming in the Year 2000*. Presented at the International TAF-seminar, Thyborøn, Denmark, May 11–13, Thyborøn Andels Fiskeindustri AMBA.
Schügerl, K. (1985) in *Comprehensive Biotechnology*, vol. 2 (Moo-Young, M., Cooney, C.L., and Humphrey, A.E., eds.), pp. 99–118, Pergamon Press, Oxford.
Sharp, D.H. (1989) *Bioprotein Manufacture—A Critical Assessment*, Halsted Press, John Wiley & Sons, New York.
Stryer, L. (1975) *Biochemistry*, W.H. Freeman and Company, San Francisco.
Uhde Brochure, LO II 19150079.
Uhde Hoechst Brochure, 'Bio-High Reactor'.
Viesturs, U.E., Sturmanis, I.A., Krikis, V.V., Propokenko, V.D., and Erickson, L.E. (1981) *Biotechnol. Bioeng.* 23, 1171–1191.
Walker, J., and Wilkinson, G.W. (1979) *Ann. N.Y. Acad. Sci.* 326, 181–191.
Wilson, B.W. (1953) *Australian J. Appl. Sci.* 4, 274.
Wilson, T.M., Nelson, P.E., Ryan, T.B., et al. (1985) *Vet. Med.* 80, 63–64.
Zhen, Y.Z. (1984) *Zhongua Zhongliu Zazhi* (in Chinese), 6, 27–29.
Zlokarnik, M. (1980) *Korrespondenz Abwasser*, 27 (3), 194–209.

CHAPTER 10

Enzymes of Industrial Potential from Methylotrophs

J.A. Duine
J.P. van Dijken

As indicated by their name, methylotrophs specialize in the metabolism of C_1 substrates. The unifying concept in metabolism is the transfer of reduction equivalents from a donor to an acceptor. Regarding the microbial world, diversity appears to exist in this unity since several pathways are available to metabolize a certain compound. Methylotrophs are no exception since a variety of unique pathways for the conversion of C_1 compounds is present. This is also expressed at the molecular level where a variety of sometimes unique enzymes and cofactors can catalyze a given reaction in these pathways.

With respect to the application of enzymes from methylotrophs, three categories can be specified: the unique enzymes directly involved in the metabolism of C_1 compounds (in this chapter the definition used is somewhat wider, including enzymes from nonmethylotrophs catalyzing the conversion of a C_1 compound or that of a C_1 moiety in a substrate); the more common enzymes operating in universal pathways; and nonmethylotrophic enzymes whose genes have been cloned and transferred to methylotrophs (see Chapter 13). The second category would be interesting if their production, stability, or substrate specificity would be more useful than that of competitors from other organisms. However, this has scarcely been explored and thus an adequate evaluation is not possible.

Most efforts have been directed towards the identification and characterization of enzymes and cofactors that are unique to C_1 metabolism. Because of the interest in production of single cell protein (SCP), in specifically hydroxylated compounds, and in biogas, attention has been focused on the enzymology involved in the dissimilation route starting with methane, the assimilation of formaldehyde, and methanogenesis. Ecological interest in the sulfur and nitrogen cycles in nature has provided information about the enzymes involved in the conversion of compounds containing a methylated nitrogen or sulfur (Cole and Ferguson 1988). However, far less is known about the enzymes involved in the degradation of synthetic C_1 chemicals (methoxylated aromatics; halogenated methane; and methylated sulfur, phosphorus, silicon, metal, and nitrogen compounds). These so-called xenobiotics are produced in large amounts and are frequently the cause of environmental pollution. Studies on their biological degradation have only just begun. As has been found for some pollutants, they are not degraded by the methylotrophs present in culture collections that were isolated on natural C_1 compounds. However, specialist microorganisms found in sewage plants using wastewater streams from the chemical industry contaminated by these compounds were able to grow on or to cometabolize these xenobiotics. Only in a few cases has the enzyme catalyzing such a special degradation step been (partly) characterized, and the potentials for application demonstrated. This approach of screening for the special methylotrophs and for enzymes with special properties has been carried out by a group at Ciba-Geigy (Ghisalba et al. 1986). Therefore, a rich arsenal of special hydrolases, lyases, and oxidoreductases involved in the degradation of these compounds is still waiting to be explored.

Application of enzymes in general has been largely restricted to those catalyzing hydrolytic or redox steps in dissimilation processes. A more recent development concerns lyases using water or ammonia as a substrate. The two main fields of applicational interest are in the development of analytical methods and the production of specialties. In this respect, enzymes from methylotrophs form no exception. For that reason, some members of the three interesting main classes, the hydrolases, the lyases, and the oxidoreductases, will be discussed. Examples will be provided in the text and in Tables 10-1 and 10-2. Although few applications for enzymes converting synthetic C_1 chemicals are in use at present, they will be briefly covered here because of their growing environmental importance. Enzymes that are more commonly used, such as alcohol oxidase and formate dehydrogenase will be treated more extensively.

TABLE 10-1 Analytical Applications

Sensor/Analyte	Enzyme	Reference
Amperometric/methanol	Methanol dehydrogenase	Higgins et al. 1984
Flow injection/dimethylformamide	Dimethylformamidase	Lüdi et al. 1988
Oxygen/ethanol	Alcohol oxidase	Verduyn et al. 1983; Belghith et al. 1987; Suye et al. 1990
Flow injection/ethanol	Alcohol oxidase	Gibson et al. 1989
Colorimetric/alcohols	Alcohol oxidase/peroxidase	Verduyn et al. 1984a; Barzana et al. 1989
Test strip/ethanol	Alcohol oxidase/peroxidase	Adams 1988; Woodward 1990
Badge/formaldehyde	Formaldehyde dehydrogenase/diaphorase	Rindt and Scholtissek 1989
pH/monomethyl sulfate	Monomethylsulfatase	Schär and Ghisalba 1985
Amperometric/CO	CO oxidase	Turner et al. 1984

TABLE 10-2 Applications in (Enantioselective) Biocatalysis

Enzyme	Product	Reference
Methanol dehydrogenase	(R)2-Octanol	Schär et al. 1985
Methanol dehydrogenase	Chloroalcohols	Janssen et al. 1985
Methanol dehydrogenase	Glyceraldehyde	Upjohn Co.
Esterase	(2R,4R)-2,4-dimethyl glutarate	Ramos Tombo et al. 1985
Esterase	(2S)-indoline-2-carboxylic acid	Ramos Tombo et al. 1987
Amidase	D- or L-Amino acids	Sanraku Ocean
Alcohol oxidase	Aldehydes	Barzana et al. 1987; Hoiberg and Hatley 1988; Duff and Murray 1989
Alcohol oxidase/catalase/aldehyde dismutase	Carboxylic acids	Kato 1986; Mizuno and Imada 1986
β-ketoester reductase	D- or L-3-Hydroxyacids	Ushio et al. 1986
s-Methane monooxygenase	Epoxides	Subramanian 1986

10.1 HYDROLASES AND LYASES

10.1.1 Carboxylic Ester Hydrolases
As part of a strategy to employ the methylotrophs isolated from wastewaters from chemical plants using C_1 compounds, a screening was made by Ramos Tombo et al. (1987) for enzymes performing enantioselective hydrolysis of esters of chiral building blocks used for the synthesis of ACE inhibitors (ACE, angiotensin-converting enzyme). The results showed that carboxylic acid ester hydrolases (EC 3.1.1.-) with opposite selectivities occur, some of them giving satisfactory kinetic resolution of the esters.

10.1.2 Sulfate Ester Hydrolases
Methylation in organic synthesis is often carried out with dimethyl sulfate. Usually, only one methyl group is transferred to the accepting group, resulting in the formation of monomethyl sulfate as a byproduct. Monomethyl sulfatase (EC 3.1.6.16) found in a *Hyphomicrobium* strain using monomethylsulfate (Ghisalba et al. 1985) hydrolyzes the substrate into methanol and sulfate. Although the enzyme could not be detached from the membranes without complete loss of activity, its suitability for determination of monomethyl sulfate was shown by incorporating membranes or whole cells of the organism in a modified pH electrode (Schär and Ghisalba 1985).

10.1.3 Phosphoric Ester Hydrolases
Organophosphorus compounds are of great economical and environmental importance. For instance, pesticides based on these structures are used world-wide in massive quantities. Special strains of *Hyphomicrobium* have been isolated which can grow on trimethyl phosphate (Ghisalba et al. 1987). Although the enzymes have not been isolated, the growth parameters indicate that phosphatases (EC 3.1.8.-) are involved in the degradation to methanol and phosphate. *O,O*-Dimethyl methylphosphonate is degraded in a similar way. Most probably, the splitting of the C-P bond in methylphosphonate is due to a lyase, because the formation of CH_4 was found in other organisms catalyzing this reaction (Wackett et al. 1987). In view of their environmental and analytical importance, it is probable that these enzymes will frequently be applied after they become available and their properties are known.

10.1.4 Amidases
N,N-Dimethylformamidase (EC 3.5.1.56) has been purified from a bacterium utilizing N,N-dimethyl formamide as a carbon, energy, and nitrogen source (Schär et al. 1986). The enzyme contains iron and an unknown prosthetic group. Its substrate specificity is rather restricted since only a few

other amides (N-ethyl formamide, N-methyl formamide, and more slowly, N,N-diethyl formamide and N,N-dimethyl acetamide) are hydrolyzed to the acid and the alkylamine. Therefore, a rather specific method for N,N-dimethyl formamide analysis was developed, making use of immobilized enzyme placed in a flow injection apparatus (Lüdi et al. 1988).

As claimed in the patent literature (Sanraku Ocean), a methylotrophic bacterium has been isolated with a D-amino-acylase which might be interesting for production of homochiral amino acids.

10.1.5 Cyanide and Cyanate Lyases

Several fungi are able to degrade cyanide (Knowles and Bunch 1986). A cyanide hydratase (EC 4.2.1.66) catalyzing the conversion of cyanide into formamide has been purified from the spores of adapted *Stemphylium loti* (Fry and Millar 1972). Cells of this fungus are used for the treatment of wastewater. (They are sold under the name of "Cyclear" by ICI.) Heavy metals inhibit the enzyme so that metal finishing wastes cannot be decontaminated with it. However, several bacteria are also known to degrade cyanide: they contain a monooxygenase that first converts cyanide into cyanate. The latter is then transformed (via carbamate) into CO_2 and NH_3 by cyanate lyase (EC 4.3.99.1). Bicarbonate plays an essential role in the mechanism of the reaction (Johnson and Anderson 1987). Apparently, these enzyme systems are less sensitive to heavy metal inhibition since wastewaters (from the electroplating industry) containing cyanide in the form of $Ni(CN)_4^{2-}$ could successfully be treated (Khedkar et al. 1989).

10.1.6 Dehalogenases

In general, the presence of a halogen substituent in a chemical structure renders such a compound recalcitrant against biological degradation. However, notably among the methylotrophs, organisms have been found which detoxify these xenobiotics either via a reductase, an oxygenase, or a halidohydrolase.

1-Haloalkane dehalogenase (EC 3.8.1.5) hydrolyzes a broad range of haloalkanes to the primary alcohol and halide ions. Such an enzyme has been found in the methylotroph *Xanthobacter autotrophicus* GJ10 (Janssen et al. 1985). Although *Arthrobacter* HA1 does not grow on C_1-compounds, bromomethane is an inducer and substrate for the enzyme (Scholtz et al. 1987b; Scholtz et al. 1988a). Lyases can also detoxify halogens. Dichloromethane dehalogenase (EC 4.5.1.3) converts dichloromethane into formaldehyde and chloride ions (Kohler-Staub and Leisinger 1985). The enzyme has been found in several *Pseudomonas* and *Hyphomicrobium* strains. As reported by Scholtz et al. (1988b), two different types exist (Keuning et al. 1985; Scholtz et al. 1987a). Gluthatione (GSH) is essential for activity as it participates in the reaction. In this respect, alkylhalidase (EC 3.8.1.1), which occurs in liver and converts bromochloromethane into formaldehyde and the halide ions, appears to be similar (Heppel and Porterfield 1948).

Several other bacteria may be able to degrade chloromethane not via a dehalogenase, but either with methane monooxygenase (Patel et al. 1982; Yokota et al. 1986; Hanson and Lipscomb 1990; Fox et al. 1990; Oldenhuis et al. 1989b) or with ammonia monooxygenase (Rasche et al. 1989). This is a cometabolic process, requiring, for example, the participation of formate (provision of reduction equivalents) and O_2 in the case of methanotrophs. Most probably, the soluble form of methane monooxygenase is involved (Tsien et al. 1989). Besides chloromethane, di- and trichloromethane were both degraded by *Methylosinus trichosporium* OB3b, but carbon tetrachloride (CCl_4) was not (Oldenhuis et al. 1989a). Although the degradation of CCl_4 has been reported for anaerobic organisms, the enzymes involved in it are unknown (Stromeyer et al. 1989). Also, a spontaneous chemical dehalogenation step occurs in this process, perhaps catalyzed by Factor 430.

Halogenated alkanes are produced in large amounts by the chemical industry. In principle, they could form attractive starting materials for the production of fine chemicals ([homochiral] alcohols and hydroxyacids) with specific enzymes. Another reason to study the biodegradation of these compounds is the fact that they cause environmental pollution. These suspected carcinogens are resistant to biodegradation in aerobic subsurface environments. This contributes to their persistence in polluted groundwaters.

10.2 OXIDOREDUCTASES

10.2.1 Alcohol Oxidoreductases

10.2.1.1 Alcohol Dehydrogenases. NAD-Dependent alcohol dehydrogenase (EC 1.1.1.1) occurs in many methylotrophic organisms where it has a function in growth on alcohols except on methanol (although the enzyme sometimes has the ability to oxidize methanol). Recently, a special NAD-dependent alcohol dehydrogenase, having a role in methanol oxidation, was found in some thermotolerant methylotrophic *Bacillus* species (Arfman et al. 1989). The inherent thermostability of the enzyme could be beneficial in situations where the labile, classical enzymes from horse liver and baker's yeast are used. NAD-Dependent alcohol dehydrogenase from methylotrophic yeasts shows the opposite enantioselectivity in 2-butanol oxidation as compared to that of baker's yeast (Hou et al. 1981b). The enzyme from a methanol-grown *Pseudomonas* species (ATCC 21439) is strictly specific for secondary alcohols (Hou et al. 1981a). Alcohol dehydrogenases and ketone reductases have been isolated from the methylotrophic yeasts *Hansenula polymorpha* and *Candida utilis*. The enzymes could have potentials for enantioselective conversions since S,S-2,3-butanediol was preferentially oxidized by 2,3-butanediol dehydrogenase, whereas the R,R-form was oxidized by dihydroxyacetone reductase, both from *Candida utilis* (Verduyn et al. 1988).

Methanol dehydrogenase (EC 1.1.99.8) occurs in all Gram-negative methylotrophs able to utilize methanol. Its application in a biosensor and in a fuel cell has been examined (Davis et al. 1983; Higgins et al. 1984). A patent has been filed on the application of the enzyme from *Methylobacterium organophilum* for the oxidation of glycerol to glyceraldehyde (Upjohn Co.). The enzyme from *Hyphomicrobium* MS 223 appeared to be enantioselective in the oxidation of 2-octanol (Schär et al. 1985). Methanol dehydrogenase is a so-called quinoprotein, that is, its cofactor is pyrroloquinoline quinone (PQQ). The large levels of enzyme synthesized and its localization in the periplasm could be responsible for the presence of substantial amounts of PQQ in the spent culture medium (van Kleef and Duine 1989). On the other hand, the Gram-positive *Nocardia* species 239 also excretes PQQ when grown on methanol, although it has no periplasm and its methanol dehydrogenase is quite different from the classical one (Duine et al. 1984). A novel tetrazolium dye-linked alcohol dehydrogenase has been discovered in this organism as well as in some other Gram-positive bacteria (van Ophem et al. 1991). Most novel cofactors discovered in methylotrophs are restricted to these organisms. However, quinoproteins have also been detected in mammals (Duine and Jongejan 1989), although it is very likely that the covalently bound cofactor is not PQQ. From recent insight into the analytical procedures originally developed for detection of covalently bound PQQ, it appears that some of these enzymes are quinoproteins because they have topaquinone as their cofactor (Janes et al. 1990), others are not but they have a reactive or modified tyrosyl residue in their protein chain (Duine 1991) (which may be the precursor of PQQ formation during application of the analytical procedures). If PQQ is a vitamin for these organisms (Killgore et al. 1989, but see also van der Meer et al. 1990a), its production by methylotrophic bacteria could form an economically attractive process. Several patents exist on this topic, and two Japanese companies already produce the compound by fermentation (Mitsubishi Gas Company; Ube Industries). It should be realized, however, that chemical synthesis (Corey and Tramontano 1981; van der Meer et al. 1990b) is an alternative which has also been realized (Fluka AG).

10.2.1.2 Alcohol Oxidase. Alcohol oxidase (EC 1.1.3.13) was discovered by Janssen and Ruelius (1968) in *Basidiomycetous* molds. A few years later, when it was found that certain yeast species are able to grow on methanol, Ogata and co-workers reported that alcohol oxidase is the key enzyme of methanol metabolism in the methylotrophic yeast *Kloeckera* sp. 2201 (Tani et al. 1972a and 1972b). Since then, many studies have appeared on methylotrophic yeasts, and in all cases the oxidation of methanol in these organisms appears to be catalyzed by an oxidase according to the following reaction:

$$CH_3OH + O_2 \rightarrow CH_2O + H_2O_2$$

Catalase is also always required to decompose the toxic hydrogen peroxide produced. The oxidase only has a physiological function during growth of yeasts on methanol. Growth on other alcohols such as ethanol and propanol proceeds via the classical alcohol dehydrogenase (EC 1.1.1.1). This may be the reason why alcohol oxidase is frequently called MOX (methanol oxidase), although it oxidizes a whole range of lower primary alcohols and formaldehyde. The name of the enzyme is abbreviated here as AOD (alcohol oxidase).

Only the AOD of methylotrophic yeasts will be discussed here. Alcohol oxidases have also been described, however, for nonmethylotrophs: for example, a secondary alcohol oxidase from *Pseudomonas* species (Morita and Watanabe 1977), an aliphatic diol oxidase from a *Flavobacterium* species (Willets 1983), fatty alcohol oxidases of n-alkane-utilizing yeasts (Kemp et al. 1990) and aryl alcohol oxidase of a ligninolytic fungus (Guillén et al. 1990). Since an excellent review has recently appeared on both fundamental and applied aspects of AOD (Woodward 1990), the reader is referred to this for more detailed information.

Properties. The property of growth on methanol is not widespread amongst yeasts, as it only occurs in the genera *Candida, Hansenula, Pichia,* and *Torulopsis*. The most detailed information on AOD is available for that purified from *Candida boidinii, Hansenula polymorpha,* and *Pichia pastoris*. Purification of the enzyme generally involves ammonium sulfate precipitation, DEAE ion exchange chromatography, and gel filtration. Using this three-step procedure, it is possible to obtain a homogeneous enzyme preparation that is devoid of catalase activity. Details on the properties of the AODs have been presented by Fujii and Tonomura (1972), Tani et al. (1972a and 1972b), Sahm and Wagner (1973), van Dijken et al. (1976b), Kato et al. (1976), Yamada et al. (1979), Couderc and Baratti (1980), Patel et al. (1981), Veenhuis et al. (1983), Ledeboer et al. (1985), Hopkins and Muller (1987), Unichika Co. Ltd. (1986), and Woodward (1990).

Regulation of Synthesis. Relevant to the applications of the enzyme is the regulation of its synthesis and that of its physiological-twin enzyme, catalase. Since H_2O_2 is a product of methanol metabolism it is not surprising that yeasts contain large amounts of catalase during growth on methanol. Both AOD and catalase are located in a special subcellular compartment, the peroxisome (Veenhuis et al. 1983). In all cases investigated, AOD appears to be present in a crystalline form. Catalase is thought to "swim" in these peroxisomal AOD crystals (Veenhuis et al. 1981). Since many applications of AOD rely on the production or detection of H_2O_2, it is evident that removal of catalase from extracts of methanol grown yeasts is an important step in the down-stream processing of the enzyme.

Provided that cultivation conditions are appropriate, AOD can make up 40% of the soluble protein of the cells (Giuseppin 1988). This is one of the rare examples of "super expression" of the synthesis of an enzyme in an organism that has not been genetically engineered. The case of "AOD superproduction" is only secondary to that of uricase (EC 1.7.3.3) in *Bacillus fastidiosus*. In this organism, as much as 50% of the soluble protein may consist of a single enzyme, namely uricase (Bongaerts 1978). The utility of the high expression rate of AOD has long been recognized by those interested in the synthesis of heterologous proteins in yeasts. Various examples are now available of high expression of foreign proteins under control of the AOD promotor in methylotrophic yeasts.

Hydrogen Peroxide Generation. Detergent manufacturers have investigated the possibility of using AOD and an alcohol as a hydrogen peroxide-generating system for the bleaching of laundry (Schreiber et al. 1977; Cox et al. 1982). This application focused on incorporation of the enzyme in liquid detergents. In powder formulations, H_2O_2 production for bleaching is accomplished by the incorporation of perborate. Perborate reacts upon contact with water to produce H_2O_2 and can therefore not be used in liquid detergents. A very elegant possibility for enzymatic generation of H_2O_2 with liquid detergents has been proposed by Unilever (Cox et al. 1982). According to the patent filing, it is possible to incorporate AOD into a liquid in the presence of high ethanol concentrations and detergents. Under these conditions the enzyme is not active, but after dilution it becomes active and generates H_2O_2.

In considering the economical feasibility of AOD in liquid detergents, it is clear that the high cost of down-stream processing of the enzyme is an important drawback. The enzymes that are presently used in biological detergents (proteases and lipases) are extracellular enzymes that can be simply collected via filtration. However, for the purification of AOD not only mechanical cell disruption is required but also complete removal of catalase is essential as it catalyzes the decomposition of H_2O_2. The physiological necessity of coordinated synthesis of AOD and catalase seems obvious: when catalase is not present, H_2O_2 intoxication can be expected to occur when cells are exposed to the substrates of AOD. A rather surprising finding was therefore that catalase-negative mutants of the methylotrophic yeast *Hansenula polymorpha* were still able to decompose H_2O_2. Cells could be grown in chemostat cultures on mixtures of glucose and H_2O_2. Not only extracellular H_2O_2 but also H_2O_2 generated intracellularly by AOD was decomposed by the catalase-negative mutant (Giuseppin et al. 1988; Verduyn et al. 1988). It appeared that the yeast contained another H_2O_2-destroying enzyme: cytochrome *c* peroxidase (CCP; EC 1.11.1.5). In cell-free extracts, CCP does not interfere with H_2O_2 production (the enzyme requires reduced cytochrome c as a cosubstrate for H_2O_2 destruction). Catalase-negative mutants are therefore an attractive option for the production of H_2O_2 via AOD and alcohols since there is no need for the expensive removal of catalase (Giuseppin et al. 1988).

Although a catalase-negative mutant of *Hansenula polymorpha* is able to survive when fed with methanol, it has only a limited capacity to decompose the H_2O_2 generated by AOD. The reason is that the cellular capacity for H_2O_2 destruction by CCP is smaller than the capacity of alcohol oxidase to generate H_2O_2. Thus, when intact cells are exposed to an excess of alcohol that saturates the alcohol oxidase, H_2O_2 is excreted in considerable amounts (Giuseppin et al. 1988). Furthermore, since CCP can easily be inactivated by simple treatments such as freeze-drying, dried whole cells of catalase-negative mutants are in principle directly applicable as an efficient H_2O_2-producing system.

Whether or not bleaching of laundry with AOD will become a reality is dependent on many factors which are beyond the scope of this chapter. H_2O_2 generation via alcohol oxidase, either with free enzyme or with whole cells, may also be applicable as an antimicrobial system, forming an alternative for the use of harsh chemicals (Woodward 1990).

Production of Aldehydes. AOD or AOD-containing cells can also be used for the conversion of alcohols into the corresponding aldehydes. Acetaldehyde, for example, is a flavor compound that, according to international regulations, has to be of biological origin in order to be added to food stuffs. Also, production of benzaldehyde with AOD from *Pichia pastoris* has been reported (Duff and Murray 1989).

Alcohol Detection. An interesting application of AOD is in the colorimetric assay of those alcohols that serve as a substrate for the enzyme, ethanol being the most important in this respect. This application was already mentioned by the discoverers of the enzyme (Janssen and Ruelius 1968). The principle of colorimetric alcohol assays with AOD was first given by Trinder (1969):

$$\text{alcohol} + O_2 \xrightarrow{\text{AOD}} \text{aldehyde} + H_2O_2$$
$$\text{reduced dye} + H_2O_2 \xrightarrow{\text{peroxidase}} \text{oxidized dye (colored)} + H_2O$$

Many variations exist with respect to the reduced dye. The authors of this chapter now have 10 years of experience with this assay, which has completely replaced the use of gas chromatography for alcohol determination in their laboratory. As described by Verduyn et al. (1984a), the assay is rapid, convenient, sensitive, and accurate. For those reasons, it may eventually also replace the classical enzymatic assay of ethanol with NAD-linked alcohol dehydrogenase.

Another analytical application of the enzyme could be its use in a biosensor. Alcohol detection with AOD in combination with an oxygen monitor has been reported by, among others, Verduyn et al. (1983), who immobilized AOD on a nylon net that covered the oxygen-permeable membrane. The electrode performed well and was used to follow the response of yeasts to exposure to pulses of sugars (Verduyn et al. 1984b). A normal oxygen probe was used as a reference electrode since the activity of the enzyme, and thus the response of the alcohol sensor, is sensitive to changes in the dissolved oxygen concentration. A similar procedure was described by Belghith et al. (1987).

Another potential application for AOD in combination with a peroxidase is its use in a spot test for alcohol in biological fluids. However, for the enzyme to be used in such a dry formulation, stability is a prerequisite. The enzyme from *Hansenula polymorpha* appears to meet these requirements, while that from *Pichia pastoris* does not (Woodward 1990). The reason for the high stability of the enzyme is not known, but probably relates to the fact that *H. polymorpha* differs from the other methylotrophic yeasts in its high temperature optimum for growth.

A new development in test strips has been made by Woodward and coworkers at the University of Leeds, U.K. (Woodward 1990). Classical test strips on which an oxidase and a peroxidase are present rely on matching the color that is produced by the joint action of the two enzymes. Woodward (1990) describes a test that is based on a threshold value, which can be preset at any value that is required. The prototype of the alcohol test consists of a card on which two reagent pads in the form of small paper discs are fixed. The discs differ in their threshold value for alcohol-dependent color development. Thus, when alcohol-containing saliva is applied, three results are possible: no color (no or only a low, acceptable, alcohol concentration); color development on one disc (medium alcohol concentration); or color development on both discs (high alcohol concentration). In view of its simplicity, this type of threshold test has the potential for commercial exploitation.

10.2.2 Aldehyde/Formate Oxidoreductases

10.2.2.1 (Form)aldehyde Dehydrogenases. Several enzymes exist which are specific for formaldehyde conversion: the widely distributed GSH- and NAD-dependent formaldehyde dehydrogenase (EC 1.2.1.1); NAD-dependent formaldehyde dehydrogenase from a *Pseudomonas putida* strain (EC 1.2.1.46); and NAD-dependent formaldehyde dehydrogenase requiring an unknown factor, present in two Gram-positive species, *Rhodococcus erythropolis* and *Nocardia* species 239 (Eggeling and Sahm 1985; van Ophem and Duine 1990). Besides these specific enzymes, several aldehyde dehydrogenases are also able to oxidize formaldehyde (although this role in vivo is sometimes questionable): isozymes from mammalian NAD-dependent aldehyde dehydrogenase (EC 1.2.1.3), such a type also occurring in *Nocardia* species 239 (van Ophem and Duine 1990); isozymes from mammalian NAD(P)-dependent aldehyde dehydrogenase (EC 1.2.1.5), such a type also present in *Hyphomicrobium* X (Poels and Duine 1989); (formal)dehyde dismutase (EC 1.2.99.4), found in a *Pseudomonas* strain (Kato et al. 1986); and dye-linked aldehyde dehydrogenase (EC 1.2.99.3), occurring in several methylotrophs.

An NAD-dependent formaldehyde dehydrogenase (unspecified) may be involved in the production of pyruvate from hydroxyacetone or methylglyoxal (Tarame et al. 1988). Formic acid has been prepared from methanol by a coupled system of alcohol oxidase, catalase, and the (form)aldehyde dismutase (Mizuno and Imada 1986). As the dismutase is in fact a "Cannizzarase" (2 aldehydes → 1 alcohol + 1 acid) and has a broad substrate specificity (Kato et al. 1984), it could be used for the interconversion of several aldehydes, acids, and primary alcohols.

The formaldehyde dehydrogenase type EC 1.2.1.46 is rather stable and adsorbed to a badge, together with diaphorase and an artificial electron acceptor, it can be used for monitoring formaldehyde in the atmosphere (Rindt and Scholtissek 1989).

10.2.2.2 Formate Dehydrogenase. NAD-Dependent formate dehydrogenase (EC 1.2.1.2) occurs in methylotrophic bacteria and yeasts (Johnson and Quayle 1964; Fujii and Tonomura 1972; Patel and Hoare 1971; Kato et al. 1974; van Dijken et al. 1976a; Schütte et al. 1976; Müller et al. 1978; Egorov et al. 1979; Hou et al. 1982; Allais et al. 1983; Avilova et al. 1985; Izumi et al. 1987). It is also found in the animal and in the plant kingdom (Ohyama and Yamazaki 1974 and 1975; Uotila and Koivusalo 1979), indicating that it is not typical for methylotrophs. It can be induced however in various microorganisms when grown on a certain substrate in the presence of formate as an additional energy source. Utilization of formate by nonmethylotrophic heterotrophs can be used to assess the upper limits of carbon conservation by these microorganisms (Gommers et al. 1988).

The formate dehydrogenase of yeasts has been the most intensively studied with respect to applications. The reader is referred to Izumi et al. (1989) who discuss the features of formate dehydrogenase of 28 methanol-utilizing yeast strains, and strains of 7 bacteria, 2 molds, and 3 nonmethylotrophic yeasts.

Applications. Formate dehydrogenase catalyzes the oxidation of formate to CO_2. The equilibrium of the reaction strongly favors CO_2 formation and NAD reduction to NADH. For that reason it is used to generate NADH from NAD, required for the conversion of α-keto acids into homochiral hydroxyacids or amino acids and (di)keto compounds into homochiral alcohols (Hummel and Kula 1989). These conversions take place in so-called enzyme-membrane reactors in which the enzymes and also their cofactors are retained. Prevention of cofactor loss is accomplished via covalent linkage of the pyridine nucleotide to a water-soluble polymer unable to pass the membrane. Attachment of NAD to polyethylene glycol (PEG) has been proven to be a good method for the regeneration of NADH with formate dehydrogenase in membrane reactors (Bückmann et al. 1981; Vasic-Racki et al. 1989; Kula and Wandrey 1987; Hummel and Kula 1989). Interestingly, the K_m of formate dehydrogenase for NAD was hardly affected by attachment of the cofactor to PEG (Hummel and Kula 1989). In this respect the enzyme showed favorable characteristics compared to, for example, glucose dehydrogenase. The latter enzyme showed a thousandfold lower affinity towards its cofactor after coupling to PEG (Hummel and Kula 1989).

10.2.2.3 β-Ketoester Reductases. Yeasts have at least two enantiospecifically different enzymes (EC 1.2.1.55; EC 1.2.1.56) for the reduction of β-keto esters (Heidlas et al. 1988). The induction of these enzymes is governed, among other factors, by the substrate used for growth. Thus it was found that methanol-grown yeast showed a stereochemical control that differed from that of glucose-grown cells (production of the D-hydroxy ester is preferred in the first case) (Ushio et al. 1986).

10.2.2.4 Carbon Monoxide Oxidoreductases. Several aerobic bacteria exist which contain a carbon monoxide oxidase (EC 1.2.3.10) catalyzing the following reaction (Meyer et al. 1986):

$$CO + H_2O + O_2 \rightarrow CO_2 + H_2O_2$$

The enzyme has been used in a biosensor for monitoring CO (Turner et al. 1984).

10.2.3 Amine Oxidoreductases

Methylamine oxidase (the copper-quinoprotein type amine oxidase (EC 1.4.3.6)) occurs in *Arthrobacter* P1 (van Iersel et al. 1986), and the same type is also present in methylotrophic yeasts (Green et al. 1983). It is very likely that this enzyme has also topaquinone (Janes et al. 1990) as a prosthetic group. Quinoprotein methylamine dehydrogenase (EC 1.4.99.3) is found in some Gram-negative bacteria grown on methylamine. The three-dimensional structure determination has been successfully carried out (Vellieux et al. 1989). The cofactor appears to consist of two tryptophans, covalently linked to each other and one of them having an o-quinone grouping (Chistoserdov et al. 1990; Ubbink et al. 1991). In contrast to alcohol oxidase and methanol dehydrogenase, the use of quinoprotein amine oxidoreductase enzymes for analytical purposes is still in its infancy. However, a fungal amine oxidase has been used in a primary amine biosensor developed for monitoring the freshness of meat and fish (Karube et al. 1980).

10.2.4 C-S Bond-Attacking Oxidoreductases

Methanethiol (methyl mercaptan) oxidase (EC 1.8.3.4) occurs in a *Hyphomicrobium* species grown on dimethyl sulfoxide (Suylen et al. 1987). The enzyme oxidizes methyl mercaptan and ethyl mercaptan to the corresponding aldehydes, sulfide, and H_2O_2. Since methylated sulfur compounds are frequently toxic and malodorous pollutants produced in oil refineries and paper industries, such an enzyme could be useful in environmental problems.

10.2.5 Monooxygenases

Methane monooxygenase occurs in methanotrophs. Two different enzymes are known: the soluble (EC 1.14.13.25) and the membrane-bound one (having different substrate and inhibitor specificities and which has been only scarcely characterized so far). The soluble one (sMMO) has been studied in a number of different laboratories, including those of Higgins, Dalton, and Hou. It consists of a flavoprotein component which transfers reduction equivalents from NADH to the hydroxylase component (the presence of a third component regulating the flow of reduction equivalents in the presence of substrate is controversial). The mechanism shows some similarity with that of cytochrome P450 (H-abstraction preceding hydroxylation) with radicals and/or carbocations as intermediates (Ruzicka et al. 1990) as deduced from the rearrangements and inversions observed with cyclohexane and cyclohexene hydroxylation (Leak and Dalton 1988).

Aromatic hydroxylation proceeds by epoxidation, followed by an NIH shift. Epoxidation of olephins is one of the attractive options. However, in studies with *Methylosinus trichosporium* (Subramanian 1986; Okura 1990), racemic products were obtained from propene and 1-butene. Since the products are not further metabolized, cometabolization of an NADH-donor like formate is required to keep the system active. Most epoxides are highly toxic for the organism so that only low levels of product are tolerated.

Monooxygenases may also participate in the degradation of methylated amines, yielding hydroxylation at either the C atom or the N atom (Large and Green 1984). The anaerobe *Eubacterium acidaminophilum* degrades methylated amines via reductases (Hormann and Andreesen 1989). Demethylation of methylated sulfur compounds also occurs with a monooxygenase. Moreover, degradation of dimethyl sulfoxide also proceeds in this way. To achieve this, the compound is first reduced to dimethyl sulfide (de Bont et al. 1981) (curiously, dimethyl sulfoxide is also a terminal electron acceptor in many bacteria [Yen and Marrs 1977]). sMMO is able to perform

O-dealkylation of aromatic ethers (Jezequel et al. 1984). In the anaerobe *Clostridium thermoaceticum*, a CO-dependent, O-demethylating enzyme system has been found (Wu et al. 1988). This organism has also a unique enzyme system to decarboxylate aromatics, providing the cell with CO_2 (Drake et al. 1989).

10.3 PERSPECTIVES

The increasing attention directed to environmental problems will stimulate development of analytical tools for monitoring methylated compounds. Enzymes occurring in methylotrophic organisms that degrade these compounds may be used for the construction of biosensors or other analytical devices. Moreover, if these enzymes can be produced on a large scale in a stable form, they could be used for the decontamination of polluted water, etc.

Another consequence of the growing awareness of environmental and health hazards caused by the deliberate use of chemicals, is the switching from the use of racemic mixtures of drugs and agrochemicals to enantiomerically pure forms (the other enantiomer is mostly inactive ballast or even harmful when it induces unwanted side effects). In view of their selectivity, enzymatic steps in the synthesis of fine chemicals or intermediates could be feasible in the production processes. Several of these compounds contain a C_1 moiety, but the unique methylotrophic enzymes have scarcely been explored for their suitability for that purpose.

Specific hydroxylation or epoxidation is one of the most difficult reactions in organic chemistry. Although asymmetric synthesis is rapidly advancing, methanotrophic bacteria may offer an alternative route if their selectivity is high enough. However, this is not the case, and at present several complications exist with the use of whole cells while the intricacy of the enzyme systems required preclude their use in vitro.

As shown in this chapter, most examples fall in the categories of the hydrolases and the oxidoreductases, the latter mainly represented by the oxidative variants (although the enzymology of methanogenesis is now rather well evolved, obviously the potentials for application of the enzymes are absent or still unexplored). Since much information is available on how to grow methylotrophs on a large scale (from SCP production), these organisms could in principle be a good source for applied enzymes. Thus, alcohol oxidase from methylotrophic yeasts can be purchased in substantial amounts from companies who supply this enzyme for use in analyses and bioconversions. However, despite the availability and potentials mentioned in this chapter, it appears that only formate dehydrogenase has attained the stage of commercial application in providing reduction equivalents for the production of L-tert-leucine (M.R. Kula, personal communication).

REFERENCES

Adams, E.C. (1988) U.S. patent 4786596A.
Allais, J.J., Louktibi, A., and Baratti, J. (1983) *Agric. Biol. Chem.* 47, 2547-2554.
Arfman, N., Watling, E.M., Clement, W., et al. (1989) *Arch. Microbiol.* 152, 280-288.
Avilova, T.V., Egorova, O.A., Ioanesyan, L.S., and Egorov, A.M. (1985) *Eur. J. Biochem.* 152, 657-662.
Barzana, E., Klibanov, A., and Karel, M. (1987) *Appl. Biochem. Biotechnol.* 15, 25-34.
Barzana, E., Klibanov, A., and Marcus, K. (1989) *Anal. Biochem.* 182, 109-115.
Belghith, H., Romette, J.L., and Thomas, D. (1987) *Biotechnol. Bioeng.* 30, 1001-1005.
Bongaerts, G.P.A. (1978) Ph.D. thesis, University of Nijmegen, Nijmegen, The Netherlands.
Bückmann, A.F., Kula, M.R., Wichmann, R., and Wandrey, C. (1981) *J. Appl. Biochem.* 3, 301-315.
Chistoserdov, A.Y., Tsygankov, Y.D., and Lidstrom, M.E. (1990) *Biochem. Biophys. Res. Commun.* 172, 211-216.
Cole, J.A., and Ferguson, S.J. (eds.) (1988) *The Nitrogen and Sulphur Cycles*, Cambridge University Press, Cambridge, UK.
Corey, E.J., and Tramontano, A. (1981) *J. Am. Chem. Soc.* 103, 5599-5600.
Couderc, C.L., and Baratti, J. (1980) *Agric. Biol. Chem.* 44, 2279-2289.
Cox, R.B., Steer, D.C., and Woodward, J.R. (1982) Unilever U.K. patent 2101167.
Davis, G., Hill, H.A.O., and Aston, W.J. (1983) *Enzyme Microbiol. Technol.* 5, 383-388.
de Bont, J.A.M., van Dijken, J.P., and Harder, W. (1981) *J. Gen. Microbiol.* 127, 315-323.
Drake, H.L., Daniel, S.C., Hsu, T., et al. (1989) in *Abstracts of the 6th International Symposium on Microbial Growth on C_1-compounds*, Göttingen, FRG.
Duff, S.J.B., and Murray, W.D. (1989) *Biotechnol. Bioeng.* 34, 153-159.
Duine, J.A. (1991) *Trends Biochem. Sci.* 16, 12.
Duine, J.A., and Jongejan, J.A. (1989) *Annu. Rev. Biochem.* 58, 403-426.
Duine, J.A., Frank, J., and Berkhout, M.P.J. (1984) *FEBS Lett.* 168, 217-221.
Eggeling, L., and Sahm, H. (1985) *Eur. J. Biochem.* 150, 129-134.
Egorov, A.M., Avilova, T.A., Dikov, M.M., et al. (1979) *Eur. J. Biochem.* 99, 569-576.
Fox, B.G., Borneman, J.G., Wacket, L.P., and Lipscomb, J.D. (1990) *Biochemistry* 29, 6419-6427.
Fry, W.E., and R.L. Millar (1972) *Archiv. Biochem. Biophys.* 151, 468-474.
Fujii, T., and Tonomura, K. (1972) *Agric. Biol. Chem.* 36, 2297-2306.
Ghisalba, O., Cevey, P., Küenzi, M., and Schär, H-P. (1985) *Conservation Recycling* 8, 47-71.
Ghisalba, O., Schär, H-P., and Ramos Tombo, G.M. (1986) in *Enzymes as Catalysts in Organic Synthesis* (Schneider, M.P., ed.), pp. 233-250, D. Reidel Publishing Co., Dordrecht, The Netherlands.
Ghisalba, O., Küenzli, M., Ramos Tombo, G.M., and Schär, H-P. (1987) *Chimia* 41, 206-215.

Gibson, T.D., Parker, S.M., and Woodward, J.R. (1989) in *Abstracts of the 6th International Symposium of Growth on C_I-compounds*, Göttingen, FRG.
Giuseppin, M.L.F. (1988) Ph.D. thesis, University of Delft, Delft, The Netherlands.
Giuseppin, M.L.F., van Eijk, H.M.J., Bos, A., Verduyn, C., and van Dijken, J.P. (1988) *Appl. Microbiol. Biotechnol.* 28, 286–292.
Gommers, P.J.F., van Schie, B.J., van Dijken, J.P., and Kuenen, J.G. (1988) *Biotechnol. Bioeng.* 32, 86–94.
Green, J., Haywood, G.W., and Large, P.J. (1983) *Biochem. J.* 211, 481–493.
Guillén, F., Martinez, A.T., and Martinez, M.J. (1990) *Appl. Microbiol. Biotechnol.* 32, 465–469.
Hanson, R.S., and Lipscomb, J.D. (1990) U.S. patent 9005708.
Heidlas, J., Engel, K-H., and Tressel, R. (1988) *Eur. J. Biochem.* 172, 633–639.
Heppel, L.A., and Porterfield, V.T. (1948) *J. Biol. Chem.* 176, 763–769.
Higgins, I.J., Aston, W.J., Best, D.J., et al. (1984) *Microbial Growth on C_I-compounds* (Crawford, R.L., and Hanson, R.S., eds.), pp. 297–305, American Society for Microbiology, Washington DC.
Hoiberg, D.A., and Hatley, G.W. (1988) U.S. patent 88/8879.
Hopkins, T.R., and Muller, F. (1987) in *Proceedings 5th International Symposium Microbial Growth on C_I Compounds* (Van Verseveld, H.W., and Duine, J.A., eds.), pp. 150–157, Martinus Nijhof Publishers, Dordrecht, The Netherlands.
Hormann, K., and Andreesen, J.R. (1989) *Arch. Microbiol.* 153, 50–59.
Hou, C.T., Patel, R., Barnabe, N., and Marczak, I. (1981a) *Eur. J. Biochem.* 119, 359–364.
Hou, C.T., Patel, R., Laskin, A.I., Barnabe, N., and Marczak, I. (1981b) *Appl. Environ. Microbiol.* 41, 829–832.
Hou, C.T., Patel, R.N., Laskin, A.I., and Barnabe, N. (1982) *Arch. Biochem. Biophys.* 216, 296–305.
Hummel, W., and Kula, M.R. (1989) *Eur. J. Biochem.* 184, 1–13.
Izumi, Y., Kanzaki, H., Monita, S., and Yamada, H. (1987) *FEMS Microbiol. Lett.* 48, 139–142.
Izumi, Y., Kanzaki, H., Morita, S., Futazuka, H., and Yamada, H. (1989) *Eur. J. Biochem.* 182, 333–341.
Janes, S.M., Mu, D., Wemmer, D., et al. (1990) *Science* 248, 981–987.
Janssen, F.W., and Ruelius, H.W. (1968) *Biochim. Biophys. Acta* 151, 330–342.
Janssen, D.B., Scheper, A., Dijkhuizen, L., and Witholt, B. (1985) *Appl. Environ. Microbiol.* 49, 673–677.
Jezequel, S.G., Kaye, B., and Higgins, I.J. (1984) *Biotechnol. Lett.* 6, 567–570.
Johnson, W.V., and Anderson, P.M. (1987) *J. Biol. Chem.* 262, 9021–9025.
Johnson, P.A., and Quayle, J.R. (1964) *Biochem. J.* 93, 281–290.
Karube, I., Satoh, I., Araki, Y., Suzuki, S., and Yamada, H. (1980) *Enzyme Microbiol. Technol.* 2, 117–120.
Kato, N. (1986) Japanese patent JP 01/86885.
Kato, N., Kano, M., Tani, Y., and Ogata, K. (1974) *Agric. Biol. Chem.* 38, 111–116.
Kato, N., Omori, Y., Tani, Y., and Ogata, K. (1976) *Eur. J. Biochem.* 64, 341–350.
Kato, N., Yamagami, T., Kitayama, Y., Shimao, M., and Sakazawa, C. (1984) *J. Biotechnol.* 1, 295–306.
Kato, N., Yamagami, T., Shimao, M., and Sakazawa, C. (1986) *Eur. J. Biochem.* 156, 59–64.
Kemp, G.D., Dickinson, F.M., and Ratledge, C. (1990) *Appl. Microbiol. Biotechnol.* 32, 461–464.

Keuning, S., Janssen, D.B., and Witholt, B. (1985) *J. Bacteriol.* 163, 635–639.
Khedkar, S., Kale, N., and Rale, V. (1989) in *Abstracts of the 6th International Symposium on Microbial Growth on C_1-compounds*, Göttingen, FRG.
Killgore, J., Smidt, C., Duich, L., et al. (1989) *Science* 245, 850–852.
Knowles, C.J., and Bunch, A.W. (1986) *Adv. Microbiol. Physiol.* 27, 73–111.
Kohler-Staub, D., and Leisinger, T. (1985) *J. Bacteriol.* 162, 676–681.
Kula, M.R., and Wandrey, C. (1987) *Methods Enzymol.* 136, 9–21.
Large, P.J., and Green, J. (1984) *Microbial Growth on C_1-compounds* (Crawford, R.L., and Hanson, R.S., eds.), pp. 155–164, American Society for Microbiology, Washington DC.
Leak, D.J., and Dalton, H. (1988) *Biocatalysis* 1, 23–36.
Ledeboer, A.M., Edens, L., Maat, J., et al. (1985) *Nucl. Acids. Res.* 13, 3063–3082.
Lüdi, H., Garn, M., Ghisalba, O., and Schär, H-P. (1988) *Int. J. Environ. Anal. Chem.* 33, 131–139.
Meyer, O., Jacobitz, S., and Krüger, B. (1986) *FEMS Microbiol. Rev.* 39, 161–179.
Mizuno, S., and Imada, Y. (1986) *Biotechnol. Lett.* 8, 79–84.
Morita, M., and Watanabe, Y. (1977) *Agric. Biol. Chem.* 41, 1535–1537.
Müller, U., Willnow, P., Ruschig, U., and Hopner, T. (1978) *Eur. J. Biochem.* 83, 485–498.
Ohyama, T., and Yamazaki, I. (1974) *J. Biochem.* (Tokyo) 75, 1257–1263.
Ohyama, T., and Yamazaki, I. (1975) *J. Biochem.* (Tokyo) 77, 845–852.
Okura, I. (1990) Japanese patent JP 02092295.
Oldenhuis, R., Vink, R., Janssen, D.B., Witholt, B. (1989a) *Appl. Environ. Microbiol.* 55, 2819–2826.
Oldenhuis, R., Vink, R., Oedzes, H., and Janssen, D.B. (1989b) in *Abstracts of the 6th International Symposium on Microbial Growth on C_1-compounds*, Göttingen, FRG.
Patel, R.N., and Hoare, D.S. (1971) *J. Bacteriol.* 107, 187–192.
Patel, R.N., Hou, C.T., Laskin, A.I., and Derelanko, P. (1981) *Arch. Biochem. Biophys.* 210, 481–488.
Patel, R.N., Hou, C.T., Laskin, A.I., and Felix, A. (1982) *Appl. Environ. Microbiol.* 44, 1130–1137.
Poels, P.A., and Duine, J.A. (1989) *Arch. Biochem. Biophys.* 271, 240–245.
Ramos Tombo, G.M., Schär, H-P., and Ghisalba, O. (1987) *Agric. Biol. Chem.* 51, 1833–1838.
Rasche, M., Hicks, R., Harding, R., Hyman, M., and Arp, D. (1989) in *Abstracts of the 6th International Symposium on Microbial Growth on C_1-compounds*, Göttingen, FRG.
Rindt, K-P. German patent DE 3720.506.
Rindt, K-P., and Scholtissek, S. (1989) in *Abstracts of the 6th International Symposium on Microbial Growth on C_1-compounds*, Göttingen, FRG.
Ruzicka, F., Huang, D.S., Donnelly, M.I., and Frey, P.A. (1990) *Biochemistry* 29, 1696–1700.
Sahm, H., and Wagner, F. (1973) *Eur. J. Biochem.* 36, 250–256.
Sanraku Ocean. Japanese patent J. 55-042534.
Schär, H-P., and Ghisalba, O. (1985) *Biotechnol. Bioeng.* 27, 897–901.
Schär, H-P., Chemla, P., Ghisalba, O. (1985) *FEMS Microbiol. Lett.* 26, 117–122.
Schär, H-P., Holzmann, W., Ramos Tombo, G.M., and Ghisalba, O. (1986) *Eur. J. Biochem.* 158, 469–475.

Scholtz, R., Leisinger, T., Suter, F., and Cook, A.M. (1987a) *J. Bacteriol.* 169, 5016–5021.
Scholtz, R., Schmuckle, A., Cook, A.M., and Leisinger, T. (1987b) *J. Gen. Microbiol.* 133, 267–274.
Scholtz, R., Messi, F., Leisinger, T., and Cook, A.M. (1988a) *Appl. Environ. Microbiol.* 54, 3034–3038.
Scholtz, R., Wackett, R.P., Egli, C., Cook, A.M., and Leisinger, T. (1988b) *J. Bacteriol.* 170, 5698–5704.
Schreiber, W., Schindler, J., and Schmid, R. (1977) Henkel GmbH. German patent 2557623.
Schütte, H., Flossdorf, J., Sahm, H., and Mula, M.R. (1976) *Eur. J. Biochem.* 62, 151–160.
Stromeyer, S., Egli, C., Tschau, T., and Leisinger, T. (1989) in *Abstracts of the 6th International Symposium on Microbial Growth on C_1-compounds*, Göttingen, FRG.
Subramanian, V. (1986) *J. Ind. Microbiol.* 1, 119–127.
Suye, S., Ogawa, A., Yokoyama, S., and Obayashi, A. (1990) *Agric. Biol. Chem.* 54, 1297–1298.
Suylen, G.M.H., Large, P.J., van Dijken, J.P., and Kuenen, J.G. (1987) *J. Gen. Microbiol.* 133, 2989–2997.
Tani, Y., Miya, T., Nishikawa, H., and Ogata, K. (1972a) *Agric. Biol. Chem.* 36, 68–75.
Tani, Y., Miya, T., and Ogata, K. (1972b) *Agric. Biol. Chem.* 36, 76–83.
Tarame, S., Sogabe, Y., and Mitsukida, N. (1988) Japanese patent 86/279426.
Trinder, P. (1969) *Anal. Clin. Biochem.* 6, 24–27.
Tsien, H.C., Brusseau, G.A., Hanson, R.S., and Wackett, L.P. (1989) *Appl. Environ. Microbiol.* 55, 3155–3161.
Turner, A.P.F., Aston, W.J., and Higgins, I.J., et al. (1984) *Anal. Chim. Acta* 163, 161–174.
Ubbink, M., Van Kleef, M.A.G., Kleinjan, D.J., et al. (1991) *J. Bacteriol.* (in press).
Unichika Co. Ltd. (1986) Japanese patent JP 217522.
Uotila, L., and Koivusalo, M. (1979) *Arch. Biochem. Biophys.* 196, 33–45.
Upjohn Co. U.S. patent 4353987.
Ushio, K., Inouye, K., Nakamura, K., Oka, S., and Ohno, A. (1986) *Tetrahedron Lett.* 27, 2657–2660.
van der Meer, R.A., Groen, B.W., Jongejan, J.A., and Duine, J.A. (1990a) *FEBS Lett.* 261, 131–134.
van der Meer, R.A., Groen, B.W., van Kleef, M.A.G., et al. (1990b) *Methods Enzymol.* 188, 260–283.
van Dijken, J.P., Oostra-Demkes, G.J., Otto, R., and Harder, W. (1976a) *Arch. Microbiol.* 111, 77–83.
van Dijken, J.P., Otto, R., and Harder, W. (1976b) *Arch. Microbiol.* 111, 137–144.
Van Iersel, J., Van der Meer, R.A., and Duine, J.A. (1986) *Eur. J. Biochem.* 161, 415–419.
van Kleef, M.A.G., and Duine, J.A. (1989) *Appl. Environ. Microbiol.* 55, 1209–1213.
Van Ophem, P.W., and Duine, J.A. (1990) *Arch. Biochem. Biophys.* 282, 248–253.
Van Ophem, P.W., Euverink, G.-J., Dijkhuizen, L., and Duine, J.A. (1991) *FEMS Microbiol. Lett.* (in press).
Väsic-Racki, D., Jonas, M., Wandrey, C., Hummel, W., and Kula, M.R. (1989) *Appl. Microbiol. Biotechnol.* 31, 215–222.

Veenhuis, M., Harder, W., Van Dijken, J.P., and Mayer, F. (1981) *Mol. Cell. Biol.* 1 (10), 949–957.
Veenhuis, M., van Dijken, J.P., and Harder, W. (1983) *Adv. Microbiol. Physiol.* 24, 2–82.
Vellieux, F.M.D., Huitema, F., Goendijk, H., et al. (1989) *EMBO J.* 8, 2171–2178.
Verduyn, C., van Dijken, J.P., and Scheffers, W.A. (1983) *Biotechnol. Bioeng.* 25, 1049–1055.
Verduyn, C., Zomerdijk, T.P.L., van Dijken, J.P., and Scheffers, W.A. (1984a) *Appl. Microbiol. Biotechnol.* 19, 181–185.
Verduyn, C., van Dijken, J.P., and Scheffers, W.A. (1984b) *J. Microbiol. Methods* 2, 15–25.
Verduyn, C., Giuseppin, M.L.F., Scheffers, W.A., and van Dijken, J.P. (1988a) *Appl. Env. Microbiol.* 54, 2086–2090.
Verduyn, C., Breedveld, G.J., Scheffers, A., and van Dijken, J.P. (1988b) *Yeast* 4, 135–142.
Wackett, L.P., Shames, S.L., Venditti, C.P., and Walsh, C.T. (1987) *J. Bacteriol.* 169, 710–717.
Willets, A. (1983) *J. Gen. Microbiol.* 129, 997–1004.
Woodward, J.R. (1990) in *Advances in Autotrophic Microbiology and One-carbon Metabolism* (Codd, G.A., Tabita, F.R., and Dijkhuizen, L., eds.) Kluwer Ac. Publishers, Dordrecht, The Netherlands (in press).
Wu, Z., Daniel, S.L., and Drake, H.L. (1988) *J. Bacteriol.* 170, 5747–5750.
Yamada, H., Sgin, K.C., Kato, N., Shimizu, S., and Tani, Y. (1979) *Agric. Biol. Chem.* 43, 877–878.
Yen, H.C., and Marrs, B. (1977) *Arch. Biochem. Biophys.* 181, 411–418.
Yokota, T., Fuse, H., Omori, T., and Minoda, Y. (1986) *Agric. Biol. Chem.* 50, 453–460.

CHAPTER 11

Production of Useful Chemicals by Methylotrophs

Yoshiki Tani

The principal feedstock in the microbial industry has been carbohydrates, mainly molasses. However, the production of molasses depends on agriculture, and the presence of nonfermentable substances in molasses necessitates extra cost in the waste treatment. *n*-Paraffins, which once were considered to be the preferred feedstock for single cell protein (SCP) production, have now become less attractive due to increases in the price of oil.

C_1 compounds having no carbon-carbon bond are utilized or synthesized by microorganisms at all oxidation levels including carbon monoxide, carbon dioxide, methane, methanol, and methylamines. Of these C_1 compounds, methanol is now considered to be preferable for the microbial industry. Methanol is water-miscible and treatment of methanol-broth waste is easier because of its high purity. Methanol utilizers, or methylotrophs, grow faster than other C_1 utilizers. Methanol is synthesized from methane, which is a major component of natural gas. The world resource of natural gas is estimated at several hundred times that of petroleum oil. The present annual world production of methanol is over 15 million tons. Recent development in C_1-synthetic chemistry offers processes in which the ethylene from petroleum is replaced by carbon monoxide or methanol as the starting material. The substitution of methanol for gasoline has been suggested as a way to reduce energy and pollution problems. Methane itself is also attractive in spite of problems such as mass transfer in an aqueous process and explosion hazards.

In the use of methanol for SCP production, the pathways for dissimilating and assimilating methanol have been extensively investigated, as summarized by Anthony (1982), Hou (1984a), and Tani (1984). So far, the existence of a variety of unique metabolic features in methylotrophs has been demonstrated. Some of them could be used to develop biochemical processes for the production of fine and commodity chemicals by methylotrophs (Tani 1985). This chapter reviews recent advances on the production of such useful chemicals by means of methylotrophic functions.

11.1 PRODUCTION OF CHEMICALS BY C_1-OXIDATIVE ENZYMES

11.1.1 Aldehydes Production by Alcohol Oxidase

In methylotrophic yeasts, methanol is oxidized to carbon dioxide through the actions of alcohol oxidase, catalase, formaldehyde dehydrogenase, S-formylglutathione hydrolase, and formate dehydrogenase. The alcohol oxidase uses molecular oxygen as the electron acceptor and, therefore, does not require regeneration or an exogenous supply of cofactor for methanol oxidation.

Methanol is mostly used to produce formaldehyde by a chemical conversion process at 500–600°C. The use of a microbial process for the conversion would require much less energy and by-product formation would be minimized.

The yeast *Candida boidinii* S2 was mutagenized to increase its alcohol oxidase activity (Tani et al. 1985a and 1985b). A mutant strain shows 2.4-times higher alcohol oxidase activity than the wild-type strain. The high activity is attributed to a quantitative change and also a change in the rate of enzyme synthesis. Formaldehyde productivity is much improved by the use of a methanol-limited chemostat culture (Sakai and Tani 1986). Cells from the chemostat culture at a dilution rate of 0.075 h^{-1} show the highest productivity, having the highest activity of alcohol oxidase and almost minimum activity of formaldehyde dehydrogenase. Formaldehyde productivity is also improved by the use of catalase-enhanced (Sakai et al. 1987b) and catabolite repression-insensitive mutants (Sakai et al. 1987a; Tani et al. 1988c).

The productivity and catalytic stability of chemostat-grown cells are improved by preincubation of the cells at 37°C (Sakai and Tani 1987a). Treatment of cells with a cationic detergent (cation M2) also improves productivity (Sakai and Tani 1988a), as the treatment causes formaldehyde and formate dehydrogenases to leak out of cells more rapidly than catalase, but there is no leakage of alcohol oxidase. After the detergent treatment, the cells are obviously damaged, and several peroxisomes fuse with each other. The detergent-treated cells could be applied for the simple process of purification and improvement of stability of alcohol oxidase to form an enzyme-azide complex (Sakai and Tani 1988b). Immobilized cells of *Pichia pastoris* are more stable to heat denaturation and retained more alcohol oxidase activity than free cells (Duff and Murray 1988a).

Acrolein is chemically produced through the direct oxidation of propylene. However, the product is difficult to recover and it is contaminated with large amounts of acetaldehyde and acetone as by-products. Acetaldehyde is used for the synthesis of a number of chemicals and is produced through ethylene oxidation. The oxidation of allyl alcohol, ethanol, or 1-propanol is performed for the production of acrolein, acetaldehyde, or propionaldehyde, respectively, with heat- and detergent-treated cells of *Candida boidinii* (Table 11–1; Sakai and Tani 1987b).

The purified alcohol oxidase shows the following relative activities towards alcohols: methanol, 100%; ethanol, 75%; 1-propanol, 30%; and allyl alcohol, 49%. However, the molar yield of acetaldehyde is the highest and that of acrolein is the lowest among the aldehydes. Also, the complete consumption of alcohol is observed on the oxidation of ethanol and 1-propanol, whereas some methanol or allyl alcohol remained, even when they are added at low concentrations. Thus, a good substrate for enzyme activity is not always a good substrate for aldehyde production, due to the inhibitory effect of aldehyde on the reaction. The optimum temperature of 4°C for aldehyde production prevents the inactivation of alcohol oxidase during this reaction. Ethanol oxidation was also examined using *Pichia pastoris* cells for the production of acetaldehyde (Duff and Murray 1988b).

TABLE 11–1 Production of Aldehydes from Alcohols by Cells of *Candida boidinii* S2

Substrate/Product	Cell Treatment	Concentration of Substrate	Concentration of Product	Yield (%)	Reaction Time (h)
Methanol/ formaldehyde	Heat	3.0 M	1.09 M	36	10
	Cation M2	3.0 M	1.16 M	39	7
		3.0 M	1.38 M	46	20
Ethanol/ acetaldehyde	Heat	0.43 M	0.43 M	99	1.5
		1.74 M	1.71 M	98	18
1-Propanol/ propionaldehyde	Heat	83 mM	83 mM	100	1.5
		333 mM	283 mM	85	19
Allyl alcohol/ acrolein	Heat	862 mM	536 mM	62	2

11.1.2 Formate Production by Formaldehyde Dismutase

Formaldehyde dismutase from *Pseudomonas putida* catalyzes the dismutation of various aldehydes, leading to the formation of equimolar amounts of the corresponding alcohols and acids (Kato et al. 1983, 1984a, 1984b, and 1986b).

Mizuno and Imada (1986) have developed an enzymatic process for the production of formate from methanol, involving the combination of alcohol oxidase and catalase for methanol oxidation and formaldehyde dismutase. In this process, methanol is first oxidized to formaldehyde, which is then converted to formate and methanol. After repetition of the sequential enzyme reactions, all the methanol added is stoichiometrically converted to formate. The reaction system does not require any electron acceptor other than oxygen. The process is improved by the immobilization of enzymes or of cells (Kato et al. 1988). Conversion of methanol to formate occurs with an immobilized enzyme system consisting of the three enzymes, and with an intact cell mixture of *Hansenula polymorpha* and *Pseudomonas putida*. The immobilized cell system can also be used for the conversion of several aliphatic alcohols, C_1 to C_4, to the corresponding acids.

11.1.3 Epoxide Production by Methane Monooxygenase

Epoxides of olefins are important for the chemical industry because of their ability to form homopolymers and copolymers. The conversion process used for producing 1,2-epoxyalkanes from 1-alkenes in *Nocardia corallina* (Furuhashi et al. 1981) and for converting α, β-halohydrins to the corresponding epoxides used by *Flavobacterium* species (Geigert et al. 1983) has been demonstrated.

The methane monooxygenase of obligate methylotrophs grown on methane is the enzyme responsible for the initial oxygenation of methane to methanol in the presence of NADH and oxygen. Because of its broad specificity, the enzyme can also oxidize various nongrowth substrates, such as substituted methane derivatives, *n*-alkanes, terminal alkenes, dimethyl ether, cyclic alkanes, and aromatic compounds.

The application of this enzyme to the epoxidation system was shown by Hou et al. (1980) and by Dalton (1980). Resting cells and cell-free extracts of *Methylococcus capsulatus, Methylosinus trichosporium,* and *Methylobacterium organophilum* oxidize C_2-C_4 1-alkenes to their corresponding 1,2-epoxides. Methanol-grown cells do not have this epoxidation activity. An advantage of this oxidation system is that the products, 1,2-epoxides, are not further metabolized and hence accumulate extracellularly. However, the requirement for a reduced cofactor, NADH, is a disadvantage of this process. Propylene oxide has been produced from propylene using a bioreactor in which the cells were adsorbed to glass beads and which was equipped to handle the gaseous substrate and product (Hou 1984b). Methanol is used for the regeneration of NADH by means of methanol dehydrogenase.

The cells also oxidize n-alkanes producing both primary and secondary alcohols, which are further oxidized and not accumulated.

11.1.4 Methyl Ketone Production by Secondary Alcohol Dehydrogenase

Methyl ketones are important industrial solvents. Methane-grown cells of methylotrophs can oxidize n-alkanes (C_3 to C_6) to their corresponding methyl ketones (Hou 1984b). The first step in the oxidation of n-alkanes to secondary alcohols is catalyzed by methane monooxygenase. An NAD-dependent secondary alcohol dehydrogenase specifically catalyzes the oxidation of secondary alcohols to methyl ketones.

Oxidation of secondary alcohols to methyl ketones also occurs in cells of methylotrophic bacteria and yeasts (Hou et al. 1979; Patel et al. 1979). The activity is found in methane-, methanol-, and methylamine-grown cells but not in succinate-grown cells. The enzyme in methane-grown cells can be used for regeneration in the epoxide production system (Hou 1984b).

11.2 PRODUCTION OF RESPIRATORY COENZYMES

11.2.1 Cytochrome *c*

Cytochrome *c* is used as a medicine for heart diseases caused by oxygen deficiency. It is usually purified from extracts of horse and cattle hearts and from yeast. Methylotrophic bacteria oxidize methanol with methanol dehydrogenase, from which electrons flow to cytochrome *c*. Therefore, methanol could be a more suitable carbon source for the production of cytochrome *c* than other carbon sources, since the growth substrate is directly connected to cytochrome *c* in methylotrophs.

An obligate methylotroph, *Methylomonas* sp. YK1, was mutagenized so as to be resistant to metabolic inhibitors related to the function of cytochrome *c* (Tani et al. 1985c). Table 11–2 lists the mutants having the highest cytochrome *c* content for each of several inhibitors. The cytochrome *c* content of the cyanide-resistant mutant *Methylomonas* sp. YK56, which was derived from strain YK8, was three times higher than that of the wild-type strain YK1. The optimization of medium composition and the feeding of methanol and peptone increases the cytochrome *c* productivity of mutant strain YK56 to about 60 mg/liter of culture medium. The maximal productivity is obtained using a jar-fermentor culture feeding methanol; 160 mg/liter is obtained after a 32-h cultivation period (Yoon and Tani 1987b). Further improvement in the amount of cytochrome *c* is attained with glycine- and glycine-analog-resistant mutants due to stimulation of the heme biosynthesis (Yoon et al. 1987).

Soluble cytochrome *c* of the wild-type strain is separated into two fractions, called cytochrome *c*-I and *c*-III, on DEAE-cellulose column chro-

TABLE 11-2 Cytochrome c Content of Inhibitor-Resistant Mutants of *Methylomonas* sp. YK1

Inhibitor	Resistant Strain	Cytochrome c Produced (mg/l of culture medium)
—	YK 1 (wild)	2.97
KCN	YK 8	6.83
Antimycin A	YK 126	5.96
Rotenone	YR 372	5.20
2,4-Dinitrophenol	YD 873	4.75
Pentachlorophenol	YP 1116	4.62
Progesterone	YPR 1385	4.95

matography. The proportion of cytochrome c-I and c-III is 2.1 and 97.9%. On the other hand, mutant strain YK56 contains three distinct forms of soluble cytochromes c: cytochrome c-II is found as a new fraction in addition to the two cytochromes of the wild-type strain. The proportion of these three soluble cytochromes c are: I, 2.4; II, 71.5; and III, 26.1%. The amount of cytochrome c-I and c-III produced is similar in both strains, while the mutant produces cytochrome c-II as its main component (Tani et al. 1986b). The properties of cytochrome c-II are clearly different from those of cytochrome c-III. This indicates that cytochrome c-II is a new product in the mutant cell. The membrane-bound cytochrome c also has two forms, which are similar but not identical to the soluble cytochromes c-II and c-III (Yoon and Tani 1987a).

The cytochrome c production of the wild-type strain and of mutant strain YK56 grown with excess methanol is higher than that with limited methanol (Yoon and Tani 1987c). The wild-type strain grown under both conditions contains two soluble cytochromes c, though the mutant strain contains three. The proportion of cytochromes c-II and c-III of the mutant change according to the culture conditions.

11.2.2 Pyrroloquinoline Quinone

In addition to the well-known pyridine nucleotide- and flavin nucleotide-dependent oxidoreductases, it has become clear that there is a third class of enzymes in which pyrroloquinoline quinone (PQQ) is involved as the coenzyme of a group of dehydrogenases and oxidases. Such a novel coenzyme was originally discovered and identified in methanol dehydrogenases of methylotrophic bacteria. PQQ has a growth-stimulating effect for various microorganisms and hence could be considered as a new vitamin.

PQQ is found in the culture medium of various microorganisms. Of these, some methylotrophs are most favorable for PQQ production, and more than 10 mg/liter of PQQ accumulates in the culture medium (Ameyama et al. 1984). The extracellular production of PQQ increases at the stationary growth phase after the carbon source is used up. PQQ preparation from methylotrophs has been commercialized in Japan.

11.2.3 Coenzyme Q_{10}

Coenzyme Q (ubiquinone) is a fat-soluble coenzyme and an essential component in the respiratory chain. Homologs with side chains of different lengths are widely distributed in nature. Among them, coenzyme Q_{10} is used to treat some form of heart disease.

The facultative methylotroph *Protomonas extorquens* has been used for the fermentative production of coenzyme Q_{10} (Urakami and Hori-Okubo 1988). The content of coenzyme Q_{10} increases as the age of the culture increases, and it reaches a maximum level during the stationary phase. The coenzyme Q_{10} content has also been increased by repeated mutagenesis to a level of approximately 3.3 mg/g dry cells (2.5 times that of the wild strain). Homologs such as coenzyme Q_9, Q_{11}, and Q_{12} are also found in increased level in some mutants. Commercial production of coenzyme Q_{10} using continuous culture techniques was begun in 1980 (Urakami et al. 1981).

11.2.4 NAD(P)H

Two dehydrogenases are needed to reduce NAD^+ in the oxidation of methanol in methylotrophs. Formate dehydrogenase is useful for the NADH-requiring system in the bioconversion process, because the substrate is fairly inexpensive and the product, carbon dioxide, does not inhibit the conversion reaction(s). Cells of a facultative methylotroph, an *Arthrobacter* species, convert NAD^+ to NADH in the presence of formate (Izumi et al. 1983). The conversion ratio and the amount of NADH produced 90% and 30 g/liter, respectively.

The oxidation of formate or hydrogen by methanogenic bacteria is mediated through F_{420} and coupled by F_{420}-$NADP^+$-oxidoreductase with the concomitant reduction of $NADP^+$. Resting cells of a methanogen produce NADPH from $NADP^+$ (Eguchi et al. 1983). The conversion ratio and the amount of NADPH produced were 60% and 6 g/liter, respectively.

11.2.5 ATP

Methylotrophic yeasts reduce NAD^+ in the oxidation pathway to supply energy for the assimilation of C_1 compounds through glyceraldehyde 3-phosphate by the xylulose monophosphate pathway. The proton motive force couples the NADH formed to the respiratory chain, while ADP is phosphorylated in the oxidative phosphorylation system.

A system for the production of ATP from AMP was first constructed using cells of *Candida boidinii* 2201 treated with Zymolyase®, a cell wall-lytic bacterial enzyme (Tani et al. 1982 and 1984a). The reaction mixture contains AMP, methanol (or formate), NAD$^+$, inorganic phosphorus, and treated cells. The yeast cells produce ATP by a sequence of reactions including phosphorylation of AMP to ADP by adenylate kinase, oxidation of methanol by formaldehyde and formate dehydrogenases to reduce NAD$^+$, and oxidative phosphorylation of ADP to ATP in the respiratory chain. This process was the first one to produce ATP through oxidative phosphorylation, as well as the first one to use a C_1 compound as the energy source.

The cell treatment was then modified in order to stabilize the production of ATP by incubating the cells with added sorbitol (Tani et al. 1984b). The sorbitol-treated cells do not lose activity even in the presence of 3 M methanol or after incubation for 36 h. Under optimal reaction conditions, the amount of ATP in the reaction mixture reaches 30 g/liter (Tani et al. 1984c).

Transmission electron microscopy showed that the high ATP-producing activity of the sorbitol-treated cells is due to the plasmolysis without destruction of essential organelles for the ATP-producing system (Yonehara and Tani 1988a). Limiting the Zn^{2+} concentration in the culture medium improves the ATP yield from AMP, the conversion rate being 92%, possibly due to repression of the enzyme AMP deaminase, for which Zn^{2+} is a cofactor and which affects the ATP-producing activity. A Zn^{2+}-limited culture provides fully active cells as to ATP production. Transmission electron microscopy shows that Zn^{2+} is a necessary factor for the formation of the peroxisomes in which alcohol oxidase and catalase are localized. Among osmo-regulatory compounds that plasmolyze the yeast cells, only sorbitol increases ATP production (Yonehara and Tani 1986). The sorbitol effect is presumably due to the increased level of NADH produced by the oxidation in parallel with the methanol oxidation.

Adenosine or adenine, which are more economical substrates than AMP, can be used as the substrate instead of AMP (Tani and Yonehara 1985). Efficient ATP production from adenosine is achieved by the addition of pyrophosphate together with Mg^{2+}. Under optimal conditions, 198 mM (100 g/liter) of ATP is produced with a conversion rate of 77% from adenosine (Yonehara and Tani 1987). ATP production from adenine takes place at a lower rate.

The appearance of the by-product IMP in the reaction mixture due to the catalysis of AMP deaminase lowers the ATP yield. Yonehara and Tani (1988b) obtained a mutant strain which is impaired with respect to its ability to grow on AMP as the sole nitrogen source. The AMP deaminase of the mutant strain exhibits the same allosteric patterns as those of the wild-type strain. Its specific activity in the cell-free extract of the mutant strain, however, is lower than that of the wild-type strain. Sorbitol-treated cells of the mutant produce ATP from AMP at a high conversion rate (95%).

The ATP production system involves NADH production in the cytosol and electron transport to the mitochondria by the glycerol phosphate shuttle mechanism to reduce FAD bound to the inner membranes. Consequently, the P/O ratio in this system is two, rather than three. The material balance of ATP production from adenosine can be summarized as the equation below:

$$\text{adenosine} + 3\text{ Pi} + \frac{3}{4}\text{ methanol} + \frac{9}{8}O_2 \rightarrow \text{ATP} + \frac{3}{4}CO_2 + \frac{9}{2}H_2O$$

Based on the free energy changes of the reactions involved in the system, the theoretical value of the overall energy efficiency in going from methanol to ATP is calculated as 14.3%. In a typical reaction mixture, the practical energy efficiency from methanol to ATP is 4.25%. Therefore, the ratio of the practical to the theoretical is 0.3 (Yonehara and Tani 1986).

ATP production can be characterized either as a biotransformation system or on the basis of energy conversion. The free energy of methanol could be employed to use methanol as an energy source in bioconversion systems for various compounds.

11.3 GLYCEROL PRODUCTION IN YEAST

There has been a great revival in interest in glycerol fermentation due to concern over diminishing petroleum reserves. The glycerol produced as a by-product in the manufacture of detergents and fatty acids does not provide a stable supply.

Methylotrophic yeasts assimilate methanol through the xylulose monophosphate pathway, in which dihydroxyacetone is phosphorylated, and resultant phosphate becomes available for the regeneration of the C_1 receptor, xylulose 5-phosphate. Dihydroxyacetone is an oxidized product of glycerol dissimilation in some microorganisms. Tani and Yamada (1987a and 1987b) examined the role of dihydroxyacetone in the assimilation of methanol and dissimilation of glycerol in methylotrophic yeasts. They found three different patterns, represented by *Candida boidinii* 2201, *Hansenula ofunaensis,* and *H. polymorpha* DL-1, with regard to the initial step of glycerol dissimilation: the phosphorylative pathway involving glycerol kinase in *C. boidinii* 2201; the oxidative pathway involving NAD^+-linked glycerol dehydrogenase in *H. ofunaensis;* and both pathways in *H. polymorpha* DL-1 (Figure 11-1).

Two types of glycerol dehydrogenase appear on DEAE-cellulose column chromatography of cell-free extracts of methylotrophic yeasts. One type, glycerol dehydrogenase-I, prefers the reduction of dihydroxyacetone at pH 6.0, and the other type, glycerol dehydrogenase-II, catalyzes the oxidation of glycerol at pH 9.0, together with the reductive activity (Yamada and Tani 1987a). Glycerol dehydrogenase-I (I-2) is a dihydroxyacetone reductase (Yamada and Tani 1987b and 1988a). Dihydroxyacetone is an intermediate in the methanol metabolism in yeast. There is good correlation between the activity of dihydroxyacetone reductase and the intracellular content of glyc-

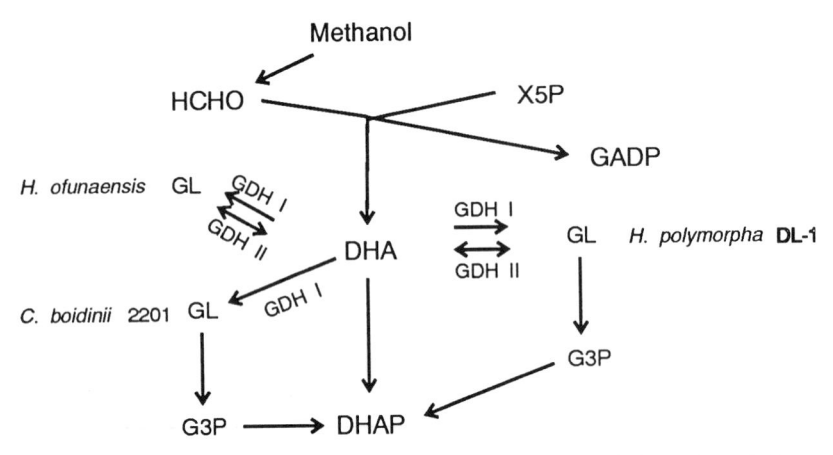

262 Production of Useful Chemicals by Methylotrophs

```
                        Methanol
                           ↙
           HCHO ←                   ← X5P
                    ↘         ↘
                               ↘ GADP
  H. ofunaensis  GL ⇌GDH I
                   GDH II          GDH I
                     ↘      DHA  ⇌        GL   H. polymorpha DL-1
                                  GDH II
                               ↓         ↓
  C. boidinii 2201  GL ← GDH I
         ↓              ↓                ↘ G3P
         ↓              ↓                  ↓
        G3P  ⟶  DHAP  ←
```

FIGURE 11-1 The relationship between the dissimilation of glycerol and the assimilation of methanol. GL, glycerol; X5P, xylulose 5-phosphate; DHA, dihydroxyacetone; GADP, glyceraldehyde 3-phosphate; G3P, glycerol 3-phosphate; DHAP, dihydroxyacetone phosphate; GDH, glycerol dehydrogenase.

erol in *H. ofunaensis* when the cells are grown on methanol. The enzyme activity increases when they are grown under low oxygen. The less-aerobic conditions after aerobic cultivation increase the glycerol synthesis. Methanol is the most effective carbon source for dihydroxyacetone reductase synthesis, and glycerol does not repress the synthesis. These observations indicate that in methanol-grown cells of *H. ofunaensis*, dihydroxyacetone reductase is responsible for growth under low oxygen by means of the reoxidation of NADH from the methanol oxidation (Yamada and Tani 1988b).

Cells of a dihydroxyacetone kinase-deficient mutant strain of *H. polymorpha* accumulate dihydroxyacetone and also glycerol from methanol (Kato et al. 1986a). Use of dihydroxyacetone reductase makes it possible to produce glycerol by growing yeast in a methanol medium, with dihydroxyacetone as the intermediate. Glycerol production using a mutant defective in glycerol catabolism may be a promising method for the development of a fermentative process for glycerol using methanol as the culture substrate.

Glycerol-negative mutants were derived by UV irradiation from *Candida boidinii* 2201 (Yamada et al. 1988). A glycerol-producing mutant, strain SK-1, is defective in glycerol-3-phosphate dehydrogenase. This confirms the involvement of the phosphorylative pathway in glycerol utilization in the wild-type strain. The extracellular accumulation of glycerol formed from methanol through dihydroxyacetone in the mutant is due to repression of further utilization.

A decrease in the glycerol in the medium is still observed in the late stage of cultivation of the mutant. This is due to the consumption of glycerol by another glycerol-3-phosphate dehydrogenase, which is NAD$^+$-dependent, after glycerol kinase-mediated phosphorylation. Two glycerol kinase inhibitors, 1-thioglycerol and glycerol-α-monochlorohydrin, increase glycerol production on its addition at the mid-logarithmic phase (Yamada et al. 1989). The addition of MnSO$_4$ at the beginning of cultivation also increases the glycerol production. Though dihydroxyacetone reductase is responsible for the generation of NAD under less-aerobic growth conditions, more glycerol is obtained on cultivation under aerobic conditions due to a higher growth rate. After rapid growth a shift of the culture from aerobic to less-aerobic conditions results in an increase in glycerol production. The respiratory inhibitors, antimycin A and 2,4-dinitrophenol, have a positive effect on glycerol production. Under optimal conditions, glycerol production reaches 1.5 g/liter.

11.4 POLYOL PRODUCTION IN YEAST

Large amounts of sorbitol are used in the food industry to keep food moist and soft, as the starting material for the synthesis of ascorbic acid, and also as a sweetening agent for diabetics. Two enzymes are known to form sorbitol by the reduction of D-fructose and D-glucose: L-iditol dehydrogenase, which catalyzes the reversible oxidation of sorbitol to fructose; and aldose reductase, which catalyzes the reduction of glucose to sorbitol.

Methylotrophic yeasts can produce the sorbitol efficiently from glucose using a resting-cell system in which methanol serves as the energy source for the regeneration of NADH (Tani and Vongsuvanlert 1987; Vongsuvanlert and Tani 1988a and 1988b). Cells of *Candida boidinii* 2201 having high sorbitol productivity are prepared by cultivating the yeast in a medium containing methanol and xylose. Sorbitol at 8.8 g/liter and 19.1 g/liter is obtained from 20 g/liter of glucose and fructose, respectively.

A sweet hexitol, iditol, appearing as colorless crystals with a melting point of 70–73°C, has not been thoroughly evaluated for practical applications, because of the lack of a practical preparation method. The chemical synthesis of iditol by the hydrogenation and acetylation of sorbose has been extensively studied. However, iditol obtained from either of these sources is contaminated with its C-2 epimer, D-glucitol (sorbitol). The low yield of the multistep chemical process results in high production costs.

Iditol is produced from sorbose by iditol dehydrogenase coupled with NADH regeneration under methanol oxidation in the resting-cell system of *Candida boidinii* 2201 (Vongsuvanlert and Tani 1988c). Maximum activities of the enzymes for iditol production are found in cells grown on a medium containing methanol and D-xylose. The highest amount of iditol, 148 g/liter (98% conversion rate), is obtained from sorbose in the presence of methanol.

Currently, xylitol, the pentitol of xylose, is frequently used as a sweetener in foods, owing to its sweetening power—two times greater than sorbitol. Xylitol is chemically produced by the hydrogenation of xylose. Recently, fermentative and biocatalytic processes for the production of xylitol have attracted much attention and been studied with bacteria and yeasts.

Xylitol is formed as a metabolic intermediary product of xylose in microorganisms in two ways: xylose is directly converted to xylitol by NADPH-dependent aldehyde reductase, or it is first isomerized by D-xylulose by xylose isomerase and then reduced to xylitol by NADH-dependent xylitol dehydrogenase.

Candida boidinii 2201 could produce xylitol and also xylulose during cultivation on a xylose medium (Vongsuvanlert and Tani 1989). A xylose (100 g/liter) medium supplemented with 2% methanol, however, gave higher amounts of xylitol (48.5 g/liter) and xylulose (3.3 g/liter).

11.5 AMINO ACID PRODUCTION

The accumulation of L-glutamate (6.8 g/liter) in the culture medium of *Methanomonas methylovora* (Oki et al. 1973) was the first report of amino acid production by methylotrophs. Subsequently, amino acid production using methylotrophs has been further developed based on data on methylotroph metabolism (Morinaga and Hirose 1984). The published reports can be classified into two groups (Tani 1985): In one, the biosynthetic pathways for L-glutamate, L-leucine, L-valine, L-tryptophan, L-phenylalanine, and L-tyrosine productions are identical to those of conventional organisms. In the other group, the biosyntheses of L-serine, L-methionine, and *O*-methyl-L-homoserine depend on the incorporation of a C_1 unit in the biosynthetic pathway.

The production of L-serine by methylotrophs (Keune et al. 1976; Tani et al. 1978; Morinaga et al. 1981a, 1981b, and 1983a; Izumi et al. 1982; Yamada et al. 1986) is one use of their unique metabolic pathway, the serine pathway. Serine hydroxymethyltransferase, which fixes formaldehyde to glycine to form L-serine, is the key enzyme in this pathway.

A glycine-resistant mutant of *Pseudomonas* MS31 stably produces L-serine from glycine (Watanabe et al. 1987a). The activity of L-serine dehydratase degrading L-serine is reduced in the mutant. A methionine-auxotrophic mutant deficient in homocysteine transmethylation activity has a serine hydroxymethyltransferase activity that is more than 1.7-fold higher than that of its parent strain (Watanabe et al. 1987b). Under optimum conditions, this mutant accumulates up to 23.9 g/liter of L-serine. The yield coefficient of L-serine from consumed glycine is 89% (mol/mol). The maximum conversion rate of added glycine (19 g/liter) to L-serine is 77% (mol/mol).

Serine production with resting cells of *Pseudomonas extorquens* NR-1 under multivariable, controlled conditions results in the highest level of L-serine production (54.5 g/liter) (Sirirote et al. 1986). Calcium-alginate-immobilized resting cells of the strain are also available for repeated use (Sirirote et al. 1988).

Somewhat higher amounts of L-serine may be produced by nonmethylotrophic microorganisms. However, the theoretical conversion rate of glycine to L-serine cannot exceed 50% in these processes, because the C_1 unit is provided by the cleavage of a second glycine molecule.

Although L-methionine is synthesized by alternate pathways in microorganisms, L-homoserine, sulfide, 5-methyltetrahydrofolate, and L-serine are essential intermediates in the biosynthesis. Thus, several steps of C_1 unit incorporation are included in the biosynthesis of L-methionine. There is, therefore, a possibility that methionine, which is now supplied by chemical synthesis, could be manufactured by a methylotrophic fermentation (Yamada et al. 1982; Morinaga et al. 1982a, 1982b, 1983b, and 1984).

Candida boidinii 2201 accumulates relatively high amount of free L-methionine (pool methionine) intracellularly (Tani et al. 1988a). An ethionine-resistant mutant strain shows an even higher level of pool methionine and, after optimization of culture conditions, 38.5 g (as dry weight) of cells and 282 mg of pool methionine per liter of culture broth are obtained after 11 days of cultivation (Lim and Tani 1988). With limited methanol in continuous cultivation, pool methionine productivity reaches a maximum value of 1.14 mg/liter/h at a dilution rate of 0.05/h. During methanol-limited growth in continuous cultivation, the pool methionine content of the mutant is about 20–35% higher than that in batch cultivation.

Productions of L-phenylalanine, L-tyrosine, and L-tryptophan by *Methylomonas methanolophila* (Suzuki et al. 1977), L-valine and L-leucine by *Methylomonas aminofaciens* (Ogata et al. 1977; Izumi et al. 1977), O-methyl-L-homoserine by facultative methylotrophs (Tanaka et al. 1980), and L-tryptophan by *Hansenula polymorpha* (Denenu and Demain 1981; Longin et al. 1982) have been reported.

11.6 POLY-β-HYDROXYBUTYRIC ACID PRODUCTION

Poly-β-hydroxybutyric acid (PHB) is a carbon-energy storage material that accumulates inside a variety of microorganisms. Solid PHB is a biodegradable thermoplastic polyester that can be utilized in ways similar to those of many conventional synthetic plastics. Therefore, the commercial use of PHB is one solution to the problem of pollution caused by nondegradable plastics. Its production is now under development, for example, by Imperial Chemical Industries Ltd., England.

PHB production by methylotrophs was first described by Braunegg et al. (1978). Large amounts of PHB are produced from methanol by means of a microcomputer-aided, fully automatic fed-batch culture technique (Suzuki et al. 1986a, 1986b, 1986c, and 1988). By controlling the feeding of ammonia, the PHB concentration in cells of *Pseudomonas extorquens* reaches 149 g/liter at 170 h and the total cell concentration is 233 g/liter. PHB yield from methanol is 0.20 (g PHB/g methanol). The degree of polymerization affects the physical characteristics of the polymer. For instance, PHB of molecular weight greater than 10^5 is not useful as a biodegradable carrier for long-term-dose medication. A change in the methanol concentration affects the average molecular weight of PHB. When the cultivation is carried out at 0.05 g/liter of methanol, the average molecular weight of PHB is above 8×10^5. At 32 g/liter of methanol, the average molecular weight of PHB is less than 0.5×10^5.

11.7 PRODUCTION OF OTHER METABOLITES

Vitamin B_{12} has a complicated corrinoid structure and can be produced by microbial processes. Methylotrophs were thought to be potentially useful as corrinoid precursors because the coenzymatic function of vitamin B_{12} may involve the transfer of a methyl group. A fed-batch culture, in which a methanol-ammonia mixture is fed in response to a direct signal of a pH change in the culture, was used to study vitamin B_{12} production by some methylotrophic bacteria (Nishio et al. 1977). A methanol-limited chemostat culture also showed an increase in intracellular content of vitamin B_{12} with increase of dilution rate (Tsuchiya et al. 1980), but the productivity of these bacteria is low—not more than 3.2 mg/liter.

The extracellular accumulation of polysaccharide in a methanol medium was reported for some methylotrophic bacteria. A polysaccharide produced by a *Pseudomonas* species contains D-allose (9.8%), which is rarely found as the sugar component in natural polysaccharides (Misaki et al. 1979). The polysaccharide from a *Hyphomicrobium* species contains a new monomethyl sugar, methyl-D-mannose (Kanamaru et al. 1982).

Pink-pigmented bacteria have frequently been isolated among the facultative methylotrophs. The pink pigment of *Protaminobacter ruber (Pseudomonas extorquens)* was shown to be a carotenoid (Sato et al. 1982; Shimizu et al. 1982).

Yeast strains belonging to the genus *Candida* are known to accumulate citric acid in a medium containing *n*-paraffins or other carbon compounds. Mutants of methylotrophic *Candida* species that are resistant to fluoroacetate were found to produce citric acid (Tani et al. 1990).

11.8 CONCLUSIONS

A variety of studies using methylotrophs for the production of useful chemicals have been reviewed in this chapter. Methanol at high concentrations inhibits the growth of microorganisms. The low optimum concentration for growth and biomass production has led to the introduction of fed-batch culture methods in which intermittent feeding of methanol is used to increase the cell concentration in a fermentor. Another disadvantage of methanol as a substrate is the extremely high energy consumption required to synthesize essential metabolites from C_1 units.

At present, commercial processes involving methylotrophs are limited to the production of SCP, coenzyme Q_{10}, PQQ, and methanol-oxidizing enzymes. However, the demand for methanol is increasing both in the chemical industry and in petroleum-substitute fuels. The interrelationship between these fields caused the drop of the price of methanol drastically in this two years, being able to compete with carbohydrate-dependent processes. The expression of heterologous genes in methylotrophic yeasts could also present attractive possibilities for the production of useful materials from methanol.

REFERENCES

Ameyama, M., Hayashi, M., Matsushita, K., Shinagawa, E., and Adachi, O. (1984) *Agric. Biol. Chem.* 48 (2), 561–565.
Anthony, C. (1982) *The Biochemistry of Methylotrophs,* Academic Press, London.
Braunegg, G., Sonnleitner, B., and Lafferty, R.M. (1978) *Eur. J. Appl. Microbiol. Biotechnol.* 6 (1), 29–37.
Dalton, H. (1980) *Adv. Appl. Microbiol.* 26, 71–87.
Denenu, E.O., and Demain, A.L. (1981) *Appl. Environ. Microbiol.* 41 (15), 1088–1096.
Duff, S.J.B., and Murray, W.D. (1988a) *Biotechnol. Bioeng.* 31 (8), 790–795.
Duff, S.J.B., and Murray, W.D. (1988b) *Biotechnol. Bioeng.* 31 (1), 44–49.
Eguchi, S.Y., Nishio, N., and Nagai, S. (1983) *Agric. Biol. Chem.* 47 (12), 2941–2943.
Furuhashi, K., Taoka, A., Uchida, S., Karube, I., and Suzuki, S. (1981) *Eur. J. Appl. Microbiol. Biotechnol.* 12 (1), 39–45.
Geigert, J., Neidleman, S.L., Liu, T.-N.E., et al. (1983) *Appl. Environ. Microbiol.* 45 (3), 1148–1149.
Hou, C.T. (1984a) in *Methylotrophs: Microbiology, Biochemistry and Genetics* (Hou, C.T., ed.), pp. 1–53, CRC Press, Boca Raton, FL.
Hou, C.T. (1984b) in *Methylotrophs: Microbiology, Biochemistry and Genetics* (Hou, C.T., ed.), pp. 145–166, CRC Press, Boca Raton, FL.
Hou, C.T., Patel, R., Laskin, A.I., Barnabe, N., and Marczak, I. (1979) *Appl. Environ. Microbiol.* 38 (1), 135–142.
Hou, C.T., Patel, R.N., and Laskin, A.I. (1980) *Adv. Appl. Microbiol.* 26, 41–69.
Izumi, Y., Asano, Y., Tani, Y., and Ogata, K. (1977) *J. Ferment. Technol.* 55 (5), 452–458.
Izumi, Y., Takizawa, M., Tani, Y., and Yamada, H. (1982) *J. Ferment. Technol.* 60 (4), 269–276.
Izumi, Y., Mishra, S.K., Ghosh, B.S., Tani, Y., and Yamada, H. (1983) *J. Ferment. Technol.* 61 (2), 135–142.

Kanamaru, K., Iwamuro, Y., Mikami, Y., Obi, Y., and Kisaki, T. (1982) *Agric. Biol. Chem.* 46 (10), 2419–2424.
Kato, N., Shirakawa, K., Kobayashi, H., and Sakazawa, C. (1983) *Agric. Biol. Chem.* 47 (1), 39–46.
Kato, N., Kobayashi, H., Shimao, M., and Sakazawa, C. (1984a) *Agric. Biol. Chem.* 48 (8), 2017–2023.
Kato, N., Yamagami, T., Kitayama, Y., Shimao, M., and Sakazawa, C. (1984b) *J. Biotechnol.* 1 (5/6), 295–306.
Kato, N., Kobayashi, H., Shimao, M., and Sakazawa, C. (1986a) *Appl. Microbiol. Biotechnol.* 23 (3/4), 180–186.
Kato, N., Yamagami, T., Shimao, M., and Sakazawa, C. (1986b) *Eur. J. Biochem.* 156 (1), 59–64.
Kato, N., Mizuno, S., Imada, Y., Shimao, M., and Sakazawa, C. (1988) *Appl. Microbiol. Biotechnol.* 27 (5/6), 567–571.
Keune, H., Sahm, H., and Wagner, F. (1976) *Eur. J. Appl. Microbiol.* 2 (3), 175–184.
Lim, W.-J., and Tani, Y. (1988) *J. Ferment. Technol.* 66 (6), 643–647.
Longin, R., Cooney, C.L., and Demain, A.L. (1982) *Appl. Biochem. Biotechnol.* 7 (4), 281–293.
Misaki, A., Tsumuraya, Y., and Kakuta, M. (1979) *Carbohyd. Res.* 75, C8.
Mizuno, S., and Imada, Y. (1986) *Biotechnol. Lett.* 8 (2), 79–84.
Morinaga, Y., and Hirose, Y. (1984) in *Methylotrophs: Biochemistry and Genetics* (Hou, C.T., ed.), pp. 107–118, CRC Press, Boca Raton, FL.
Morinaga, Y., Yamanaka, S., and Takinami, K. (1981a) *Agric. Biol. Chem.* 45 (6), 1419–1424.
Morinaga, Y., Yamanaka, S., and Takinami, K. (1981b) *Agric. Biol. Chem.* 45 (6), 1425–1430.
Morinaga, Y., Tani, Y., and Yamada, H. (1982a) *Agric. Biol. Chem.* 46 (1), 57–63.
Morinaga, Y., Tani, Y., and Yamada, H. (1982b) *Agric. Biol. Chem.* 46 (2), 473–480.
Morinaga, Y., Yamanaka, S., and Takinami, K. (1983a) *Agric. Biol. Chem.* 47 (9), 2113–2114.
Morinaga, Y., Tani, Y., and Yamada, H. (1983b) *Agric. Biol. Chem.* 47 (12), 2855–2860.
Morinaga, Y., Tani, Y., and Yamada, H. (1984) *Agric. Biol. Chem.* 48 (1), 143–148.
Nishio, N., Tsuchiya, Y., Hayashi, M., and Nagai, S. (1977) *J. Ferment. Technol.* 55 (2), 151–155.
Ogata, K., Izumi, Y., Kawamori, M., Asano, Y., and Tani, Y. (1977) *J. Ferment. Technol.* 55 (5), 444–451.
Oki, T., Kitai, A., Kouno, K., and Ozaki, A. (1973) *J. Gen. Appl. Microbiol.* 19 (1), 79–83.
Patel, R.N., Hou, C.T., Laskin, A.I., Derelanko, P., and Felix, A. (1979) *Appl. Environ. Microbiol.* 38 (2), 219–223.
Sakai, Y., and Tani, Y. (1986) *Agric. Biol. Chem.* 50 (10), 2615–2620.
Sakai, Y., and Tani, Y. (1987a) *J. Ferment. Technol.* 65 (4), 489–491.
Sakai, Y., and Tani, Y. (1987b) *Agric. Biol. Chem.* 51 (9), 2617–2620.
Sakai, Y., and Tani, Y. (1988a) *Appl. Environ. Microbiol.* 54 (2), 485–489.
Sakai, Y., and Tani, Y. (1988b) *Agric. Biol. Chem.* 51 (1), 227–233.
Sakai, Y., Sawai, T., and Tani, Y. (1987a) *Appl. Environ. Microbiol.* 53 (8), 1812–1818.

Sakai, Y., Tamura, K., and Tani, Y. (1987b) *Agric. Biol. Chem.* 51 (8), 2177-2184.
Sato, K., Mizutani, T., Hiraoka, M., and Shimizu, S. (1982) *J. Ferment. Technol.* 60 (2), 111-115.
Shimizu, S., Sato, K., Hiraoka, M., Yamashita, F., and Kobayashi, Y. (1982) *J. Ferment. Technol.* 60 (2), 163-166.
Sirirote, P., Yamane, T., and Shimizu, S. (1986) *J. Ferment. Technol.* 64 (5), 389-396.
Sirirote, P., Yamane, T., and Shimizu, S. (1988) *J. Ferment. Technol.* 66 (3), 291-297.
Suzuki, M., Berglund, A., Unden, A., and Heden, C.-G. (1977) *J. Ferment. Technol.* 55 (5), 466-475.
Suzuki, T., Yamane, T., and Shimizu, S. (1986a) *Appl. Microbiol. Biotechnol.* 23 (5), 322-329.
Suzuki, T., Yamane, T., and Shimizu, S. (1986b) *Appl. Microbiol. Biotechnol.* 24 (5), 366-369.
Suzuki, T., Yamane, T., and Shimizu, S. (1986c) *Appl. Microbiol. Biotechnol.* 24 (5), 370-374.
Suzuki, T., Deguchi, H., Yamane, T., Shimizu, S., and Gekko, K. (1988) *Appl. Microbiol. Biotechnol.* 27 (5/6), 487-491.
Tanaka, Y., Araki, K., and Nakayama, K. (1980) *Biotechnol. Lett.* 2 (2), 67-74.
Tani, Y. (1984) in *Methylotrophs: Microbiology, Biochemistry and Genetics* (Hou, C.T., ed.), pp. 55-85, CRC Press, Boca Raton, FL.
Tani, Y. (1985) in *Biotechnology and Genetic Engineering Reviews*, vol. 3 (Russell, G.E., ed.), pp. 111-135, Intercept Ltd., Newcastle upon Tyne, England.
Tani, Y., and Vongsuvanlert, V. (1987) *J. Ferment. Technol.* 65 (4), 405-411.
Tani, Y., and Yamada, K. (1987a) *FEMS Microbiol. Lett.* 40 (2/3), 151-153.
Tani, Y., and Yamada, K. (1987b) *Agric. Biol. Chem.* 51 (7), 1927-1933.
Tani, Y., and Yonehara, T. (1985) *Agric. Biol. Chem.* 49 (3), 637-642.
Tani, Y., Kanagawa, T., Hanpongkittikun, A., Ogata, K., and Yamada, H. (1978) *Agric. Biol. Chem.* 42 (12), 2275-2279.
Tani, Y., Mitani, Y., and Yamada, H. (1982) *Agric. Biol. Chem.* 46 (4), 1097-1099.
Tani, Y., Mitani, Y., and Yamada, H. (1984a) *Agric. Biol. Chem.* 48 (2), 431-437.
Tani, Y., Mitani, Y., and Yamada, H. (1984b) *J. Ferment. Technol.* 62 (1), 99-101.
Tani, Y., Yonehara, T., Mitani, Y., and Yamada, H. (1984c) *J. Biotechnol.* 1 (1), 119-127.
Tani, Y., Sakai, Y., and Yamada, H. (1985a) *Agric. Biol. Chem.* 49 (9), 2699-2706.
Tani, Y., Sakai, Y., and Yamada, H. (1985b) *J. Ferment. Technol.* 63 (5), 443-449.
Tani, Y., Yoon, B.D., and Yamada, H. (1985c) *Agric. Biol. Chem.* 49 (8), 2385-2391.
Tani, Y., Lim, W.-J., and Yang, H.-C. (1988a) *J. Ferment. Technol.* 66 (2), 153-158.
Tani, Y., Yoon, B.D., and Yamada, H. (1988b) *Agric. Biol. Chem.* 50 (10), 2545-2552.
Tani, Y., Sawai, T., and Sakai, Y. (1988c) *Biotechnol. Bioeng.* 32 (9), 1165-1169.
Tani, Y., Sakai, Y., and Cho, S.-G. (1990) *Appl. Microbiol. Biotechnol.* 34(1), 5-9.
Tsuchiya, Y., Nishio, N., and Nagai, S. (1980) *J. Ferment. Technol.* 58 (5), 485-487.
Urakami, T., and Hori-Okubo, M. (1988) *J. Ferment. Technol.* 66 (3), 323-332.
Urakami, T., Terao, I., and Nagai, I. (1981) in *Microbial Growth on C_1-Compounds* (Dalton, H., ed.), pp. 349-359, Heyden & Sons Ltd., London.
Vongsuvanlert, V., and Tani, Y. (1988a) *Agric. Biol. Chem.* 52 (2), 405-411.

Vongsuvanlert, V., and Tani, Y. (1988b) *Agric. Biol. Chem.* 52 (7), 1817–1824.
Vongsuvanlert, V., and Tani, Y. (1988c) *J. Ferment. Technol.* 66 (5), 517–523.
Vongsuvanlert, V., and Tani, Y. (1989) *J. Ferment. Bioeng.* 67 (1), 35–39.
Watanabe, M., Morinaga, Y., Takenouchi, T., and Enei, H. (1987a) *J. Ferment. Technol.* 65 (5), 563–567.
Watanabe, M., Morinaga, Y., and Enei, H. (1987b) *J. Ferment. Technol.* 65 (6), 617–620.
Yamada, H., Morinaga, Y., and Tani, Y. (1982) *Agric. Biol. Chem.* 46 (1), 47–55.
Yamada, H., Miyazaki, S.S., and Izumi, Y. (1986) *Agric. Biol. Chem.* 50 (1), 17–21.
Yamada, K., and Tani, Y. (1987a) *Agric. Biol. Chem.* 51 (9), 2401–2407.
Yamada, K., and Tani, Y. (1987b) *Agric. Biol. Chem.* 51 (9), 2629–2631.
Yamada, K., and Tani, Y. (1988a) *Agric. Biol. Chem.* 52 (3), 711–719.
Yamada, K., and Tani, Y. (1988b) *Agric. Biol. Chem.* 52 (8), 1951–1956.
Yamada, K., Kuwae, S., and Tani, Y. (1988) *Agric. Biol. Chem.* 52 (8), 1945–1949.
Yamada, K., Kuwae, S., and Tani, Y. (1989) *Agric. Biol. Chem.* 53 (2), 541–543.
Yonehara, T., and Tani, Y. (1986) *Agric. Biol. Chem.* 50 (4), 899–905.
Yonehara, T., and Tani, Y. (1987) *J. Ferment. Technol.* 65 (3), 255–260.
Yonehara, T., and Tani, Y. (1988a) *Agric. Biol. Chem.* 52 (4), 909–914.
Yonehara, T., and Tani, Y. (1988b) *J. Ferment. Technol.* 66 (5), 591–594.
Yoon, B.D., and Tani, Y. (1987a) *Agric. Biol. Chem.* 51 (2), 339–347.
Yoon, B.D., and Tani, Y. (1987b) *J. Ferment. Technol.* 65 (3), 261–266.
Yoon, B.D., and Tani, Y. (1987c) *J. Ferment. Technol.* 65 (6), 621–628.
Yoon, B.D., Ueno, M., and Tani, Y. (1987) *J. Ferment. Technol.* 65 (6), 629–634.

PART IV

Genetics

CHAPTER 12

Molecular Genetics of Methylotrophic Bacteria

Mary E. Lidstrom
Yuri D. Tsygankov

The development of viable commercial processes using bacteria often requires genetic information and genetic manipulations. Growing interest in the commercial applications of methylotrophic bacteria has resulted in increased research on the genetics and molecular biology of these organisms, and substantial progress in these areas has been made in the past few years. Genetic studies in the methylotrophs have proven difficult in the past, due to the lack of efficient gene transfer systems. However, the application of broad-host-range cloning vectors to genetic studies in methylotrophs in the past 10 years has proven extremely successful and has provided the technical means for continued studies. A number of recent reviews cover the genetics and molecular biology of methylotrophic bacteria (Holloway 1984; De Vries 1986; De Vries et al. 1990; Lidstrom 1990; Lidstrom and Stirling 1990) and the reader is referred to these for details not covered in this manuscript. This review will focus on those aspects of genetics in methylotrophic bacteria with particular importance to commercial applications.

Yuri D. Tsygankov acknowledges the excellent work of A.Y. Chistoserdov, S.M. Kazakova, I.G. Serebrijski, M.V. Gomelski, L.V. Kletsova, G.N. Marchenko, and S.Y. Shilova. Mary E. Lidstrom acknowledges the support of the Office of Naval Research, the Department of Energy, and the National Institutes of Health.

12.1 GENERAL MOLECULAR GENETICS

12.1.1 Recombination and *recA*

The development of a conventional genetic recombination system in methylotrophic bacteria requires the construction of *recA* mutants. These are important for studies involving plasmids carrying fragments of chromosomal DNA and for enhancing the stability of recombinant plasmids encoding eukaryotic and prokaryotic foreign genes.

The *recA* gene product is involved in general processes of DNA metabolism: general recombination, DNA repair, and expression of SOS functions. For *Escherichia coli* it has been demonstrated that the *recA* protein combines DNA-dependent ATPase activity and protease activity, which cleaves and inactivates both the *lexA*-encoded repressor of the SOS genes and repressors for some phages (Lanzov 1985; Little and Mount 1982; Walker 1984). A protein analogous to *recA* has been shown to exist in a variety of bacterial strains other than *E. coli* (Better and Helinski 1983; Keener et al. 1984; Ohman et al. 1985; Sano and Kageyama 1987; Zaitzev et al. 1986; Goldberg and Mekalanos 1986; Hamood et al. 1986; Paul et al. 1986; Koomey and Falkow 1987; Resnick and Nelson 1988; Ramesar et al. 1988), including the obligate methylotrophic species *Methylophilus methylotrophus* (Finch et al. 1986) and *Methylobacillus flagellatum* (Gomelsky et al. 1989).

In the methylotrophs noted above, the genes for chicken ovalbumin, mouse dihydrofolate reductase (Hennam et al. 1982), interferon α1 (de Maeyer et al. 1982), and interferon αF (Chistoserdov et al. 1987) are deleted from recombinant plasmids over the course of several generations. It has been suggested that plasmid instability could be minimized by the use of *recA*-like hosts (Powell and Byrom 1982). *M. methylotrophus* and *M. flagellatum recA* genes have been cloned, and they restore UV and mitomycin C resistance to *recA* mutants of *E. coli* (Finch et al. 1986; Gomelsky et al. 1989). In both cases, hybridization analysis demonstrated sequence similarity between the *recA* genes from the methylotrophic bacterium and a fragment of *E. coli* DNA that contains the *recA* gene.

The product of the *M. methylotrophus recA* gene was identified in minicells as a 36-kDa protein (Finch et al. 1986), somewhat smaller than that of the 37.8-kDa *E. coli* RecA protein (Sancar et al. 1980). The *M. methylotrophus* RecA protein is able to substitute for the *E. coli* RecA protein in vivo in catalyzing the inactivation of the LexA repressor of the SOS genes. However, synthesis of the *M. methylotrophus* RecA protein does not appear to be inducible. It has been suggested that the *M. methylotrophus recA* promoter lacks an appropriate operator for the *E. coli* LexA protein (Finch et al. 1986). In P1 transduction experiments, it was demonstrated that the *M. flagellatum recA* gene product restored approximately 50% of the recombination defect in the *E. coli recA* cells. Although no phages are known for *M. flagellatum*, its RecA protein catalyzes lambda cI-repressor cleavage under UV-light-stimulated lambda prophage induction in *E. coli recA* strains containing the hybrid plasmid with the *M. flagellatum recA* gene (Gomelsky et al. 1989). Such interspecies complementation provides further evidence that the function of the RecA protein has been highly conserved in a number of prokaryotes.

Ruvkun and Ausubel (1981) have described a general method of site-directed mutagenesis in prokaryotes which involves mutagenizing cloned fragments in *E. coli* with insertion of a marker and then replacing the wild-type sequences in the original organism with the mutant sequences by homologous recombination (marker exchange). This method was used in generating *recA* mutants of *M. flagellatum* (Gomelsky et al. 1989). These *recA* mutants were more sensitive to UV irradiation than the wild-type strain, and the level of recombination was more than 100 times lower than in the wild-type strain. *M. flagellatum recA* mutants might, therefore, be used as a host for the stable maintenance of plasmids in large-scale fermentation cultures.

The nucleotide sequence of the *M. flagellatum recA* gene and the deduced protein sequence has been determined (Gomelsky et al. 1989). The coding part of the gene consists of 1,032 base pairs (bp) specifying 344 amino acids of which 70% are identical to the corresponding residues of the *E. coli* RecA protein and 73% to the *Pseudomonas aeruginosa* RecA protein. Taking into consideration conserved substitutions, the level of homology is even higher. The essential differences are in the C-terminal sequences and in the 32–38 amino acid regions, which may take part in the control of protease activity by interaction with the carboxy terminus (Sano and Kageyama 1987). The molecular weight of the *M. flagellatum* RecA protein is 36.9 kDa. Unlike all the known sequenced bacterial *recA* genes, the approximately 300-bp region preceding the start codon for the *M. flagellatum recA* gene has no SOS box (the LexA repressor binding site), which presumably explains the constitutive expression of the gene in *E. coli* independent of DNA damage.

12.1.2 Promoters

Information on promoters in methylotrophic bacteria has only recently begun to appear, and, as yet, no definitive analyses have been carried out. However, the results obtained so far suggest that promoter structure may differ from that in *Escherichia coli* for some genes but may be similar for others. In *Methylomonas clara*, several putative promoter regions have been identified on a small cryptic plasmid but only one of these (pL) was found to be active in *M. clara* (Table 12-1; Metzler et al. 1988). This promoter was also active in *E. coli*, and it showed significant similarity to the *E. coli* σ^{70} promoter consensus sequence (Table 12-1). The gene(s) controlled by this promoter is not known.

TABLE 12-1 Putative Promoter Sequences in Some Methylotrophic Bacteria

Organism or plasmid and gene	Sequence −35	Separation (bp)	−10	Distance to Transcriptional Start Site (bp)	Reference[1]
Methylomonas clara cryptic plasmid pL	TTGCGT	16	TAAAAT	7	1
moxF gene					
Methylobacterium organophilum XX	AAAGACA	18	TAGAAA	5	2
Methylobacterium extorquens AM1	AAAGACA	18	TAGAAA	6	3
Methylomonas albus BG8	AAAGGAA	16	CGGAAA	no data	4
Methylomonas sp. A4	AAATCCA	17	AAGAAA	no data	5
dcmA gene					
Methylobacterium sp. DM4	CTTGACA	18	TAGAAC	5	6
E. coli (consensus)[2]	TTGACA	15–19	TATAAT	5–9	
Methylotrophs (putative consensus)[2]	**AAAGACA**	16–18	**TAGAAA**	5–7	

[1] References: 1, Metzler et al. 1988; 2, Machlin and Hanson 1988; 3, Anderson et al. 1990; 4, Kuhn and Lidstrom, unpublished; 5, Waechter-Brulla and Lidstrom, unpublished; 6, LaRoche and Leisinger 1990.
[2] The nucleotides that are shown in bold face are present in 80% or more of the known sequences.

Transcriptional start sites have been mapped for three genes, each in a different *Methylobacterium* species, and sequence data upstream from *moxF* are available for two methanotrophs (Table 12-1). A strongly conserved sequence is present at the −10 and −35 position for the *Methylobacterium* genes, and a similar sequence is present between 70 and 200 bp upstream of the other genes shown in Table 12-1. This may represent the promoter sequence for these genes, and a putative methylotroph consensus sequence is noted in Table 12-1. In none of these cases has a construction with a reporter gene been used to demonstrate promoter activity in vivo, and no mutagenesis studies have been carried out to define the important sequences.

In *Paracoccus denitrificans*, a GC-rich sequence (GCGGCGGC) quite different from the A-rich putative consensus sequence shown in Table 12-1 has been implicated in promoter activity (De Vries et al. 1990). A similar sequence is present upstream from the *moxF* gene of this organism (Harms et al. 1987), but the transcriptional start site has not yet been mapped to confirm the role of this sequence. In this case, a fusion to *lacZ* has been constructed and used to show that this region does contain a promoter active in *P. denitrificans* (Harms et al. 1989b). Sequences similar to this GC-rich stretch are present upstream of the *moxF* genes of *Methylobacterium organophilum* XX (Machlin and Hanson 1988) and *M. extorquens* AM1 (Anderson et al. 1990). However, these are between 300–600 nucleotides upstream of the transcriptional start sites, and so their significance is unknown.

Some foreign promoters are expressed in methylotrophs. Many (but not all) drug resistance genes on cloning vectors are expressed (see De Vries et al. 1990 for a comprehensive list), and the *tac* promoter has been shown to be active in *Methylobacillus flagellatum* (Chistoserdov et al. 1987) and *Methylophilus methylotrophus* (Byrom 1984). These data suggest that most methylotrophs contain an equivalent to the σ^{70} RNA polymerase subunit found in *E. coli*.

12.1.3 Codon Usage

A few genes from methylotrophs have now been sequenced, providing information on codon usage for the proteins encoded by these genes. Table 12-2 shows codon usage for four *mox* genes from *Methylobacterium extorquens* AM1, and similar data have been obtained for genes from *M. organophilum* XX (Machlin and Hanson 1988) and *Paracoccus denitrificans* (Harms et al. 1987). In other bacteria with high G+C ratios, a strong bias exists in the third position for G and C (West and Iglewski 1988), and this is also true in the *Methylobacterium* and *Paracoccus* strains, which have G+C ratios of 65–69 mol%. In these bacteria, several codons involving A or U in the third position are used rarely or not at all. Very little information is available concerning codon usage in the methylotrophs with low G+C ratios. Table 12-3 lists results for the *recA* gene of *Methylobacillus flagellatum* (M.V. Gomelsky and Y.D. Tsygankov, unpublished observations),

Molecular Genetics of Methylotrophic Bacteria

TABLE 12-2 Codon Usage for Four *mox* Genes of *Methylobacterium extorquens* AM1

Codon	Amino Acid	F	J	G	I	Total	% Synonymous	Codon	Amino Acid	F	J	G	I	Total	% Synonymous
UUU	F	1	1	0	0	2	6	ACU	T	0	0	0	1	1	1
UUC	F	20	5	5	3	33	94	ACC	T	24	11	8	3	46	58
UUA	L	0	0	0	0	0	0	ACA	T	1	1	0	0	2	2
UUG	L	1	1	0	0	2	2	ACG	T	17	6	8	0	31	39
CUU	L	4	0	2	0	6	6	GCU	A	2	3	0	0	5	4
CUC	L	26	10	10	3	49	49	GCC	A	35	28	11	11	85	68
CUA	L	0	0	1	0	1	1	GCA	A	2	2	1	0	5	4
CUG	L	19	13	7	2	41	42	GCG	A	15	9	3	3	30	24
AUU	I	0	0	2	0	2	5	UAU	Y	4	1	1	1	7	14
AUC	I	21	9	5	5	40	95	UAC	Y	28	8	5	2	43	86
AUA	I	0	0	0	0	0	0	CAU	H	1	0	0	0	1	6
AUG	M	18	10	5	2	35	100	CAC	H	9	1	4	1	15	94
GUU	V	0	0	0	0	0	0	CAA	Q	1	1	2	0	4	11
GUC	V	21	11	3	3	38	49	CAG	Q	15	8	7	2	32	89
GUA	V	0	0	1	0	1	2	AAU	N	1	1	3	1	6	11
GUG	V	18	15	5	0	38	49	AAC	N	32	4	8	4	48	89

12.1 General Molecular Genetics

UCU	S	0	0	0	0	0	AAA	K	1	4	0	0	5	5
UCC	S	6	4	4	3	17	32	AAG	K	44	18	13	16	91	95
UCA	S	1	0	0	0	1	2	GAU	D	14	7	1	3	25	27
UCG	S	14	8	2	1	25	47	GAC	D	37	17	10	2	66	73
CCU	P	0	0	0	0	0	0	GAA	E	4	1	2	2	9	17
CCC	P	10	11	4	4	29	42	GAG	E	15	15	9	5	44	83
CCA	P	0	0	0	0	0	0	UGU	C	1	0	0	0	1	8
CCG	P	23	7	8	2	40	58	UGC	C	3	3	4	2	12	92
UGG	W	19	2	3	1	25	100	AGU	S	0	0	1	0	1	2
UAA	Ter	1	0	0	0	1	25	AGC	S	4	5	0	0	9	17
UAG	Ter	0	0	0	0	0	0	AGA	R	0	0	0	0	0	0
UGA	Ter	0	1	1	1	3	75	AGG	R	0	1	0	0	1	4
CGU	R	4	2	0	1	7	15	GGU	G	0	0	0	0	0	11
CGC	R	10	13	4	1	28	62	GGC	G	60	12	20	6	98	79
CGA	R	0	2	0	0	2	4	GGA	G	1	2	1	0	4	3
CGG	R	3	3	1	0	7	15	GGG	G	1	5	3	0	9	7

Source: From Anderson et al. 1990; Nunn and Anthony 1988; and Nunn et al. 1989.

TABLE 12-3 Codon Usage for the *recA* Gene of *Methylobacillus flagellatum*

Codon	Amino Acid	No.	Codon	Amino Acid	No.	Codon	Amino Acid	No.	Codon	Amino Acid	No.
UUU	F	1	UCU	S	3	UAU	Y	1	UGU	C	0
UUC	F	7	UCC	S	9	UAC	Y	5	UGC	C	0
UUA	L	0	UCA	S	1	UAA	Ter	0	UGA	Ter	0
UUG	L	4	UCG	S	5	UAG	Ter	0	UGG	W	1
CUU	L	1	CCU	P	1	CAU	H	1	CGU	R	1
CUC	L	10	CCC	P	3	CAC	H	3	CGC	R	13
CUA	L	2	CCA	P	0	CAA	Q	3	CGA	R	0
CUG	L	14	CCG	P	6	CAG	Q	9	CGG	R	1
AUU	I	3	ACU	T	0	AAU	N	4	AGU	S	0
AUC	I	25	ACC	T	12	AAC	N	10	AGC	S	3
AUA	I	0	ACA	T	1	AAA	K	4	AGA	R	1
AUG	M	11	ACG	T	6	AAG	K	18	AGG	R	1
GUU	V	2	GCU	A	3	GAU	D	5	GGU	G	4
GUC	V	11	GCC	A	16	GAC	D	15	GGC	G	28
GUA	V	2	GCA	A	7	GAA	E	16	GGA	G	1
GUG	V	7	GCG	A	10	GAG	E	12	GGG	G	1

Source: M.V. Gomelsky and Y.D. Tsygankov, Unpublished results.

showing that, as expected, the strong bias for the third position found in the high G+C strains does not operate for this coding region in this strain.

12.2 MUTAGENESIS

Some chemical mutagens are not effective for the isolation of mutants in methylotrophs, for reasons that are still not entirely clear. The names and the characteristics of mutants isolated in methylotrophs and some reasons for difficulties in mutagenesis have been reviewed by Holloway (1984), De Vries (1986), and De Vries et al. (1990), and will not be repeated here. Although a number of mutagenesis techniques are being used for mutant selection in both facultative and obligate methylotrophs, to date there is little consensus of results which can be applied to the group as a whole.

12.2.1 Chemical Mutagenesis

It has been suggested that one of the common reasons preventing effective mutant isolation in methylotrophs is the practice of selecting auxotrophic mutants on complete media (Whitta et al. 1985; Tsygankov and Kazakova 1987). It was demonstrated with *Methylobacillus flagellatum* that most of the amino acids and their mixtures have a strong inhibitory effect on auxotrophic strains, which results in the reduction of growth rate and plating efficiency. The plating efficiency of some auxotrophs on complete medium was 5-10 times lower than on minimal medium with respective supplements. Polyauxotrophic mutants which were selected on minimal medium were fully incapable of growth on complete medium (Tsygankov and Kazakova 1987).

The above data suggest that the isolation of auxotrophs in arbitrary methylotrophic bacteria could be made easier by searching for only one requirement at a time and plating the mutagenized cells on minimal medium supplemented with individual amino acids, vitamins, or nucleotides. This strategy must involve the analysis of the inhibitory effect of nutrient supplements on the growth of the wild-type strain and, in particular, of the mutant strains. This approach has been successfully used for isolation of auxotrophs of the obligate methylotroph *M. flagellatum*, the restricted facultative methylotroph organism W3A1 and the methanotroph *Methylomonas methanica* (Tsygankov and Kazakova 1987).

12.2.2 Transposon Mutagenesis

Another technique which has proven to be successful for the isolation of C_1 metabolic and auxotrophic mutants in a variety of methylotrophs is that of transposon mutagenesis. Transposable drug resistance elements (Tn) are useful for a variety of genetic manipulations in bacteria. Transposons can be inserted into the chromosome or into plasmids, and such insertions are usually stable once formed. Many transposons carry one or more drug-resistance determinants, and this is particularly useful for chromosome mutations caused by insertion as it provides a positive selection for mutant phenotypes that are otherwise unselectable. This requires the use of a vector system that will carry the transposon into the cell but which will not itself survive as an independent replicon (suicide plasmid). To date, naturally occurring transposons have not been identified in methylotrophic bacteria.

Transposon mutagenesis with a suicide delivery plasmid has been used to obtain auxotrophs and C_1 metabolism mutants in facultative and obligate methylotrophs. Using the plasmid pMO75 (R91-5::Tn5), Whitta et al. (1985) have developed a procedure for transposon mutagenesis using the well-characterized facultative methylotroph *Methylobacterium extorquens* AM1. Plasmid pMO75 has a wide host range of conjugation but a narrow host range of replication. In matings between *Pseudomonas aeruginosa* PAO (pMO75) and *M. extorquens* AM1, transconjugants were found at low frequency which expressed the kanamycin resistance carried by the transposon. A total of 5,070 kanamycin-resistant transconjugant colonies were examined. Of these, six were mutants in methanol utilization (C_1 utilization, cou) and six were auxotrophs. One of these mutants, cou-4, is similar to the previously described, nitrosoguanidine-induced, mutant strain PCT48, and it was suggested that they are blocked in the conversion of acetyl-CoA to glyoxylate. Of the six auxotrophs isolated, three required thiamin and two required tyrosine. The overall frequency of mutant isolation following Tn5 mutagenesis in *M. extorquens* AM1 was 0.3%.

In *M. flagellatum*, plasmid pMD100 has been used successfully for transposon mutagenesis. Plasmid pMD100 is a hybrid plasmid containing a ColE1 replicon and RP4 conjugation genes, which enable the plasmid to have a broad host range but to replicate in only a limited range of species, not including methylotrophs; it also carries transposon Tn501 (mercury-resistance, Hg^r; Ely 1984). Plasmid pMD100 was transferred to *M. flagellatum* by mating, and cells were plated onto minimal medium containing a mixture of three amino acids which do not inhibit the growth of *M. flagellatum*. From 204 Hg^r transconjugant colonies obtained on media supplemented with methionine, histidine, and isoleucine, four (2%) were auxotrophs requiring histidine and methionine simultaneously for growth. From two separate matings, about 600 Hg^r colonies selected on media containing phenylalanine, tryptophan, and tyrosine were obtained and examined. Of these, three also carried mutations in phenylalanine biosynthesis, giving an overall mutant recovery of 0.5% (Y.D. Tsygankov, unpublished observations).

With the transposon Tn5-132 loaded onto the suicide plasmid pRK2013, several auxotrophic mutants (*leu*, *ilv*, and *pur*) of *Hyphomicrobium facilis* B-552 were obtained (Gliesche and Hirsch 1989). However, when a similar system with Tn5 was used in *M. methylotrophus*, no auxotrophs were isolated among thousands of Km^r colonies, even though Tn5 was able to insert into the chromosome and express the kanamycin-resistance determinant (Bohanon et al. 1988). This was shown to be due to the instability of the insertions in this organism (see below).

The site-directed mutagenesis technique (marker exchange) described above has been successfully used to generate both auxotrophic and metabolic mutants in methylotrophs, using Tn insertions in cloned genes. These include *nifD* (nitrogen-fixation) mutants in *Methylosinus* sp. 6 (Toukdarian and Lidstrom 1984), cytochrome *c*552 mutants in *Paracoccus denitrificans* (van Spanning et al. 1989), *glnA* (glutamine synthetase) mutants in *Methylococcus capsulatus* (Davidson and Murrell 1989), and mutants of pyrroloquinoline quinone (PQQ) biosynthesis in *Methylobacterium organophilum* DSM760 (Biville et al. 1989). Tryptophan auxotrophs of the obligate methylotroph *Methylophilus methylotrophus* have also been obtained with a marker exchange technique in which the cloned *trpE* gene encoding anthranilate synthase, mutagenized with Tn5, was exchanged with the intact chromosomal gene. The selected auxotrophic mutants were unstable because of the unstable nature of Tn5 insertions in the *M. methylotrophus* chromosome. If the transposon gene function of the Tn5 inserted into the *trpE* gene was inactivated before being mobilized into *M. methylotrophus*, stable Km^r tryptophan auxotrophs were obtained by double-crossover homologous recombination with the chromosome (Bohanon et al. 1988).

12.3 SURROGATE AND CLASSICAL GENETICS

12.3.1 Genetic Exchange Systems

An effective genetic exchange system is of paramount importance for studying genome organization and mapping the genes responsible for C$_1$ functions in methylotrophic bacteria. No high-frequency genetic transformation systems are available for methylotrophs, although much effort has been expended in attempts to develop them. However, low-frequency systems have been reported for *Methylococcus capsulatus* (Williams and Bainbridge 1971) and *Methylobacterium organophilum* (O'Connor et al. 1977) but these required high levels of DNA and were unreliable. Reports of transformation in *Paracoccus denitrificans* have not been repeated (De Vries et al. 1990).

No generalized transduction systems are in use in methylotrophs. One report of their use has been made for *Xanthobacter autotrophicus* (Wilke and Schlegel 1979), but no chromosomal mapping has been carried out.

Conjugational gene transfer systems using broad-host-range plasmids have been much more successful (De Vries et al. 1990; see next section). In terms of investigating the genetic constitution of the chromosome, conjugation is undoubtedly the most useful system. Most conjugation systems are physiologically similar, and two separate components can be recognized: mating pair formation with a potential recipient cell; and conjugal DNA metabolism that subsequently transfers and replicates the plasmid DNA from the *oriT* site. Despite this overall similarity, numerous distinct conjugation systems have been identified.

12.3.2 Conjugational Mapping

For many years, one of the major limitations for genetic analysis of methylotrophs was the lack of known indigenous chromosome-mobilizing plasmids that would allow gene transfer studies. No narrow-host-range sex factor comparable to the F factor has been detected in methylotrophs. This difficulty has been overcome by transferring broad-host-range plasmids with chromosome-mobilizing ability (Cma) to methylotrophic bacteria. To date, the *Pseudomonas aeruginosa* drug-resistance plasmid R68.45 and its derivatives have been the plasmids of choice for obtaining gene transfer in methylotrophs. Plasmid R68.45 promotes nonpolarized transfer of the chromosome from multiple origins. The recombinants of R68.45 matings have been shown to inherit relatively short segments of the donor chromosome, usually not exceeding 15 min (Haas and Holloway 1976). Plasmid R68.45 can mobilize the chromosome of *M. extorquens* AM1, and Goodwin and co-workers have constructed a preliminary linkage map on the basis of data from three- and four-factor crosses (Tatra and Goodwin 1983 and 1985). However, the precision of chromosomal mapping with IncP1 plasmids is uncertain, since some bacteria carry out a retrotransfer (shuttle transfer) during which the chromosomal markers can be transferred both ways: from donor to the recipient and vice-versa (Mergeay et al. 1987). Kletsova and Tsygankov (1990) have demonstrated that the retrotransfer phenomenon occurs in *M. flagellatum*. Analysis of the progeny obtained from R68.45- or pULB113-mediated matings revealed that the share of retrotransconjugants formed by retrotransfer can vary from 0 to 100%, depending on the specific donor-recipient pair and mating conditions.

12.3.3 Complementation Mapping

R68.45 and pULB113 can also form prime plasmids in which a segment of bacterial chromosome is inserted into the plasmid DNA. Prime plasmids carrying segments of the bacterial genome retain the conjugational broad-host-range of the IncP1 plasmids and, when they are transferred to an appropriate *E. coli* or *P. aeruginosa* strain, they can complement a specific range of auxotrophic markers. This provides a means of studying the expression of selected bacterial functions and can be used as an effective approach for mapping gene distribution. This alternative mapping system was called complementation mapping (Moore et al. 1983). The coinheritance and cotransfer of plasmid and chromosomal markers indicates that they are located on the same piece of DNA. Moreover, if a given prime plasmid can complement more than one function in heterologous hosts, this is evidence that genes responsible for each of these functions are on the same fragment of bacterial chromosome and hence are linked. The chief advantage of this mapping technique is that it is not necessary to obtain mutants of the bacterium being mapped, and the prime plasmids can be constructed from the wild-type strain.

The complementation mapping procedure was used to construct a preliminary chromosome map of the obligate methylotrophs *M. methylotrophus* and *Methylophilus viscogenes*, both organisms which do not have the range of suitable auxotrophic mutants adequate for an effective mapping system. Holloway and co-workers have utilized the pattern of complementation of *P. aeruginosa* and *E. coli* auxotrophs by prime plasmids and cosmid clones carrying genome segments of *M. methylotrophus* and *M. viscogenes* to construct complementation maps for both of these organisms (Table 12-4; Moore et al. 1983; Lyon et al. 1988).

In order to isolate and characterize genome fragments of *M. flagellatum*, complementation mapping was also used by Tsygankov et al. (1990). The *M. flagellatum* genes were cloned within pULB113 (RP4::miniMucts) by complementing auxotrophic mutations of *E. coli* and *P. aeruginosa*. Plasmid pULB113, like plasmid pMO172 used for linkage analysis of *M. methylotrophus* and *M. viscogenes*, is efficiently transferred to *E. coli* and *P. aeruginosa*. The R-prime plasmids formed were analyzed by complementation of auxotrophic mutations in *E. coli*, *P. aeruginosa*, and *M. flagellatum*. The merodiploid state created by the introduction of R-primes into *M. flagellatum* is unstable, so R-primes were maintained in *E. coli*. R-prime plasmids carrying fragments of *M. flagellatum* DNA complemented 33 of the 49 tested *E. coli* auxotrophic mutations, 17 of the 25 tested *M. flagellatum* mutations, and 3 of the 4 tested *P. aeruginosa* mutations (Tsygankov et al. 1990). The composition of *M. flagellatum* genes in the linkage groups determined from the data of heterologous and homologous crosses coincide. These R-primes are especially useful as a first step in cloning the genes for C_1 functions individually by the complementation of methanol-minus mutants in methylotrophic bacteria.

The organization of the *M. flagellatum* genome is similar to the genomes of the other obligate methylotrophs *M. methylotrophus* and *M. viscogenes*. In those cases in which overlapping of linkage groups was observed, the gene order in all three strains was similar (Table 12-4). Despite the fact that the level of DNA-DNA homology between *M. flagellatum* and *M. methylotrophus* was less than 10% (Govorukhina et al. 1987) and that there is a very low DNA-DNA homology level between *M. viscogenes* and *M. methylotrophus* (Lyon et al. 1988), the order of the genes in the known genome section of these methylotrophs is highly conserved.

12.3.4 Hfr-Like Mapping

Although over 30 genes of *M. methylotrophus* and *M. viscogenes* have been mapped in seven and six linkage groups, respectively, by the complementation mapping technique discussed above, the precise location of all these

TABLE 12-4 Genetic Linkage Groups in *M. flagellatum* (M.f), *M. methylotrophus* (M.m), and *M. viscogenes* (M.v)

Group	Strain	Markers[1]
1	M.v	CtrpF CargG CargF Cleu10 CtrpA Cmet9011
	M.m[2]	Cpur66 CtrpF CargG CargF Cleu10 CpheA CtrpA CpurF Cmet9011
	M.f	Ctrp45 trp3 CargG nal1 leuB pheA CtrpA CpurF met4
2	M.v	CpyrE CtrpE CtrpD CtrpC CmetA Cmet28 Cphe3
	M.m	CpyrE CtrpE CtrpD CtrpC CmetA Cmet28
	M.f	CpyrE trpE Ctrp(KS-30) met5 Cmet28 pur2
3	M.v	ChisI ChisIII ChisIIA ChisIIB
	M.m	
	M.f	ChisB ChisC ChisF Chis68 Chis4 hisB his3 his4 his6
4	M.v	CargA CargB
	M.m	
	M.f	CargA ala1
5	M.v	CpyrB CproC
	M.m[2]	CpyrB CproC Cpur66
	M.f	CpurL CthyA CthrC CthrA2 Cstr31 CthrB CpurE CargH CargE CargB Cthi1 CproA bioB CpyrB CproC CilvC
6	M.v	CproB Cpur136
	M.m	Cleu8 CproB Cpur136
	M.f	
7	M.v	CpyrD ChisV Cleu8
	M.m	
	M.f	
8	M.v	CpyrF Cphe2
	M.m	
	M.f	
9	M.v	
	M.m	CmetE CmetC CmetB Cthr48
	M.f	

[1] The marker abbreviations are explained in the caption of Figure 12–1 (see later).
[2] Group 1 and group 5 overlap at the Cpur66 marker.

genes on the chromosome has not yet been determined. The chief limitation has been the absence of a single plasmid-mediated high-frequency-of-recombination system that was applicable for mapping genes around the chromosome precisely by time units in interrupted mating crosses. The value of a genetic map based on time units is well illustrated by the map constructed for *Enterobacteriaceae* species and *Pseudomonas aeruginosa*.

One of the main approaches for constructing Hfr-like donor strains, applicable for many nonenteric bacteria, consists of using plasmids which have a wide host range of conjugation but a narrow host range of replication. Selection for antibiotic resistance results in strains containing these plasmids, integrated into the chromosome. One such plasmid, pAS8-121 is a tetracycline-sensitive derivative of the hybrid plasmid pAS8 (RP4::ColE1, fused at their EcoRI sites) carrying a Tn7 insertion in a gene necessary for RP4-mediated replication (Sakanyan et al. 1978). It is conjugation proficient, transferring between *E. coli* strains with high frequencies, and replicates via the ColE1 replicon system, so it is a suicide plasmid in nonenteric bacteria. Plasmid pAS8-121 may be suitable as a general-purpose carrier plasmid for introducing transposons into the genomes of diverse bacteria (Sato et al. 1981), and it has been successfully used for generating a series of Hfr donors in the phototrophic nitrogen-fixing bacterium *Rhodobacter sphaeroides* (Kameneva et al. 1986).

Plasmid pAS8-121 has been used by Serebrijski et al. (1989) to construct Hfr-like donors of *M. flagellatum*. Among 17 independently isolated tranconjugants of *M. flagellatum* harboring drug-resistance markers of pAS8-121 only one, designated HfrA, mobilized the chromosome in a polarized manner. It has been shown that this stable Hfr-like donor is the result of a rearrangement of the integrated plasmid DNA, involving the elimination of plasmid ColE1 and the Ap[r] marker. The rest of the tranconjugants tested were able to transfer each of the selected chromosomal markers with similar frequencies, which is characteristic for broad-host-range Cma-type plasmids of the IncP1 group.

The Hfr-like donor constructed has only one origin of transfer, at a site which has been arbitrarily designated 0 min on the *M. flagellatum* chromosome. This donor has been shown to be satisfactory for both plate mating and interrupted matings in the region of 0 to 40 min, but for more distal markers, interrupted matings give too few recombinants for accurate data to be obtained. Donor *M. flagellatum* HfrA was used to construct a time of entry map of the chromosome. This was achieved by determining the time of entry of six randomly dispersed markers, four of which are present in the individual linkage groups established by the R-prime plasmid mapping (Table 12-4). Among the six markers localized in interrupted crosses, five are transferred within 30 min of crossing. Figure 12-1 shows a preliminary linear map of the *M. flagellatum* chromosome constructed on the basis of time-of-marker entry combined with the established group of linked genes. Genetic circularity for *M. flagellatum* has not yet been demonstrated.

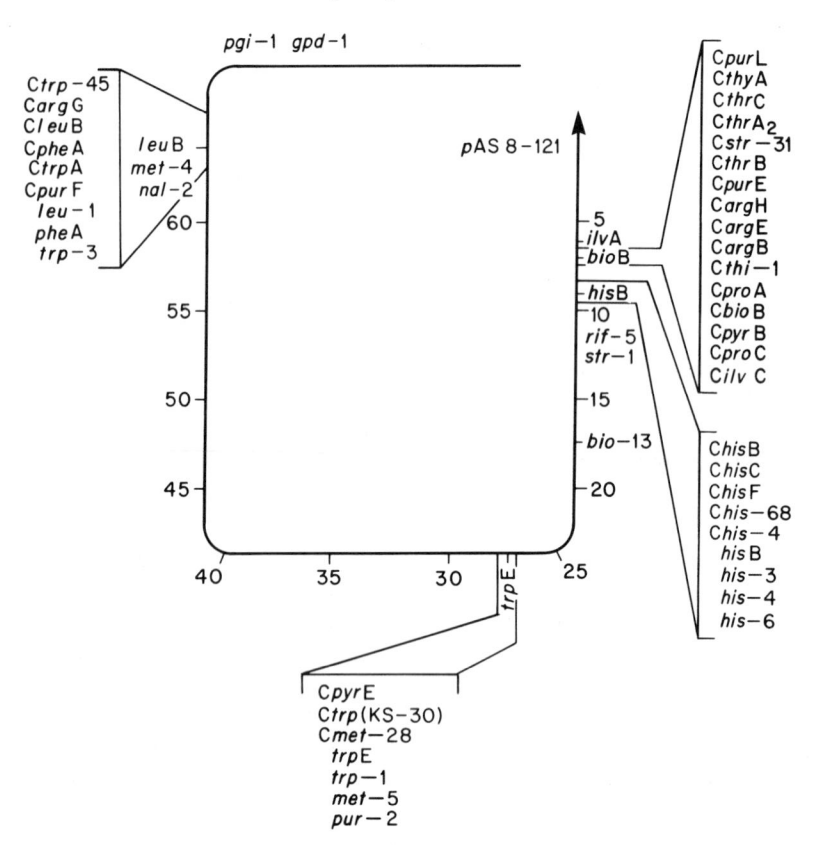

288 Molecular Genetics of Methylotrophic Bacteria

FIGURE 12-1 Chromosome map of *M. flagellatum*, derived from the data of Serebrijski et al. (1989); Kletsova and Tsygankov (1990); and Tsygankov et al. (1990). The arrowhead points to the chromosome transfer origin (zero time) of plasmid pAS8-121 integrated into the chromosome. Markers whose location is indicated by a short bar joining the locus designation to the map were located by interrupted matings. The numbers outside the rectangle (the chromosome) indicate the time scale (in min). The prefix "C" indicates clusters of closely linked markers identified by complementation with R-prime plasmids in heterospecific matings. Genes without additional symbols were identified by complementation with *M. flagellatum* auxotrophic mutations. A square bracket indicates that the precise order of the genes in the linkage group is unknown.

About half of all the mapped biosynthetic *M. flagellatum* genes are localized on a small segment of the genetic map between 6 and 10 min. This result is consistent with the idea of the existence of auxotrophic-rich regions on chromosomes of obligate methylotrophs (Lyon et al. 1988), similar to that found in *Pseudomonas* (Holloway and Morgan 1986).

12.4 GENE CLONING IN THE METHYLOTROPHS

12.4.1 Vectors

The application of gene cloning techniques has been quite successful in the Gram-negative methylotrophs, mainly due to the availability of a number of broad-host-range cloning vectors that transfer to and replicate in most Gram-negative methylotrophs. These vectors are usually derived from plasmids of the incompatibility groups IncPI, IncQ, and IncW, and they include vectors for cosmid cloning, subcloning, and expression, and for constructing translational and transcriptional fusions. Extensive lists of these vectors have been published recently (Schmidhauser et al. 1986; De Vries et al. 1990; Lidstrom and Stirling 1990) and will not be repeated here. In the Gram-positive methylotrophs, cloning vectors are not yet available.

12.4.2 Transfer Systems

In most Gram-negative methylotrophs, the only high-frequency genetic transfer system available for plasmids is conjugation and all of the generally used broad-host-range vectors are transferrable to methylotrophs by conjugation, usually from *E. coli* hosts (De Vries et al. 1990; Lidstrom and Stirling 1990). The frequencies of transfer vary from 10^{-1} to 10^{-7} per recipient, and in some cases low frequencies can be increased by the isolation of mutants deficient in restriction enzyme systems (De Vries 1986). Specific examples of gene cloning and characterization are presented in the next sections.

12.5 REGULATION AND GENETICS OF AMINO ACID BIOSYNTHESIS

12.5.1 Threonine in *Methylobacillus flagellatum*

There is no reason to believe that the organization of genes involved in pathways of amino acid biosynthesis is basically different in methylotrophs than in the other bacterial genera which have been extensively studied. However, little information is available for methylotrophs on the genetic organization and regulation of amino acid biosynthesis despite the commercial interest in using such organisms for amino acid production. The threonine biosynthesis genes of *M. flagellatum* have been used in Y.D. Tsygankov's laboratory as a convenient model for such studies, which are described below.

12.5.1.1 Biochemistry and Regulation.
M. flagellatum possesses a specific active transport system for L-threonine which is also specific for L-serine. This system is capable of carrying out the energy-dependent transport of

threonine against a 100-fold concentration gradient. This threonine carrier functions asymmetrically, that is, it mediates the transport of the substrate into the cell, but not out of the cell. Intracellular ^{14}C-threonine did not undergo significant degradation (Bourd et al. 1989).

The key enzymes of threonine biosynthesis have been detected in whole cell extracts of *M. flagellatum*. The first enzyme of the pathway for the biosynthesis of the aspartate-family amino acids, aspartate kinase, is formed at more than 80% of the level found in *E. coli*. Aspartate kinase was completely inhibited by threonine or lysine at 5 mM, and at 1 mM, the corresponding values for aspartate kinase activity were 5 and 14%, respectively. In contrast, aspartate kinase activity was not inhibited by methionine at concentrations up to 5 mM, but methionine at a concentration of 1 mM caused a 1.4-fold increase in enzyme activity. Aspartate kinase activity was completely inhibited when a mixture of threonine and lysine was added at a concentration of 1 mM each. Addition of 1 mM threonine or lysine to the crude extract of a mutant of *M. flagellatum* resistant to 20 mg threonine per ml did not alter the activity of the aspartate kinase, indicating that this enzyme is now insensitive to the feedback inhibition (V.D. Tsygankov, unpublished observations).

The activity of homoserine kinase, the product of *thrB*, was also detected in *M. flagellatum* cell extracts. At a final threonine concentration of 5 mM, the enzyme activity was reduced by 55–65%. Other aspartate-family amino acids did not alter the homoserine kinase activity.

Partially purified enzyme preparations of *M. flagellatum* have been examined for the presence of homoserine dehydrogenase. Both NAD$^+$- and NAD(P)$^+$-dependent activities were detected. The level of NAD$^+$-dependent activity was significantly higher than for NAD(P)$^+$-dependent activity and was not inhibited in the presence of threonine or methionine. The NAD(P)$^+$-dependent activity was reduced 30–50% when threonine or methionine was added at a concentration of 5 mM.

These data suggest that the control of threonine biosynthesis in *M. flagellatum* is effected at the enzyme level, with the first enzyme of threonine biosynthesis pathway being regulated by feedback inhibition. To date there is no experimental evidence that the threonine biosynthetic pathway is regulated at the level of repression.

12.5.1.2 Genetics. Genes corresponding to threonine biosynthesis were located in the 6- to 9-min region of the *M. flagellatum* chromosome map (see Figure 12–1). Three of these genes were carried by an individual prime plasmid and hence were linked on the *M. flagellatum* genome.

Threonine biosynthesis genes of *M. flagellatum* were identified by complementation of *E. coli* Thr mutants after transfer of a genomic clone bank constructed in the broad host range vectors pAYC31 or pSL5. Complementation data suggested that a cloned 10.6-kb *M. flagellatum* DNA fragment encoded the structural gene for aspartate kinase. The recipient *E. coli* strain was deficient in *thrA*1, *metLM*, and *lysC*, genes encoding isoenzymes of aspartate kinases, which are involved in the biosynthesis of threonine, methionine, and lysine, respectively. Hybrid plasmids complemented mutations in these aspartate kinases but not any other threonine biosynthesis mutants. Whole cell extracts of the complemented *E. coli* strain produced aspartate kinase with the regulation pattern of the *M. flagellatum* enzyme.

Two of the threonine biosynthesis genes, *thrA*2, encoding homoserine dehydrogenase and *thrC*, encoding threonine synthetase, were identified by complementation of appropriate mutations of *E. coli* or *P. aeruginosa* strains. Both mutations were complemented by a plasmid containing an 8.8-kb insert. In maxicells, the hybrid plasmid containing the 8.8-kb fragment reproducibly expressed two polypeptides of 54 and 40 kDa. A subcloned 2.4-kb fragment was found to complement *thrC* mutations in both orientations. A plasmid which only contains the *thrC* gene and does not complement *thrA*2 mutations directs the synthesis of the 54-kDa protein. A tightly linked 4.1-kb fragment subcloned in both orientations specifies the synthesis of the 40-kDa protein, the product of the *thrA*2 gene. These results indicate that the expression of *thrA*2 and *thrC* genes may not share a common transcriptional control mechanism. No genetic evidence is available at this time to suggest that a regulatory locus governs the coordinate expression of these two genes.

A *M. flagellatum* structural gene encoding homoserine kinase activity (*thrB*) was localized to a 1.1-kb fragment by mutant complementation. This fragment directs the synthesis of a 30-kDa polypeptide. The *thrB* gene is not closely linked to the previously described threonine biosynthesis genes. Plasmids carrying an insertion of 15–20 kb complemented *thrB* mutations but not other *thr* mutations.

12.5.2 Arginine and Serine in *M. methylotrophus*

To overcome the difficulties of isolating auxotrophic mutants in obligate methylotrophs required for distinguishing alternate pathways of amino acid biosynthesis, Kearney and Holloway (1987) examined *M. methylotrophus* cell extracts for the presence of several key enzymes in order to study the regulation and routes of arginine and serine biosyntheses. The specific activities of N-acetylornithine glutamate transacetylase, N-acetylglutamate kinase, and ornithine transcarbamylase are present, while acetylornithinase activity is absent in *M. methylotrophus* cell extracts. Thus *M. methylotrophus* could be expected to synthesize arginine via the N-acetylornithine glutamate transacetylase pathway, as the fluorescent pseudomonads do. The first enzyme of the arginine biosynthesis pathway, N-acetylglutamate kinase, was completely inhibited when the final arginine concentration was 10 mM or greater. In contrast to other bacteria, arginine biosynthetic enzymes in

M. methylotrophus are not repressed by arginine, even though arginine is transported into the cells.

It has been suggested that synthesis of serine in *M. methylotrophus* proceeds via the phosphorylated intermediate hydroxypyruvate phosphate. Two enzymes involved in this pathway, 3-phosphoglycerate dehydrogenase and phosphoserine phosphatase, were detected in *M. methylotrophus* cell extracts (Kearney and Holloway 1987). The first enzyme of serine biosynthesis (3-phosphoglycerate dehydrogenase) was reduced 50% at a serine concentration of 1 mM. In contrast, the level of phosphoserine phosphatase was not changed when the final serine concentration was 1 mM. It is likely that serine biosynthetic enzymes are not repressed.

12.5.3 Aromatic Amino Acids in Facultative and Obligate Methylotrophs

In all organisms studied thus far, the biosynthesis of all three aromatic amino acids has been shown to be a highly complex and diverse sequence of reactions and is known to generate a variety of significant metabolites. The first step involves the condensation of phosphoenolpyruvate and erythrose-4-phosphate, is catalyzed by 3-deoxy-D-arabinoheptulose-7-phosphate (DAHP)-synthase. Seven reactions of the shikimate pathway lead to the synthesis of the branch point compound chorismate from which all of the aromatic amino acids and aromatic vitamins are derived (Pittard and Gibson 1970).

Numerous studies have been done on aromatic amino acid synthesis in the genera *Escherichia, Salmonella, Bacillus, Corynebacterium*, and *Pseudomonas*; however, only a few reports describe the regulation of this pathway in methylotrophic bacteria. The biosynthesis of aromatic amino acids was studied in the facultative methylotrophic bacterium *Nocardia* sp. 239 (Boer et al. 1985, 1986). Recently, the regulation of the aromatic amino acid biosynthetic pathway was studied in a facultative serine cycle methylotrophic strain, "*Pseudomonas*" (*Methylobacterium*) sp. M (Maksimova et al. 1986). It has been shown that the control of aromatic pathways occurs both at the level of feedback inhibition by aromatic amino acids and by repression of the genes encoding the enzyme synthesis. The activity of DAHP synthase was inhibited more than 50% by 0.5 mM tyrosine, whereas other metabolites (tryptophan, anthranilate, phenyl pyruvate) had less effect (25–30% inhibition by 1.0 mM). The regulation of synthesis of DAHP synthase was examined in amino acid auxotrophs which were grown in excess or in limiting amounts of the required amino acid. The synthesis of DAHP synthase was repressed by tyrosine and phenylalanine, and the level of enzyme activity increased threefold under conditions of limiting amino acid, compared to growth in its excess (Olechnovich et al. 1987).

12.5 Regulation and Genetics of Amino Acid Biosynthesis

Metabolic options for phenylalanine and tyrosine biosynthesis in microorganisms include the use of the phenylpyruvate and *p*-hydroxyphenylpyruvate routes, or the arogenate route, or both simultaneously (Byng and Jensen 1983). It has been shown that *Pseudomonas* sp. M lacks the arogenate route and the synthesis of both amino acids takes place via phenylpyruvate and *p*-hydroxyphenylpyruvate. This bacterium possesses the following enzymes for phenylalanine and tyrosine biosynthesis: chorismate mutase, prephenate dehydratase, and prephenate dehydrogenase (Olechnovich et al. 1987). Moreover, the presence of phenylalanine hydroxylase, which catalyzes the conversion of phenylalanine to tyrosine, has been demonstrated. Neither phenylalanine nor tyrosine affected the synthesis of chorismate mutase and prephenate dehydratase. Prephenate dehydratase activity was inhibited by both phenylalanine and tryptophan, and prephenate dehydrogenase synthesis was also repressed by tryptophan. Phenylalanine hydroxylase was induced by phenylalanine. Thus, it has been shown that tryptophan takes part in the regulation of tyrosine and phenylalanine biosynthesis.

Tryptophan biosynthesis genes in *Pseudomonas* sp. M are regulated by repression of the *trpE, trpD,* and *trpC* genes by tryptophan. It was also shown that the *trpE, trpD,* and *trpC* genes are derepressed noncoordinately. The *trpF* gene product was synthesized constitutively. The *trpA* and *trpB* genes are inducible by indole-3-glycerophosphate.

Anthranilate synthase and tryptophan synthase were sensitive to feedback inhibition. The tryptophan concentrations giving 50% inhibition were estimated to be 9 mM and 1 mM, respectively. All 5-DL-methyltryptophan (5MT)-resistant mutants of *Pseudomonas* sp. M demonstrate reduced tryptophan synthase feedback inhibition of about 4- to 11-fold (Olechnovich et al. 1986; Maksimova et al. 1986). Therefore, in *Pseudomonas* sp. M, the biosynthesis of the aromatic amino acids is controlled by feedback inhibition of enzymes and repression of their synthesis, as in the fluorescent pseudomonads.

The repression of DAHP synthase was recently found in *M. flagellatum* (Maksimova, unpublished observations). It has been demonstrated that in this strain the enzyme activity increased 15- to 20-fold under conditions of limitation with aromatic amino acids. DAHP synthase consists of three isoenzymes, and each one is subjected to allosteric inhibition with the corresponding amino acid. Activity of DAHP synthase was inhibited more than 50% by 7 mM phenylalanine, whereas other metabolites had less effect (50% inhibition by 1 mM).

E. coli mutants defective in *trpA, trpB, tyrA,* and *pheA* were complemented with a gene library of *M. flagellatum* DNA constructed in the plasmid pMYF131 (Planutene and Tsygankov, unpublished observations). Of these two genes, *trpA* and *trpB,* encoding the alpha and beta subunits of tryptophan synthase, are closely linked on a subcloned fragment of approximately 10 kb.

12.6 GENES INVOLVED IN C₁ OXIDATION AND ASSIMILATION

12.6.1 Methane Oxidation

Methane is oxidized to methanol in the methanotrophic bacteria by the enzyme methane monooxygenase (Anthony 1986). As noted in Chapter 5, two forms of the enzyme are known to exist at least in some strains, a soluble form (sMMO) and a membrane-bound or particulate form (pMMO). The pMMO appears to be present in all methanotrophs, but it has not been well characterized. The sMMO has been purified from three strains and has been characterized in some detail. Five polypeptides are known to be present in the protein: three (alpha, beta, and gamma) in component A, the hydroxylase; one in component C, the reductase; and one in component B, whose function is not yet clear (Anthony 1986). N-Terminal amino acid sequence data from purified proteins have been used to design oligonucleotide probes for identifying cloned sMMO genes. In *Methylococcus capsulatus* Bath, a probe for the gamma subunit of component A has been used to identify a clone containing *mmoZ* (encoding the gamma subunit). The insert in this clone has been sequenced, and other sMMO genes were found to be clustered on this fragment including *mmoX* (alpha subunit), *mmoY* (beta subunit), *mmoC* (component C protein), and *mmoB* (component B protein), along with one open reading frame (ORF Y) of unknown function (Mullens and Dalton 1987; Stainthorpe et al. 1989; Stainthorpe et al. 1990; Pilkington et al. 1990). The order is: *mmoX—mmoY—mmoB—mmoZ—*ORF Y*—mmoC*.

At least one sMMO gene has been cloned from *Methylosinus trichosporium* OB3b, a type II methanotroph. In this case, an immunological assay was used to detect the B component from expression clones analyzed in *E. coli*, and putative clones were identified by hybridization to a probe designed from N-terminal amino acid sequence data (Wattenberg et al. 1990). The identity of the hybridizing region was confirmed by sequencing the B component gene (*mmoB*).

12.6.2 Methanol Oxidation Genes

12.6.2.1 Genetic Maps. In Gram-negative methylotrophs, methanol is oxidized to formaldehyde by a periplasmic system consisting of the enzyme methanol dehydrogenase (MeDH) and a specific cytochrome *c*, often termed cytochrome c_L (cyt c_L) (Anthony 1986; also see Chapter 5). MeDH uses the cofactor pyrroloquinoline quinone (PQQ) and contains two subunits of approximately 60 and 10 kDa, usually in an $\alpha_2\beta_2$ structure (Anthony 1986). Genetic analysis of this system in *Methylobacterium* strains has shown it to be quite complex. In *M. organophilum* XX, at least 13 complementation classes are involved (Machlin et al. 1988; Bastien et al. 1989), while in *M. extorquens* AM1, at least 20 complementation classes (called *mox* genes) are involved (Lidstrom 1990; Lee and Lidstrom, unpublished observations). These classes have been identified by a combination of mutant complementation and expression studies. Although not all are known to encode separate polypeptides, they clearly reflect functional classes. For convenience here, they will be designated as genes. A particularly valuable approach in analyzing this system has been interspecies complementation, which has allowed the rapid identification of genes in three different *Methylobacterium* strains, *M. organophilum* XX (Bastien et al. 1989), *M. organophilum* DSM760 (Biville et al. 1989), and *M. extorquens* AM1 (Lidstrom 1990), as well as three methanotrophic strains, *Methylomonas albus* BG8 (Stephens et al. 1988), *Methylosporovibrio methanica* 81Z (Bastien et al. 1989), and *Methylomonas* sp. A4 (Waechter-Brulla and Lidstrom, unpublished observations). Figure 12–2 shows a map of the known C_1 genes, including the *mox* genes in these strains. The order of the fragments shown is not known, although for *M. extorquens* AM1 the order shown is consistent with data from R68.45-mediated crosses (Tatra and Goodwin 1983 and 1985). The known and potential similarities of the genetic maps in these diverse organisms is striking, although some differences exist. Mutants in methanol oxidation have also been isolated in one RuMP cycle obligate methylotroph, *Methylophilus methylotrophus* (see section 12.6.4) and were placed into 5 groups, based on complementation and biochemical analyses (Dawson and Goodwin 1990).

12.6.2.2 Proposed Functions. The large number of genes identified so far that are required to convert methanol to formaldehyde indicates that this system must be complex. The functions of these genes are only beginning to be determined, but the available information is summarized in Table 12–5. In *M. extorquens* AM1, besides the structural genes for the MeDH (*moxF* and *moxI*) and the cyt c_L (*moxG*), three genes (*moxA, K,* and *L*; formerly called *moxA*1, *A*2, and *A*3) appear to be involved in apoprotein/PQQ assembly or processing, and seven genes (*moxP, C, T, V, H, U,* and *O*) appear to be required for PQQ synthesis or transport to the periplasm. The functions of the other genes are unknown, but many of the mutant phenotypes are pleiotropic, and so they may be involved in regulation, processing, or stability, or in unknown components of methanol oxidation in vivo. Although less phenotypic information is available for the other two strains, their isofunctional genes, as determined by interspecies mutant complementation, are noted in Table 12–5.

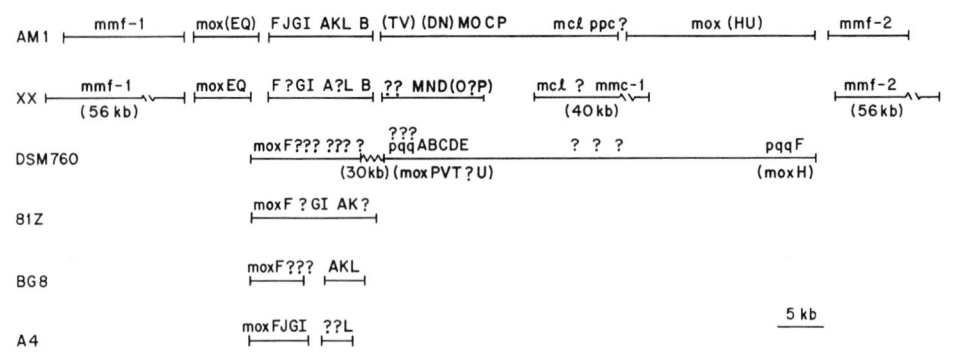

296 Molecular Genetics of Methylotrophic Bacteria

```
AM1  |--mmf-1--|  |--mox(EQ)--|  |--FJGI AKL B--|  |--(TV)(DN) MO CP--|  |--mcℓ ppc ?--|  |--mox (HU)--|  |--mmf-2--|

XX   |--mmf-1--|~\~|--moxEQ--|  |--F?GI A?L B--|  |--?? MND(O?P)--|  |--mcℓ ? mmc-1--|~\~  |--mmf-2--|~\~
       (56 kb)                                                              (40 kb)              (56 kb)

                                            ???
DSM760           |--moxF??? ???  ?  pqqABCDE--|   |--? ? ?--|         |--pqqF--|
                      (30kb) (moxPVT?U)                                (moxH)

81Z              |--moxF ? GI AK?--|

BG8              |--moxF???--| |--AKL--|

                                                                       5 kb
A4               |--moxFJGI--| |--??L--|
```

FIGURE 12-2 Organization of C_1 genes in six strains of methylotrophic bacteria. Vertical bars denote separate fragments with unknown distances between them. The order of the fragments shown is not known, but for *M. extorquens* AM1, it is consistent with R68.45-mediated chromosomal mapping data. The order of separate fragments in the other strains is unknown, but the fragments are placed in the same order as in AM1 for consistency. Parentheses around letters indicate unknown gene order. For *M. organophilum* XX, the numbers underneath designate the known size of the fragment; and for *M. organophilum* DSM760, the number underneath designates a DNA fragment with no known C_1 genes. The letters underneath the *pqq* genes designate the equivalent AM1 *mox* genes. Question marks indicate there are no data for those genes. The *mox* and *pqq* genes are as noted in Table 12-5. Abbreviations: mmf-1, acetyl-CoA oxidation to glyoxylate; mcl, malyl-CoA lyase; ppc, PEP carboxylase; mmc-1, pleiotrophy for C_1 oxidation functions, mmf-2, glycerate kinase; AM1, *Methylobacterium extorquens* AM1; XX, *M. organophilum* XX; DSM760, *M. organophilum* DSM760; 81Z, *Methylosporovibrio methanica* 81Z; BG8, *Methylomonas albus* BG8; A4, *Methylomonas* sp. A4.

The molecular basis of regulation for this system has not yet been determined, but in both *M. organophilum* XX and *M. extorquens* AM1, transcriptional start site mapping has shown that the *moxF* gene is transcriptionally regulated in response to C_1 compounds (Machlin and Hanson 1988; Anderson 1988). Therefore, one or more of the unknown *mox* genes may be involved in transcriptional regulation.

12.6.2.3 Sequence Comparisons. The genes for the alpha subunit of the MeDH have been sequenced from *Paracoccus denitrificans* (Harms et al. 1987), *M. organophilum* XX (Machlin and Hanson 1988), and *M. extorquens* AM1 (Anderson et al. 1990). Figure 12-3 shows a comparison of the amino acid sequences deduced for these *moxF* genes and a gene for a PQQ-linked alcohol dehydrogenase from *Acetobacter aceti* (Inoue et al. 1989). The sequences for the two *Methylobacterium* strains are almost identical (96% similarity), the *P. denitrificans* sequence diverges somewhat more (80% similarity), and the *A. aceti* sequence is even more divergent (30% simi-

TABLE 12-5 Properties of the Methanol Oxidation Genes in Three *Methylobacterium* Strains.

Gene Designation[1]			Approximate Polypeptide Size (kDa)	Proposed Function[2]
AM1	XX	DSM760		
moxF	VC	*moxF*	60	MeDH alpha subunit
moxJ			30	?
moxG	VB3		19	cyt c_L
moxI	VB2		10	MeDH beta subunit
moxA	VB1			PQQ/apoprotein assembly or modification
moxK				PQQ/apoprotein assembly or modification
moxL	VA2		19	PQQ/apoprotein assembly or modification
moxB	VA1			Regulation
moxP	VID	*pqqA*		PQQ synthesis or transport
moxV		*pqqB*		PQQ synthesis or transport
moxT		*pqqC*		PQQ synthesis or transport
		pqqD		PQQ synthesis or transport
moxU		*pqqE*		PQQ synthesis or transport
moxH		*pqqF*		PQQ synthesis or transport
moxC				PQQ synthesis or transport
moxO	VID			PQQ synthesis or transport
moxM	VIA			?
moxN	VIB			?
moxD	VIC			?
moxQ	VIIA			?
moxE	VIIB			?
moxR				?
moxS				?

[1] AM1, *M. extorquens* AM1; XX, *M. organophilum* XX; DSM760, *M. organophilum* DSM760. The blank spaces indicate that no data are available.
[2] The MeDH subunits and cyt c_L have been definitely identified. The other functions are only proposed, based on available phenotypic data.
Sources: Data from Nunn and Lidstrom 1986a and 1986b; Anderson and Lidstrom 1988; Lidstrom 1990; Machlin et al. 1988; Bastien et al. 1989; Biville et al. 1989.

larity). Comparisons of N-terminal regions of *moxF* sequences from two type I methanotrophs, *M. albus* BG8 and *Methylomonas* A4, show high similarity (75–80%; Kuhn, Waechter-Brulla, and Lidstrom, unpublished observations) suggesting that this gene has been highly conserved in diverse methylotrophs.

The only other available sequence comparisons for C_1 genes are for *moxJ* from *M. extorquens* AM1 (Anderson et al. 1990) and ORF 2 from *P. denitrificans* (Harms et al. 1989a). ORF 2 is located immediately down-

```
                  10          20          30          40         50
AaADH    MIRPASAKRRSLLGILAAGTICAAALPYAAVFARADGQGNTGEAIIHADD
PdMDH    MNR-NTPKARGASS-LAMAMAMGLAVITTA-PATANDQ-----LVELAKD
XXMDH    MSRFVTSVSALAM--LALAPAA----LSS--VAYANDK-----LVELSKS
AM1MDH   MSRFVTSVSALAM--LALAPAA----LSS--GAYANDK-----LVELSKS

              60          70          80          90         100         110
HPENWLSYGRTYSEQRYSPLDQINRSNVGDIKLLGYYTLDTNRGQEATPLVVDGIMYATT
-PANWVMIGRDYNAQNYSEMTDINKENVKQLRPAWSFSTGVLHGHEGTPLVVGDRMFIHT
-DDNWVMPGKNYDSNNYSELKQVNKSNVKQLRPAWTFSTGLLNGHEGAPLVVDGKMYVHT
-DDNWVMPGKNYDSNNFSDLKQINKGNVKQLRPAWTFSTGLLNGHEGAPLVVDGKMYIHT

             120         130         140         150         160
NWSKME-ALD-AATGKLLWQYDPKVPGNIADKGCCDIVNRGAGYWNG------KVFWGTF
FFPNITFALDLNEPGKILWQNKPKQNPTARIVACCDVVNRGLAYWPGDDQVKPLIFRTQL
SFPNNTFALDLDDPGHILWQDKPKQNPAARAVACCDLVNRGLAYWPGDGKTPALILKTQL
SFPNNTFALGLDDPGTILWQDKPKQNPAARAVACCDLVNRGLAYWPGDGKTPALILKTQL

            170         180         190         200         210         220
DGRLVAADAKTGKKVWAVNTIPADASLGKQRSYTVDGAVRVAKGLVLIGNGAEFGARGF
DGHIVAMDAETGETRWIMEN--SDIKVG----STLTIAPYVIKDLVLVGSSGAELGVRGY
DRHVVALNAETGETVWKVEN--SDIKVG----STLTIAPYVVKDKVIIGSSGAELGVRGY
DGNMAALNAETGETVWKVEN--SDIKVG----STLTIAPYVVKDKVIIGSSGAELGVRGY

           230         240         250         260         270
VSAFDAETGKLKWR-FYIVFNNK----NEPDHAASDNILMNKAYKTWGPKGAWVRQGGGG
VTAYDVKSGEMRWRAFATGPDEELLLAEDFNAPNPHYGQKNLGLETWEG-DAW--KIGGG
LTAYDVKTGQVWRAYATGPDKDLLLADDFNVKNAHYGQKGLGTATWEG-DAW--KIGGG
LTAYDVKTGEQVWRAYATGPDKDLLLASDFNIKNPHYGQKGLGTGTWEG-DAW--KIGGG

280         290         300         310         320         330
TVWDSLVYDFVSDLIYLAVGNGSPWNYKYRSEGIGSNLFLGSIVALKPETGEYVWHFQAT
TNWGWYAYDPEVDLFYYGSGNPAPWNETMRP---GDNKWTMAIWGREATTGEAKFAYQKT
TNWGWYAYDPGTNLIYFGTGNPAPWNETMRP---GDNKWTMTIFGRDADTGEAKFGYQKT
TNWGWYAYDPGTNLIYFGTGNPAPWNETMRP---GDNKWTMTIFGRDADTGEAKFGYQKT

340         350         360         370         380         390
PMDQWDYTSVQQIMTLDMPVK-GEMRHVIVHAPKNGFFYVLDAKTGEFLS-GKNYVYQNW
PHDEWDYAGVNVMMLSEQEDKQGQMRKLLTHPDRNGIVYTLDRTNGDLISADKMDDTVNW
PHDEWDYAGVNVMMPSEQKDKDGKTRKLLTHPDRNGIVYTLDRTDGALVSANKLDDTVNV
PHDEWDYAGVNVMMLSEQKDKDGKARKLLTHPDRNGIVYTLDRTDGALVSANKLDDTVNV

400         410         420         430         440         450
ANGLDPLTGRFMYNFDGLYTLNGKFWYGIFGPLGAHNFMAMAYSFKTHLVYIPAHQTPFG
VKEMQLDTGLPVRDPEFGTRMDHKARDICPSAMGYHNQGHDSYDPERKVFMLGINHICMD
FKTVDLKTGQPVRDPEYGTRMDHLAKDVCPSAMGYHNQGHDSYDPKRELFFMGINHICMD
FKSVDLKTGQPVRDPEYGTRMDHLAKDICPSAMGYHNQGHDSYDPKRELFFMGINHICMD

460         470         480         490         500         510
YKNQVGGFKPHADSWNVGLDMTKNGLFDTPEARTAYIKDLHGWLLAWDPVKMETVW-KID
WEPFMLPYR--AGQFFVGATLIMYPGPKATAER-AG----AGQIKAYDAISGEMKWEKME
WEPFMLPYR--AGQFFVGATLNMYPGPKGDRQNYEG----LGQIKAYNAITGSYKWEKME
WEPFMLPYR--AGQFFVGATLNMYPGPKGDRQNYEG----LGQIKAYNAITGDYKWEKME
```

Continued

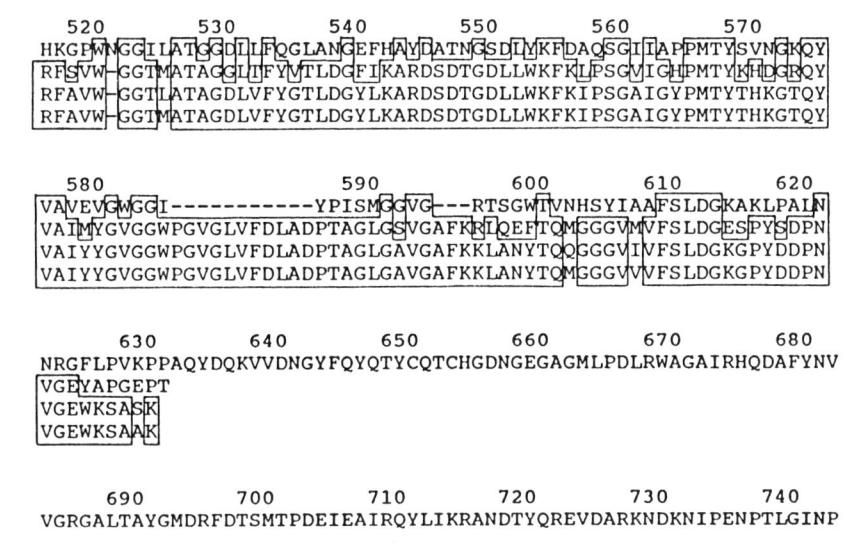

```
        520        530        540        550        560        570
HKGFWNGGILATGGDLLFQGLANGEFHAYDATNGSDLMKFDAQSGIIAPPMTYSVNGKQY
RFSVW-GGTMATAGGLIFYMTLDGFIKARDSDTGDLLWKFKLPSGVIGHPMTYKHDGRQY
RFAVW-GGTLATAGDLVFYGTLDGYLKARDSDTGDLLWKFKIPSGAIGYPMTYTHKGTQY
RFAVW-GGTMATAGDLVFYGTLDGYLKARDSDTGDLLWKFKIPSGAIGYPMTYTHKGTQY

        580        590        600        610        620
VAVEVGWGGI-----------YPISMGGVG---RTSGWTVNHSYIAAFSLDGKAKLPALN
VAIMYGVGGWPGVGLVFDLADPTAGLGSVGAFKRLQEFTQMGGGVMVFSLDGESPYSDPN
VAIYYGVGGWPGVGLVFDLADPTAGLGAVGAFKKLANYTQQGGGVIVFSLDGKGPYDDPN
VAIYYGVGGWPGVGLVFDLADPTAGLGAVGAFKKLANYTQMGGGVVVFSLDGKGPYDDPN

        630        640        650        660        670        680
NRGFLPVKPPAQYDQKVVDNGYFQYQTYCQTCHGDNGEGAGMLPDLRWAGAIRHQDAFYNV
VGEYAPGEPT
VGEWKSASK
VGEWKSAAK

        690        700        710        720        730        740
VGRGALTAYGMDRFDTSMTPDEIEAIRQYLIKRANDTYQREVDARKNDKNIPENPTLGINP
```

FIGURE 12-3 Comparison of the amino acid sequences deduced from nucleotide sequences for four PQQ-linked proteins. Boxed areas show identical amino acids. Abbreviations: AaADH, PQQ-linked alcohol dehydrogenase from *Acetobacter aceti*; Pd MDH, methanol dehydrogenase from *Paracoccus denitrificans*; XX MDH, methanol dehydrogenase from *M. organophilum* XX; AM1 MDH, methanol dehydrogenase from *M. extorquens* AM1.

stream of *moxF* and therefore is a likely candidate for the *moxJ* homolog in that organism. Sequence comparisons show only 13% overall identity, but an internal segment of approximately 150 amino acids shows over 50% similarity with no gaps, suggesting that these two genes are equivalent (Anderson et al. 1990).

12.6.3 Serine Pathway Genes

The organization of genes specific for the serine pathway of formaldehyde assimilation have been studied in *M. organophilum* XX and *M. extorquens* AM1. In *M. organophilum* XX, linkage analysis with pleiotropic methanol utilization mutants suggested that at least three separate linkage groups were present (O'Connor and Hanson 1978; O'Connor 1981). These findings were confirmed by later mapping studies involving cosmid complementation that showed the presence of three separate linkage groups involving serine pathway genes, which were in turn separate from three other linkage groups involving *mox* genes (Figure 12-2; Allen and Hanson 1985; Machlin et al. 1988).

In *M. extorquens* AM1, two serine pathway genes (*ppc*, PEP carboxylase; and *mcl*, malyl-CoA lyase) have been shown to be linked to several *mox* genes (Figure 12-2; Fulton et al. 1984; Lidstrom et al. 1987), and two others (encoding glycerate kinase and a gene involved in oxidation of acetyl-CoA to glyoxylate) have been shown to be on separate, unlinked fragments (Stone and Goodwin 1989). As noted previously, the order of separate fragments for *M. extorquens* AM1 and *M. organophilum* XX is not known.

No information is yet available concerning the molecular mechanism for the regulation of serine pathway genes, but enzyme activity data suggest they are regulated by the presence of C_1 compounds in a system separate from the regulation of the oxidation genes (McNerney and O'Connor 1980; O'Connor, 1981).

12.6.4 Ribulose Monophosphate Cycle

The enzymes involved in the ribulose monophosphate (RuMP) cycle for formaldehyde assimilation in obligate methylotrophs have been investigated in some detail (Anthony 1982). However, studies on the genetic regulation of the RuMP cycle enzymes have been less successful, since they require the isolation of mutants that block separate stages of formaldehyde oxidation and assimilation.

Kletsova et al. (1988) have described a general approach for the isolation of mutants of C_1 metabolism in *M. flagellatum*. This strain uses the 2-keto, 3-deoxy, 6-phosphogluconate aldolase/transaldolase (KDPGA/TA) variant of the RuMP cycle (Anthony 1982). Since C_1 substrates are the only source of carbon and energy for this bacterium, mutations in C_1 metabolism genes are lethal. Therefore, the isolation of mutants defective in RuMP cycle enzyme activities requires the selection of conditionally lethal (temperature sensitive, *ts*) lesions. More than 4,000 random *ts* mutants were isolated in *M. flagellatum*, and 500 were identified that were unable to grow at high temperature in the presence of exogenous nutrient supplements. Each of these was tested for enzyme activities in cell-free extracts at permissive and nonpermissive temperatures. This analysis led to the identification of mutants defective in the genes *pgi* and *gpd*, encoding phosphoglucoisomerase and glucose 6-phosphate dehydrogenase, respectively, two enzymes of the RuMP cycle. Mutations that were *pgi*- and *gpd*-minus have complete growth inhibition at nonpermissive temperatures, which further confirms the existence of only one pathway for the incorporation of formaldehyde carbon into the cellular material of *M. flagellatum*.

The isolation of C_1 metabolism mutants allowed mapping of the genes responsible for key enzymes of the RuMP cycle. For mapping *pgi* and *gpd* genes in *M. flagellatum*, plasmids R68.45 and pULB113, and Hfr-type donor-mediated crosses were used (Kletsova and Tsygankov 1990). The *pgi* and *gpd* genes were both linked to methionine and leucine auxotrophy and nalidixic acid resistance markers, which all map in the late region of the *M. flagellatum* chromosome (Figure 12-1).

A similar approach was used by Hutchinson and Goodwin (1989) and Dawson and Goodwin (1990) for the isolation of auxotrophs and mutants unable to utilize methanol and/or methylamine at the nonpermissive temperature in *M. methylotrophus*. The biochemical characterization of *ts* mutants which grew normally on methanol and methylamine at 30°C but grew only on methylamine at 37°C, showed that these mutants are defective in methanol oxidation. These mutants synthesized the alpha subunit of methanol dehydrogenase when grown at 30°C on methylamine, but not when grown on this substrate at 37°C.

REFERENCES

Allen, L.N., and Hanson, R.S. (1985) *J. Bacteriol.* 161, 955–962.
Anderson, D.J. (1988) Ph.D. thesis, University of Washington, Seattle, WA.
Anderson, D.J., and Lidstrom, M.E. (1988) *J. Bacteriol.* 170, 2254–2262.
Anderson, D.J., Morris, C.J., Nunn, D.N., Anthony, C., and Lidstrom, M.E. (1990) *Gene* 90, 173–176.
Anthony, C. (1982) *The Biochemistry of Methylotrophs*, Academic Press, London.
Anthony, C. (1986) *Adv. Microb. Physiol.* 27, 113–210.
Bastien, C., Machlin, S., Zhang, Y., Donaldson, K., and Hanson, R. (1989) *Appl. Env. Microbiol.* 55, 3124–3130.
Better, M., and Helinski, R. (1983) *J. Bacteriol.* 155, 311–316.
Biville, F., Turlin, E., and Gasser, F. (1989) *J. Gen. Microbiol.* 135, 2917–2930.
Boer, L., Clement, W., Dijkhuizen, L., and Harder, W. (1985) *Antonie van Leewenhoek* 51, 566–567.
Boer, L., Vrijbloed, W., and Dijkhuizen, L. (1986) *Abstracts 5th International Symposium on Microbial Growth on C_1 Compounds* (Duine, J.A., and van Verseveld, H.W., eds.), Groningen, Netherlands.
Bohanon, M., Bastien, C., Yoshida, R., and Hanson, R. (1988) *Appl. Environ. Microbiol.* 54, 271–273.
Bourd, G.I., Malin, G.M., and Tsygankov, Y.D. (1989) *FEMS Microbiol. Lett.* 65, 109–112.
Byng, G.S., and Jensen, R.A. (1983) in *Isoenzymes*, vol. 8 (Ratazzi, M.C., Scandalios, J.G., and Witt, S., eds.), pp. 115–140, Liss, New York.
Byrom, D. (1984) in *Microbial Growth on C_1 Compounds* (Crawford, R.L., and Hanson, R.S., eds.), pp. 221–223, American Society for Microbiology, Washington, DC.
Chistoserdov, Y.A., Eremashvili, M.R., Mashko, S.V., et al. (1987) *Molekularnaya genetika, microbiologia i virusologia (USSR)* 8, 36–41.
Davidson, S., and Murrell, J.C. (1989) *Abstracts 6th International Symposium on Microbial Growth on C_1 Compounds*, Göttingen, FRG.
Dawson, A., and Goodwin, P. (1990) *J. Gen. Microbiol.* 136, 1373–1380.
de Maeyer, E., Skup, D., Prasad, K.S.N., et al. (1982) *Proc. Natl. Acad. Sci. USA* 79, 4256.
De Vries, G. (1986) *FEMS Microbiol. Rev.* 39 (3), 235–258.
De Vries, G., Kues, U., and Stahl, U. (1990) *FEMS Microbiol. Rev.* 75, 57–102.
Ely, B. (1984) *Mol. Gen. Genet.* 200, 402–404.

Finch, P., Brough, C., and Emmerson, P. (1986) *Gene* 44 (1), 47-53.
Fulton, G., Nunn, D., and Lidstrom, M. (1984) *J. Bacteriol.* 160, 718-723.
Gliesche, C.G., and Hirsch, P. (1989) *Abstracts 6th International Symposium on Microbial Growth on C_1 Compounds*, Göttingen, FRG.
Goldberg, I., and Mekalanos, J.J. (1986) *J. Bacteriol.* 165, 715-722.
Gomelsky, M.V., Gak, E.R., and Tsygankov, Y.D. (1989) *Abstracts 6th International Symposium on Microbial Growth on C_1 Compounds*, Göttingen, FRG.
Govorukhina, N.I., Kletsova, L.V., Tsygankov, Y.D., Trotsenko, Y.A., and Netrusov, A.I. (1987) *Mikrobiologiya (USSR)* 56, 849-854.
Haas, D., and Holloway, B.W. (1976) *Mol. Gen. Genet.* 144, 243-253.
Hamood, A.N., Pettis, G.S., Parker, C.D., and McIntosh, M.A. (1986) *J. Bacteriol.* 167, 375-378.
Harms, N., De Vries, G., Maurer, K., Hoogendijk, J., and Stouthamer, A. (1987) *J. Bacteriol.* 169, 3969-3975.
Harms, N., Van Spanning, R., Oltmann, F., and Stouthamer, A. (1989a) *Antonie van Leewenhoek* 56, 47-50.
Harms, N., Van Spanning, R., Oltmann, F., and Stouthamer, A. (1989b) *Abstracts 6th International Symposium on Growth on C_1 Compounds*, Göttingen, FRG.
Hennam, J.F., Cunningham, A.E., Sharpe, G.S., and Atherton, K.T. (1982) *Nature* 297, 80-82.
Holloway, B. (1984) in *Methylotrophs: Microbiology, Biochemistry and Genetics* (Hou, C.T., ed.), pp. 87-106, CRC Press, Boca Raton, FL.
Holloway, B.W., and Morgan, A.F. (1986) *Ann. Rev. Microbiol.* 40, 79-105.
Hutchinson, N., and Goodwin, P.M. (1989) *Abstracts 6th International Symposium on Microbial Growth on C_1 Compounds*, Göttingen, FRG.
Inoue, T., Sunagawa, M., Mori, A., et al. (1989) *J. Bacteriol.* 171, 3115-3122.
Kameneva, S.V., Polivtseva, T.P., Balanina, N.V., and Schestakov, S.V. (1986) *Genetica (USSR)* 11, 2664-2673.
Kearney, P., and Holloway, B.W. (1987) *FEMS Microbiol. Lett.* 44, 7-11.
Keener, S.L., McNamee, K.P., and McEntee, K. (1984) *J. Bacteriol.* 160, 153-160.
Kletsova, L.V., and Tsygankov, Y.D. (1990) *Arch. Microbiol.* 153, 139-145.
Kletsova, L., Chibisova, E., and Tsygankov, Y. (1988) *Arch. Microbiol.* 149, 441-446.
Koomey, M., and Falkow, S. (1987) *J. Bacteriol.* 169, 790-795.
Lanzov, V.A. (1985) *Genetica (USSR)* 21, 1413-1427.
LaRoche, S.D., and Leisinger, T. (1990) *J. Bacteriol.* 172, 164-171.
Lidstrom, M.E. (1990) *FEMS Microbiol. Rev.* 87, 431-436.
Lidstrom, M.E., and Stirling, D. (1990) *Ann. Rev. Microbiol.* 44, 27-57.
Lidstrom, M.E., Nunn, D.N., Anderson, D.J., Stephens, R.L., and Haygood, M.G. (1987) in *Microbial Growth on C_1 Compounds* (Van Verseveld, H., and Duine, H., eds.), pp. 248-254, Martinus Nijhoff, Amsterdam.
Little, J.W., and Mount, D.W. (1982) *Cell* 29, 11-22.
Lyon, B., Kearney, P., Sinclair, M., and Holloway, B. (1988) *J. Gen. Microbiol.* 134, 123-132.
Machlin, S., and Hanson, R. (1988) *J. Bacteriol.* 170, 4739-4747.
Machlin, S., Tam, P., Bastien, C., and Hanson, R. (1988) *J. Bacteriol.* 170, 141-148.
Maksimova, N.P., Olechnovich, I.N., and Fomichev, Y.K. (1986) *Genetica (USSR)* 21, 194-199.
McNerney, T., and O'Connor, M. (1980) *Appl. Environ. Microbiol.* 40, 370-375.

Mergeay, M., Lejeune, P., Sadouk, A., Gerits, J., and Fabry, L. (1987) *Mol. Gen. Genet.* 209, 61–70.
Metzler, T., Marquardt, R., Prave, P., and Winnacker, E.-L. (1988) *Mol. Gen. Genet.* 211, 210–214.
Moore, A.T., Nayudu, M., and Holloway, B.W. (1983) *J. Gen. Microbiol.* 129, 785–799.
Mullens, I., and Dalton, H. (1987) *Biotech.* 5, 490–493.
Nunn, D., and Anthony, C. (1988) *Biochem. J.* 256, 673–676.
Nunn, D., and Lidstrom, M.E. (1986a) *J. Bacteriol.* 166, 581–590.
Nunn, D., and Lidstrom, M.E. (1986b) *J. Bacteriol.* 166, 591–598.
Nunn, D., Day, D., and Anthony, C. (1989) *Biochem. J.* 260, 857–862.
O'Connor, M. (1981) in *Microbial Growth on C_1 Compounds* (Dalton, H., ed.), pp. 294–300, Heyden, London.
O'Connor, M., and Hanson, R.S. (1978) *J. Gen. Microbiol.* 104, 105–111.
O'Connor, M., Wopat, A., and Hanson, R. (1977) *J. Gen. Microbiol.* 98, 265–272.
Ohman, D.E., West, M.A., Flynn, J.L., and Goldberg, J.B. (1985) *J. Bacteriol.* 162, 1068–1074.
Olechnovich, I.N., Maksimova, N.P., and Fomichev, Y.K. (1986) *Mol. Genet. Microbiol. Virusol. (USSR)* 12, 34–36.
Olechnovich, I.N., Maksimova, N.P., and Fomichev, Y.K. (1987) *Genetica (USSR)* 23, 414–420.
Paul, K., Grosh, S., and Das, J. (1986) *Mol. Gen. Genet.* 203, 58–63.
Pilkington, S.J., Salmond, G.P.C., Murrell, J.C., and Dalton, H. (1990) *FEMS Microbiol. Lett.* 72, 345–348.
Pittard, J., and Gibson, F. (1970) in *Current Topics in Cellular Regulation*, vol. 2 (Horecker, B.L., and Stardtman, E.R., eds.), pp. 29–63, Academic Press, Inc., New York.
Powell, K.A., and Byrom, D. (1982) in *Genetics of Industrial Microorganisms* (Ikeda, Y., and Beppu, T., eds.), p. 345, Kodansha, Tokyo.
Ramesar, R.S., Woods, D.R., and Rawlings, D.E. (1988) *J. Gen. Microbiol.* 134, 1141–1146.
Resnik, D., and Nelson, D.R. (1988) *J. Bacteriol.* 170, 48–55.
Ruvkun, G., and Ausubel, F. (1981) *Nature* 298, 85–88.
Sakanyan, V.A., Yakubov, L.Z., Alikhanian, S.I., and Stepanov, A.I. (1978) *Mol. Gen. Genet.* 165, 331–341.
Sancar, A., Sachelek, C., Konigsberg, W., and Rupp, W.D. (1980) *Proc. Natl. Acad. Sci. USA* 77, 2611–2615.
Sano, Yu., and Kageyama, M. (1987) *Mol. Gen. Genet.* 208, 412–419.
Sato, M., Staskawicz, B.J., Panopoulos, N.J., Peter, S., and Honma, M. (1981) *Plasmid* 6, 325.
Schmidhauser, T.J., Ditta, G., and Helinski, D.R. (1986) in *The Vectors, Cloning Vectors and their Uses* (Rodriguez, R.C., and Denhardt, D.T., eds.), pp. 345–387, Butterworths, Boston.
Serebrijski, I.G., Kazakova, S.M., and Tsygankov, Y.D. (1989) *FEMS Microbiol. Lett.* 59, 203–206.
Stainthorpe, A., Murrell, J., Salmond, G., Dalton, H., and Lees, V. (1989) *Arch. Microbiol.* 152, 154–159.
Stainthorpe, A., Lees, V., Salmond, G., Dalton, H., and Murrell, J. (1990) *Gene* 91, 27–34.

Stephens, R., Haygood, M., and Lidstrom, M. (1988) *J. Bacteriol.* 170, 2063–2069.
Stone, S., and Goodwin, P. (1989) *J. Gen. Microbiol.* 135, 227–235.
Tatra, P., and Goodwin, P. (1983) *J. Gen. Microbiol.* 129, 2629–2634.
Tatra, P., and Goodwin, P. (1985) *Arch. Microbiol.* 143 (2), 169–178.
Toukdarian, A., and Lidstrom, M. (1984) *J. Bacteriol.* 157, 979–983.
Tsygankov, Y.D., and Kazakova, S.M. (1987) *Arch. Microbiol.* 149, 112–119.
Tsygankov, Y.D., Kazakova, S.M., and Serebrijski, I.G. (1990) *J. Bacteriol.* 172, 2747–2754.
Van Spanning, R.J.M., Wansell, C., Harms, N., Oltmann, L.F., and Stouthamer, A.H. (1989) *Abstracts 6th International Symposium on Microbial Growth on C_1 Compounds*, Göttingen, FRG.
Walker, G.A. (1984) *Microbiol. Rev.* 48, 60–93.
Wattenberg, E.V., Tsien, H.C., and Hanson, R.S. (1990) Abstracts Annual Meeting American Society for Microbiology, Anaheim, CA.
West, S., and Iglewski, B. (1988) *Nucl. Acid Res.* 16, 9323–9330.
Whitta, S., Sinclair, M., and Holloway, B. (1985) *J. Gen. Microbiol.* 131 (6), 1547–1551.
Wilke, D., and Schlegel, H.G. (1979) *J. Gen. Microbiol.* 115, 403–410.
Williams, E., and Bainbridge, B.W. (1971) *J. Appl. Bacteriol.* 34, 685–689.
Zaitzev, E.N., Zaitzeva, E.M., Bakhlanova, I.V., et al. (1986) *Genetica (USSR)* 22, 2721–2727.

CHAPTER 13

Heterologous Gene Expression in Methylotrophic Yeast

Juerg F. Tschopp
James M. Cregg

The *Pichia pastoris* host strain employed for heterologous gene expression was originally developed during the 1970s by Phillips Petroleum Co. for the efficient utilization of methanol for the production of single cell protein (SCP) (Wegner 1983). As a result of the fermentation development studies required for SCP production, the growth characteristics of *P. pastoris* are well defined. The organism is routinely propagated in continuous culture at densities of approximately 130 gm dry weight per liter for periods of up to 30 days in 1,500-liter fermentors. The culture medium is inexpensive and defined, consisting of a pure carbon source (methanol), biotin, salts, trace elements, and water, and is free from undefined ingredients which can be potential sources of pyrogens and toxins. Because *P. pastoris* cultures are normally grown in a medium containing methanol at a pH below the op-

---

The authors wish to acknowledge that research described in this manuscript was performed by a large number of scientists. We recognize the contributions of Geneva R. Davis, Robert Siegel, Cathy Stillman, Russell Brierley, Patty Koutz, Gonul Velicelebi, Sally Provov, Michael Akong, Mary Ellen Digan, Stephen Lair, William Craig, Greg Holtz, Peter Kellaris, Anita Melton, Rossanne Kosson, Tom Gingeras, Rich Buckholz, Lynn Grinna, and Gregory Thill. The authors also wish to thank Michael Harpold and Jean Sartor for editorial comments and Sue Perry for the preparation of this manuscript.

timum of most microorganisms, *P. pastoris* cultures are less susceptible than most to contamination. Studies to date suggest that *P. pastoris* SCP is non-pathogenic and devoid of toxins and pyrogens.

The biochemical pathway for methanol utilization in *P. pastoris* is well defined and appears to be similar to the one used in all methylotrophic yeasts (Anthony 1982; Veenhuis et al. 1983; Veenhuis and Harder 1987; Harder and Veenhuis 1989). Critical to the *P. pastoris* expression system is the observation that many of the methanol pathway enzymes are present at higher levels in methanol-grown cells than in cells grown on other carbon sources. An example of this apparent regulation by methanol in *P. pastoris* is the enzyme alcohol oxidase (AOX), which is not present in glucose or glycerol-grown cells, but constitutes up to 30% of the total soluble protein in methanol-grown cells (Couderc and Baratti 1980). Concomitant with the massive synthesis of AOX, the peroxisomes in these yeasts grow in number and in size (Veenhuis et al. 1983). The advent of recombinant DNA technology and efficient methods for yeast transformation created the possibility in which the high-cell density fermentation technology for *P. pastoris* could be combined with the *AOX*-promoter-based expression system for the production of large quantities of foreign proteins at reasonable costs.

The *P. pastoris* heterologous gene expression system was also developed in response to problems previously encountered using bakers' yeast, *Saccharomyces cerevisiae,* for heterologous expression. One major source of these problems in *S. cerevisiae* is the promoters selected for expression. The promoters are either weak, not well regulated, or not easily regulated in fermentor cultures. As a result, expression strains which look promising in the laboratory are not productive when moved to commercial scale (Shuster 1987). Other problems with *S. cerevisiae* are associated with secretion. Foreign proteins secreted from *S. cerevisiae* are often at concentrations too low to be commercially attractive (Kingsman et al. 1987). In addition, transit through the *S. cerevisiae* secretory pathway can result in the addition of very long oligosaccharide chains (Ballou 1982; MacKay 1987; Moonen et al. 1987; Schultz et al. 1987; Van Arsdell et al. 1987). Possibly as a consequence of this hyperglycosylation, many foreign proteins secreted from *S. cerevisiae* cannot pass through the cell wall and remain trapped in the periplasmic space. Using the *P. pastoris* expression system, however, foreign proteins are secreted into the medium at exceptionally high levels. Furthermore, it was discovered that in *P. pastoris,* these proteins undergo significantly less carbohydrate addition than in *S. cerevisiae* (Tschopp et al. 1987b; Grinna and Tschopp 1989). Thus, the prospects for the successful commercial production of a variety of foreign proteins in *P. pastoris* appear to be bright (Cregg et al. 1989b).

The purpose of this review is to present knowledge relevant to the *P. pastoris* expression system, as well as the history of the early work on developing the *P. pastoris* system. The discussion of foreign gene expression focuses on vector and host strain construction technology, selected examples of foreign gene expression, post-translational modification of secreted proteins, and strategies for the production of foreign proteins in fermentor cultures.

13.1 VECTORS AND HOSTS USED FOR HETEROLOGOUS GENE EXPRESSION

13.1.1 Isolation and Structure of Methanol-Regulated Genes

The primary element in any expression system is the promoter employed to drive heterologous gene expression. Ideally, the promoter should meet three requirements. The first and most obvious requirement is that it should be efficiently transcribed so that a high level of expression is possible. The second requirement is that the promoter should be tightly regulated so that foreign protein production strains can sometimes be in an "expression-off" mode, since strains that express products constitutively may be at a selective disadvantage relative to nonexpressing strains. This disadvantage is often not obvious during shake flask culture experiments; however, the increased number of generations required in a commercial-scale process may allow nonexpressing strains to take over and reduce the yield of the expressing cell population. The third and perhaps the least-recognized requirement is that it should be possible to induce promoter expression cheaply and easily in large-volume, high-density, fermentor cultures, using existing fermentation technology.

A unique attraction of the *P. pastoris* expression system is its promoter, which is derived from the *P. pastoris* alcohol oxidase 1 gene (*AOX1*) and fully meets the above requirements. The *AOX1* gene was isolated from a cDNA library of RNA from methanol-grown *P. pastoris* cells by plus-minus screening for clones which hybridize only to RNAs from methanol-grown, and not ethanol-grown, cells (Ellis et al. 1985). *AOX-1*-encoding cDNAs were identified by comparing the predicted amino acid sequence of the 5' end of these cDNAs with the known N-terminal sequence of the AOX protein (Ellis et al. 1985). A genomic clone of *AOX1* which included its promoter was isolated and characterized. Northern blots showed a large amount of *AOX1* message in methanol-grown cells, but no detectable message in glucose-grown cells. The pattern of *AOX1* regulation was most consistent with the existence of catabolite repression-derepression and methanol-induction mechanisms. A second functional alcohol oxidase gene (*AOX2*) has also been isolated (Cregg and Madden 1987). The protein-coding portions of the two *AOX* genes are greater than 90% homologous at both the DNA and predicted amino acid sequence levels (Koutz et al. 1989). However, no sequence homology exists outside the coding portion of the genes in either the 5' or 3' direction. Gene disruption studies show that *AOX2* is the source of only a small fraction of the alcohol oxidase activity in methanol-grown cells of *P. pastoris* (Cregg and Madden 1987; Cregg et al. 1989a). This appears to be primarily a consequence of the difference in the transcriptional strengths of their promoters, since the steady-state level of *AOX1* message is much higher than that of *AOX2* and the protein-coding portion of *AOX2* can functionally substitute for that of *AOX1* when placed under the control of the *AOX1* regulatory sequences (Cregg et al. 1989a). The *AOX2* promoter, due to its lesser transcriptional strength, is not routinely used for heterologous gene expression. Other methanol-regulated genes have been isolated and partially characterized as well (Ellis et al. 1985). The promoter of the dihydroxyacetonesynthetase gene (*DAS*), the gene for the first enzyme in the methanol assimilatory pathway, has been used in heterologous gene expression and found to be comparable in strength to the *AOX1* promoter (Tschopp et al. 1987a; G. Thill, personal communication). Like *AOX1*, the *DAS* promoter is regulated by methanol induction and catabolite repression mechanisms. Interestingly, unlike *AOX1*, expression under typical derepressing conditions was not observed.

13.1.2 Transformation and Maintenance of Heterologous Genes in Yeast Host Cells

A second element in any expression system is the ability to transfer and maintain DNA sequences of interest in the cell. Transformation methods of *P. pastoris* are similar to those for *Saccharomyces cerevisiae* and are based on auxotrophic mutant hosts and vectors which contain complementary biosynthetic genes (Cregg et al. 1985). Vectors are transformed into *P. pastoris*, using modified versions of either the spheroplast method or the whole-cell, lithium salt method (Hinnen et al. 1978; Ito et al. 1983; Cregg et al. 1985). Transformation frequencies can be as high as 10^5 transformants per μg vector DNA for the spheroplast method and 10^3 transformants per μg vector DNA for the whole-cell method, but depend strongly on the type of vector employed.

To maximize the stability of heterologous gene expression strains, expression cassettes are integrated into the *P. pastoris* genome. As observed in *S. cerevisiae*, DNA sequences integrate into the *P. pastoris* genome by homologous recombination between sequences shared by the vector and the host (Cregg et al. 1985 and 1989a; Cregg and Madden 1987 and 1989). Thus vectors can be directed to integrate at preselected sites in the *P. pastoris* genome. Most expression hosts are created using a modified version of the one-step gene replacement method developed for *S. cerevisiae* (Rothstein 1983), and the universal *P. pastoris* expression plasmid pAO804 can be used for this purpose (Figure 13–1A). Heterologous gene sequences are modified using site-directed mutagenesis to add *EcoR* I sites 5' and 3' of the translation start and stop sequences, respectively, before insertion into vector pAO804. In addition to the heterologous gene (Y) expression cassette (5'-AOX1p-Y-3'-AOXt), the plasmid is composed of pBR322 sequences, which allow the

13.1 Vectors and Hosts Used for Heterologous Gene Expression

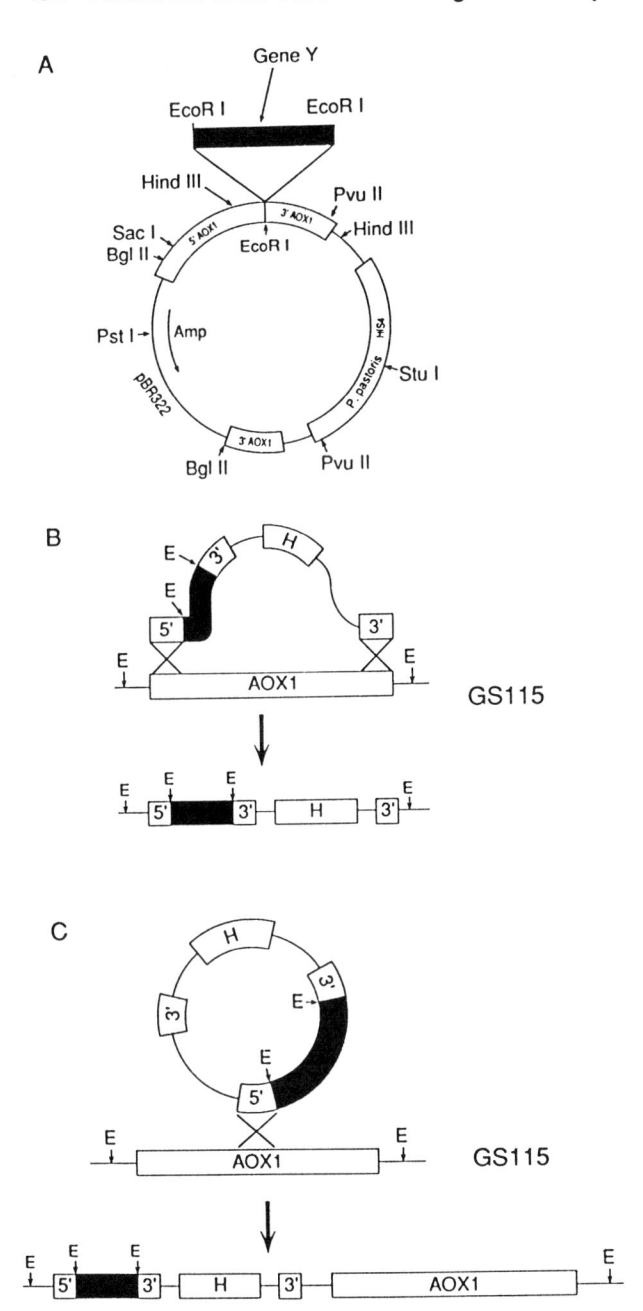

FIGURE 13-1 Technique for the construction of a *P. pastoris* expression host. (A) Restriction map of expression vector pAO804 with inserted heterologous gene Y. (B) Diagram of events leading to the replacement of the *AOX1* gene with the *Bgl* II fragment from expression vector. (C) Diagram of events leading to additive integration of the expression vector (the lengths of the gene fragments are not to scale).

propagation of the plasmid in *Escherichia coli,* the *PHIS4* gene, which provides a means of selecting transformants in *P. pastoris* strain GS115 (*his4*), and a fragment containing sequences 3' of the *AOX1* gene, which, together with the *AOX1* promoter fragment, direct a linear *Bgl* II fragment from the plasmid into the *AOX1* locus. Proper insertion of the *Bgl* II fragment results in the deletion of the entire *AOX1* gene (Figure 13-1B). The same plasmid can be digested at a single restriction site in *PHIS4* or *AOX1* to direct the plasmid to recombine by addition-type integration at either loci without disruption of *AOX1* (Figure 13-1C). This results in a host fully competent for methanol growth (Mut⁺) as opposed to methanol-defective (Mut⁻) growth after gene disruption. *P. pastoris* strains which are defective in *AOX1* grow very slowly on methanol, but are still metabolically active and often express heterologous genes at higher levels than do corresponding Mut⁺ strains, particularly in shake flask cultures (Cregg et al. 1989b). Expression cassettes which are integrated are quite stable even in large volume and high-density cultures (Cregg et al. 1987).

13.1.3 Fermentation and Scale-up

For the production of foreign proteins in fermentor cultures of *P. pastoris*, two culture methods have been devised, both of which are two-stage schemes. One employs a host which is partially defective in methanol utilization and a fed-batch fermentation process. The other utilizes a host with wild-type ability to utilize methanol and a continuous-culture process. In the first stage of both schemes, expression hosts are cultured in defined minimal medium on glycerol as the carbon source. The final cell density of the culture is preselected by the initial concentration of glycerol in the medium. On glycerol, cell growth is rapid (generation time, 3.0 h), but heterologous gene expression is repressed. Thus, a large mass of cells can be generated without significant selection for nonexpressing mutant strains.

Upon deletion of glycerol, fermentor cultures are ready to enter the second, or production, stage. Although both fermentation schemes initiate production by the addition of methanol to the culture, it is at this point that the schemes begin to differ. In the fed-batch fermentation method with *AOX1*-defective hosts, methanol is continually added to maintain a concentration of about 0.4%. These cultures grow slowly on methanol, since they are reliant on the weaker *AOX2* gene for the oxidation of methanol. However, they are still metabolically active and able to induce high levels of heterologous protein synthesis. Most cultures are harvested after less than two generations on methanol, near the point at which foreign protein accumulation begins to level off. This fed-batch fermentation method has been used to produce a number of foreign proteins, including β-galactosidase (Cregg and Madden 1988), hepatitis B surface antigen (Cregg et al. 1987), tumor necrosis factor (Sreekrishna et al. 1989), and invertase (Tschopp et al. 1987b). With each of these *P. pastoris* strains, this growth and production

13.1 Vectors and Hosts Used for Heterologous Gene Expression 311

regimen resulted in consistent yields of foreign protein and scaled up easily from shake flask to high-density fermentor cultures (Cregg et al. 1989b; Siegel and Brierley 1989).

The second and more recently developed fermentation scheme utilizes a *P. pastoris* host that is wild type for methanol utilization and a production phase in which cultures are fed methanol at a growth-limiting rate (Digan et al. 1989). The generation of biomass on glycerol is followed by a two-part production stage. In the first part, methanol is added in a fed-batch mode to the culture. During this period, foreign protein production is induced, the culture density is further increased, and the cells are acclimated to growth on methanol. The second part is a continuous-culture mode in which medium is added at a growth-limiting rate and the fermentation broth is harvested at an equal rate. The growth rate optimal for maintaining foreign protein production at the highest rate was determined empirically by changing the feed rate and monitoring the effect on volumetric productivity (grams of product per liter of culture per hour).

The fed-batch and continuous-culture fermentation schemes have been compared for the production of bovine lysozyme (Digan et al. 1989; Siegel and Brierley 1989). Both schemes result in lysozyme concentrations of approximately 300 mg/liter of medium in cultures at a cell density of about 100 gm of cells (dry weight) per liter. However, the volumetric productivity of the continuous culture scheme is approximately five times that of the fed-batch scheme employed with the *AOX1*-defective host. At 75 h after the addition of methanol, a mutant host grown in a batch culture of about 10 liters at a density of 75 gm (dry weight) per liter yields about 2 gm of lysozyme, whereas a similar-volume culture of the wild-type host grown in a continuous-culture fermentor at a slightly higher average density (100 gm dry weight per liter) generates approximately 8 gm of lysozyme by 75 h after the methanol shift (Digan et al. 1989).

Both of these approaches yield substantially higher levels of foreign protein than do methods which involve the growth of wild-type expression strains on methanol at non-growth-limiting rates. An explanation for the success of the schemes has not been proven. It is suggested that the key in both approaches is that the cells experience carbon-limited growth conditions. With the *AOX1*-defective host, this condition results from the mutation which reduces the ability of the strain to utilize methanol. In the continuous-culture method, cells fully able to utilize methanol are fed methanol at a slow rate. It is known that methylotrophic yeasts grown in methanol-limited cultures contain alcohol oxidase levels which are significantly higher than those of yeasts grown in the presence of excess methanol (Veenhuis et al. 1983). This observation suggests that *AOX1* expression may be substantially repressed in methanol-excess cultures of *P. pastoris*.

The existence of at least two highly productive fermentation methods for *P. pastoris* provides the potential manufacturer of a foreign protein with a choice. Continuous culturing holds the potential of superior productivity, while batch culturing with *AOX1*-defective hosts is technically easier to perform and has considerable tolerance for variations in and perturbances of the growth conditions.

13.2 EXPRESSION OF HETEROLOGOUS PROTEINS IN THE CYTOPLASM

13.2.1 *E. coli* β-Galactosidase

Escherichia coli β-galactosidase was the first heterologous protein expressed and studied in a methylotrophic yeast (Tschopp et al. 1987a). The protein was chosen as a model system to examine the regulation and relative strength of the *AOX1* and *DAS* promoters when forming gene fusions between these control regions and the *E. coli* β-galactosidase gene. The *Pichia pastoris* strain used in the experiment contains a single integrated copy of an *AOX1-lacZ* fusion vector (Tschopp et al. 1987a; Cregg and Madden 1988) and produces a protein composed of the first 15 amino acids of alcohol oxidase fused to the ninth amino acid of β-galactosidase. As observed with alcohol oxidase, no β-galactosidase is expressed in extracts prepared from cells grown on either glucose or glycerol. However, in methanol-grown cells, alcohol oxidase and β-galactosidase are the most prominent protein species in the extracts. These experiments provided the first insight into *AOX1* promoter regulation. The regulation of expression was found to rely on both catabolite repression-derepression and methanol-induction mechanisms (Tschopp et al. 1987a). Under the best-known derepressing conditions (carbon starvation), expression levels reached only 2-4% of the maximal levels observed in methanol.

The level of β-galactosidase synthesized in *P. pastoris* is comparable to that observed in the best *E. coli* strains. Differences, however, in β-galactosidase expression levels among *P. pastoris* hosts exist. In an *AOX1*-defective host, approximately 20% of the total soluble protein can be β-galactosidase, while *AOX1* wild-type strains only produce about 5% (Table 13-1) (Tschopp et al. 1987a; Cregg and Madden 1988). After the shift to methanol, AOX activity and mRNA in the wild-type strain increased and reached a peak at approximately 15 h before dropping fivefold to a constant level. The relatively sharp rise and fall in *AOX* message and activity levels correspond to the time when the culture initiated growth on methanol, and thus this peak may be the consequence of temporary overexpression of AOX during a period of metabolic adjustment to the new carbon source. Interestingly, β-galactosidase activity levels in this wild-type *AOX1-lacZ* fusion strain did not display a similar kinetics. These differences in the relative kinetics and expression levels of enzymes from the same promoter have been observed in many instances of heterologous gene expression in general and are thought to be due to differences in message stability or translatability, or to the potential existence of regulatory sequences in the protein-

TABLE 13-1 Expression Levels of Foreign Proteins Produced in *P. pastoris* Hosts

Expression Mode	*Host AOX1 Type*[1]	*Protein Expressed*	*Concentration (g/liter culture)*[2]	*% of Total Protein*[3]
Cytoplasmic	Def	β-Galactosidase	ND	20
	WT	β-Galactosidase	ND	5
	Def	Tumor necrosis factor	8	25
	Def	Hepatitis B surface antigen	0.3	3
	WT	Hepatitis B surface antigen	ND	0.3
	Def	Streptokinase	0.08	<1
Secreted	Def	Invertase	2–3	80–90
	Def	Bovine lysozyme	0.2–0.3	80
	WT	Bovine lysozyme	0.2–0.3	80

[1] Def, *AOX1*-defective strain; WT, wild-type strain.
[2] ND, not determined.
[3] For cytoplasmically expressed proteins, the numbers represent the percentage of total soluble protein. For secreted proteins the number is the percentage of total protein in the culture medium.

coding portion of the *AOX1* gene. The *AOX1* defective strain which expresses AOX from the transcriptionally less efficient *AOX2* gene, contained approximately one-third of the AOX activity of the wild-type strain. On the other hand, β-galactosidase activity in this host increased after 50 h in methanol, to a level four times that of the wild-type strain.

As yet there is no explanation for the enhancement of *AOX1*-promoter transcription in an *AOX1*-defective background. It is possible, however, that the effect is the result of increased derepression of the *AOX1* promoter in mutant *P. pastoris* cells over that in wild-type cells (Cregg et al. 1989b). Two observations support this notion that *AOX1* may be partially catabolite-repressed (i.e., repressed by a metabolite produced by methanol utilization) during rapid growth of wild-type cells on methanol. First, in wild-type cells at early times after the introduction of methanol, AOX activity rises to a level that is four to five times higher than in *P. pastoris* cells actively growing on methanol. Second, when methylotrophic yeast cells are fed methanol at growth-limiting rates, AOX levels are significantly higher than in cells fed methanol at non-growth-limiting rates (Veenhuis et al. 1983). Both of these observations suggest that the *AOX1* promoter is capable of higher transcriptional levels than those observed in cultures grown in standard methanol excess (non-growth-rate-limiting). To limit the growth rate of *P. pastoris* on methanol and thereby attempt to obtain induced and "fully" derepressed expression of the *AOX1* promoter, we took advantage of the presence of the "weaker" *AOX2* gene to limit availability of AOX protein.

13.2.2 Hepatitis B Surface Antigen

The surface antigen gene of the hepatitis B virus encodes a hydrophobic protein (HBsAg) of approximately 24 kDa. Approximately 100 monomers of HBsAg assemble into an antigenic particle of approximately 2.2×10^3 kDa, which is 22 nm in diameter (Tiollais et al. 1981). This assembly does not occur in bacteria but does, to a limited extent, occur in *S. cerevisiae*. For expression in *P. pastoris*, a vector was constructed in which an HBsAg gene fragment composed only of protein-coding sequences was placed between the *AOX1* promoter and the *AOX1* transcriptional terminator (Figure 13-1A). The expression cassette was integrated into the *AOX1* locus by both an "addition-type" integration and by replacement of the *AOX1* gene. A comparison of the two strains revealed that the *AOX1* wild-type strain produced only 10% as much HBsAg as the *AOX1*-defective strain. The slower-growing *AOX1*-defective strain was much more productive than the wild-type expression strain. Batch fermentations ran between 150–200 h and yielded 22-nm-particle levels (by Ausria assay) of 2–3% soluble protein (Cregg et al. 1987). Volumetric yields were routinely between 400–600 mg per liter (Table 13-1). These represent significant increases over expression results reported using *S. cerevisiae*-based production systems, which produce less than 1 mg HBsAg/liter (Hitzeman et al. 1983). Fermentation scale-up from one liter to 240 liters at various cell densities was achieved in a short period of development time without loss in HBsAg productivity and was extremely reproducible and easy to control. The molecular weight by SDS-PAGE of the recombinant HBsAg was identical to native HBsAg isolated from serum of hepatitis B carriers (Table 13-2). All of the detectable 24-kDa monomer appeared to be particle associated (Table 13-2). The majority of the particle sediments on a 20–80% sucrose equilibrium gradient at a density similar to the normal, blood-borne, 22-nm particle with a density

TABLE 13-2 Size and Post-Translational Modifications of Proteins Produced in *P. pastoris* Hosts

Protein	Molecular Weight (kDa)	N-Terminal Processing[1]	N-Linked Glycosylation	Biological Activity
β-Galactosidase	~110	ND	None	Active
Tumor necrosis factor	17.4	No methionine	None	$1-2 \times 10^7$ U/mg
Hepatitis B surface antigen	24	No methionine	None	22-nm particle; immunogenic
Streptokinase	47	ND	None	Active
Invertase	85–90	Accurate	31–36%; Man_{8-14}[2]	3,000 U/mg
Bovine lysozyme	14	Accurate	None	Active

[1] ND, not determined.
[2] Man_{8-14}, 8–14 mannose residues per oligosaccharide chain.

of 1.16 g/ml (Cregg et al. 1987). This density is characteristic of a high-molecular-weight protein-lipid complex containing approximately 30% lipid. The high degree of particle formation contrasts with the situation found in *S. cerevisiae* where monomer production is high relative to particle formation (Hitzeman et al. 1983).

13.2.3 Streptokinase

Streptokinase is a potent plasminogen activator with widespread clinical use as a thrombolytic agent (Martin 1982). It is naturally produced and secreted by various strains of hemolytic streptococci from which secretion into the external medium is directed by a 26-amino acid signal peptide that is cleaved during the secretion process. The protein has a molecular weight of about 47 kDa (Table 13-2) and was found to be composed of 415 amino acid residues (Jackson et al. 1986). Recombinant DNA-based expression in *E. coli* and *Streptococcus sanquis* has been achieved with moderate success (Malke et al. 1984; Malke et al. 1985; Jackson et al. 1986). As an alternative to bacterial systems, streptokinase has been expressed in *P. pastoris* (Hagenson et al. 1989). The DNA encoding the native streptokinase protein, without the signal sequence, was inserted into a *P. pastoris* expression vector and then inserted into the *P. pastoris* host by "additive" integration (Figure 13-1C). Streptokinase production was rapidly induced upon switching to methanol medium and leveled off in about 14 h. Thereafter, the amount of streptokinase per volume was relatively constant. *P. pastoris* cells were grown to a moderate cell density of 46 g/liter on limiting methanol-feed. The amount of streptokinase expressed during continuous fermentation was determined to be about 77 mg/liter of fermentor volume (Table 13-1). During fermentation the streptokinase gene was found to be stably maintained without adverse effects on cell growth or viability. Cell lysates from producing cells activate plasminogen, as determined by casein degradation on agar plates and by hydrolysis of a synthetic ester substrate. A prominent protein species of 47 kDa was detected by SDS-PAGE and immunoblots and was identical to the size of that reported for streptokinase secreted from *Streptococcus equisimilis*.

The streptokinase expression was not optimized for high-level expression. Nevertheless, the *Pichia* system should facilitate basic studies of streptokinase functional domains and plasminogen interactions. Furthermore, this expression system may serve as a basis for the development of genetically modified streptokinase molecules with improved clinical performance.

13.2.4 Tumor Necrosis Factor

Tumor necrosis factor (TNF) is an antitumor protein secreted by macrophages (Carswell et al. 1975; Pennica et al. 1984). The mature form of TNF is a nonglycosylated protein consisting of 157 amino acid residues with a molecular weight of 17.4 kDa (Table 13-2). For expression in *P. pastoris*, the TNF gene was inserted between the 5' and 3' regulatory sequences of *AOX1* in expression vector pAO804 (Sreekrishna et al. 1988). The resulting expression vector was inserted in the *P. pastoris* host by the *AOX1* gene displacement method (Figure 13-1B). These transformants induced by methanol for TNF expression showed approximately 30% of their soluble protein as TNF in cell-free lysates (Sreekrishna et al. 1989). The TNF expression level was significantly higher than that obtained with an *E. coli* expression system (Sreekrishna et al. 1989). Expression of TNF in batch fermentations under high-cell density conditions (100 g cell dry weight) reached 6–10 g/liter within 48 h of methanol exposure and had no apparent toxic effects on *P. pastoris*. TNF from *P. pastoris* extracts was purified to greater than 95% purity in a single step by passing the extract through an immuno affinity column containing anti-TNF monoclonal antibodies (Sreekrishna et al. 1989). The recombinant TNF had the expected N-terminal amino acid sequence as reported for natural human TNF. In particular, it was evident that the N-terminal methionine was properly processed. *E. coli*-derived TNF, purified in a similar fashion, contained N-terminal methionine on about 25% of the molecules, suggesting that *E. coli* is less efficient in removing the N-terminal methionine than *P. pastoris*. TNF contained in *P. pastoris* lysates was biologically active as determined by its cytotoxic effect on murine L-929 fibroblast cells (Table 13-2). The specific activity of TNF purified from *P. pastoris* ($1-2 \times 10^7$ U/mg) was identical to that isolated from *E. coli* (Sreekrishna et al. 1989). Most importantly, unlike in *E. coli*, the TNF produced in yeast was completely soluble and could be readily extracted from mechanically disrupted cells under nondenaturing conditions.

The level of TNF expression is remarkable and is the highest level ever reported for expression of a cytoplasmic protein. At about 30% of the soluble protein, the expression level of TNF is about the same as that of alcohol oxidase in wild-type *P. pastoris*. Taking into account that the monomeric size for TNF is about four times smaller than that of alcohol oxidase, the molar amount of TNF is in fact four times greater. The *P. pastoris* TNF expression level is also higher than the highest reported level for TNF expression in *E. coli* (Yamada et al. 1985). Similar cytoplasmic expression levels in *S. cerevisiae* have been described for only a few proteins, namely for intracellular expression of hepatitis B core antigen present as 28-nm particles (Kniskern et al. 1986) and human superoxide dismutase (N-acetylated protein) (Hallewell et al. 1987).

13.3 EXPRESSION AND SECRETION OF HETEROLOGOUS PROTEINS

The majority of proteins produced via recombinant DNA technology are naturally occurring secretory proteins or cell-surface proteins, such as viral glycoproteins. A choice therefore exists as to whether to produce such heterologous proteins as cytoplasmic proteins or to synthesize them by transit through the secretory pathway into the growth medium. There are several advantages to producing heterologous proteins as secreted proteins: These include the higher degree of initial purity and the ease of purification of the protein from the medium, the increased attainment of the correct conformation of the protein, and the increase in ability to attain the proper posttranslational modification of the protein.

For secretion of heterologous proteins, the *Pichia* system has several advantages over other expression systems. *P. pastoris* cells grow in a protein-free and nutrient-free, defined minimal medium and, therefore, only proteins normally secreted by *Pichia* contribute to the background of other proteins in the culture medium. This situation contrasts favorably with production in rich medium based on yeast extract or peptone, or with serum-based medium, such as that used for the growth of mammalian cells. Furthermore, levels of endogenous protein secretion are quite low in *P. pastoris*. This characteristic differs from bacterial and fungal hosts, such as *Bacillus subtilis* and *Aspergillus,* which secrete large amounts of endogenous proteins, particularly proteases which may degrade protein products. Despite or perhaps because of this low level of endogenous protein secretion, *P. pastoris* cells have the capacity to secrete large quantities of proteins. Efficient secretion of proteins from *Pichia* holds the promise of yielding a product of high purity secreted into a simple protein-poor medium.

13.3.1 Invertase from *Saccharomyces cerevisiae*

Invertase was chosen as a model system to examine the secretion mechanism and post-translational modification in *P. pastoris*. This enzyme has been extensively studied and has been used as a model protein to study secretion and post-translational events in *S. cerevisiae* (Schekman and Novick 1982). Knowledge of invertase expression in *P. pastoris* allows direct comparison of the two different yeasts as hosts for secretion of heterologous proteins.

Invertase is a secreted molecule residing in the periplasmic space of *S. cerevisiae*. It is synthesized as a precursor with a 19-amino acid signal sequence and is secreted as a glycoprotein containing, on average, 10 N-linked oligosaccharide moieties (Trimble and Maley 1977). The nonglycosylated polypeptide has a molecular mass of 60 kDa (Perlman et al. 1982). In *Saccharomyces,* the protein exists as a heterogeneously glycosylated species of 110–180 kDa. The coding sequence for *S. cerevisiae* invertase, including its signal sequence, was inserted behind the *AOX1* promoter. When this expression vector was integrated into the *Pichia* genome in a *AOX1*-deficient host and these cells were cultured in methanol, invertase was secreted into the growth medium. During growth in methanol, active invertase accumulated to a level of 2–3 g/liter of culture (Tschopp et al. 1987b). Virtually all the invertase activity was associated with the medium ($\sim$80% of total) and periplasm ($\sim$20% of total), indicating that this heterologous protein can efficiently traverse the secretory pathway. High-level expression and efficient secretion of invertase resulted in a high degree of initial purity in the fermentation medium, which was estimated at greater than 80% of total medium protein (Tschopp et al. 1987b). Purification of the molecule to homogeneity with a minimal number of steps and a high level of recovery was readily achieved.

A large difference in apparent molecular size was observed between the invertase molecules secreted by *S. cerevisiae* and those secreted by *P. pastoris*. The difference in size was due to oligosaccharide addition, since enzymatic removal of oligosaccharide from invertase secreted by both yeasts produced deglycosylated protein of the same size, approximately 60 kDa (Tschopp et al. 1987b). As determined by SDS-PAGE, glycosylated and enzymatically deglycosylated invertase differed in size by 20-25 kDa in *P. pastoris* (Table 13-2). If distributed evenly over 10 glycosylation sites used by *Saccharomyces*, this mass of carbohydrate would indicate an average of 8-12 mannose residues per oligosaccharide chain. In contrast, *Saccharomyces* invertase contains high-mannose-type oligosaccharides, which are extensively modified by the addition of 50 or more mannose residues per chain (Ballou 1982). Thus, the average size of oligosaccharide chains present on the molecules made in *P. pastoris* was considerably smaller than those on invertase from *Saccharomyces*. The specific activity of invertase purified from the medium of *Pichia* transformants was approximately 3,000 U/mg of protein, a value comparable to that obtained with invertase from *Saccharomyces* (Table 13-2) (Trimble and Maley 1977). This suggests that the tertiary structure of the protein has been maintained during secretion from *Pichia*, and that the novel form of glycosylation was compatible with full enzymatic activity. The N-terminal protein sequence was identical to the sequence of the enzyme from *Saccharomyces*. Thus, signal peptide recognition and removal were accurate in *P. pastoris* for this protein.

13.3.2 Bovine Lysozyme

Bovine stomach lysozyme is an approximately 14-kDa nonglycosylated protein containing four putative disulfide bonds (Table 13-2) (Jolles and Jolles 1984). The gene encoding bovine lysozyme, along with its signal peptide-coding sequence, was isolated as a cDNA clone, and inserted into the generalized *Pichia* expression vector pAO804 (Figure 13-1A) (Digan et al. 1989). A bovine lysozyme-expressing, *AOX1*-deficient host synthesized approximately 200-250 mg/liter of bovine lysozyme as a secreted product over the course of 180 h when grown at a cell density of 130 g/liter of dry weight (Table 13-1) (Digan et al. 1989). Greater than 90% of the total bovine lysozyme synthesized is present in the medium. Moreover, bovine lysozyme is the major component of the medium. It can be purified to homogeneity quickly using a two-step procedure involving a concentration step and ion-exchange chromatography. The N-terminal sequence and specific activity of purified bovine lysozyme secreted from *P. pastoris* cells are identical to those of the native enzyme and, as expected, the protein was not glycosylated (Table 13-2). Bovine lysozyme was the first mammalian protein produced by continuous, high-cell density fermentation from a recombinant microorganism.

13.4 INTRACELLULAR TRANSPORT AND POST-TRANSLATIONAL MODIFICATIONS

Results from the expression of invertase and bovine lysozyme demonstrate that for the secretion of these proteins, the signal sequence native to the protein is satisfactory. Both invertase and bovine lysozyme are efficiently secreted using their own signal sequences (Tschopp et al. 1987b; Digan et al. 1988 and 1989). When secreted from *P. pastoris*, the native foreign proteins have an amino-terminal sequence identical to that of the protein obtained from the native host. Thus, for many proteins secreted from *P. pastoris*, it is not necessary to replace foreign signal sequences with a signal sequence derived from a *P. pastoris*-secreted protein.

The glycosylation of proteins in *P. pastoris* is similar in several respects to that observed in *S. cerevisiae*. In both organisms the majority of the oligosaccharide chains added to proteins as they pass through the secretory pathway are the N-asparagine-linked, high-mannose type (Reddy et al. 1988; Ziegler et al. 1988). However, there is at least one major difference; invertase secreted from *P. pastoris* is clearly smaller and more homogeneous in size than invertase secreted from *S. cerevisiae* (Tschopp et al. 1987b). Whereas invertase from *S. cerevisiae* ranges in size from 100 to 140 kDa, the same protein secreted from *P. pastoris* migrates at about 85 to 90 kDa (Table 13-2). An analysis of these invertases shows that their N-terminal sequences and mobility on SDS-PAGE after deglycosylation are identical (Tschopp et al. 1987b) and that each contains an average of approximately 10 oligosaccharide chains per molecule (Grinna and Tschopp 1989). The difference in size is due to shorter average length of these chains (Grinna and Tschopp 1989). For *P. pastoris* invertase, approximately 85% of the oligosaccharide chains are in the size range of $Man_{8-14}GlcNAc_2$ (Man, mannose; GlcNAc, N-acetylglucoseamine), whereas only about 20% of the *S. cerevisiae* chains are in this size range (Reddy et al. 1988). Further analysis of *P. pastoris* oligosaccharides obtained from invertase and from total endogenous glycoproteins reveals that *P. pastoris* can extend mannose chains, but even the longest of these (approximately 30 mannose residues per chain; Grinna and Tschopp 1989) is significantly shorter than the structures typically found on *S. cerevisiae* glycoproteins (50 to 150 mannose residues per chain; Ballou 1982).

13.5 DISCUSSION

The *P. pastoris* expression system appears to be ideally suited for the production of a variety of proteins on a commercial scale. The essential elements of the system include the ability to integrate foreign DNA into the *P. pastoris* genome in a controlled manner and the *AOX1* promoter, which is tightly regulated and efficiently transcribed. The unusual strength of the *AOX1* promoter results in high expression levels, even though many expression strains contain only a single integrated copy of a foreign gene expression cassette. Integrated cassettes are stable and not subject to the problems associated with *S. cerevisiae* plasmid-based expression hosts. Another important aspect of the system is that reliable methods for the growth of recombinant expression hosts in high-density, large-volume fermentors have been developed.

The outlook for the secretion of heterologous gene products, using the *P. pastoris* expression system, appears to be particularly promising. The secreted products are found in the culture medium and are not retained intracellularly or in the periplasm. The concentration of the product in the medium is high, both in an absolute sense and relative to endogenous secreted proteins. As a result, the foreign protein products appear to be virtually the only protein in culture broth, a clear advantage in reducing the time and steps required for processing and purification.

The *P. pastoris* expression system should be particularly advantageous for manufacture of recombinant proteins which are required in large quantities. The yield of product is high, the cost of the growth medium is low, and the transition from shake flask to high-density, large-volume fermentors is consistently smooth. These properties might prove to be useful for the production of enzymes which perform commercially interesting reactions or which have superior kinetic properties, pH optima, or thermal stability characteristics.

REFERENCES

Anthony, C. (1982) in *The Biochemistry of Methylotrophs* (Anthony, C., ed.), pp. 269–295, New Academic Press, Inc., New York.

Ballou, C.E. (1982) in *The Molecular Biology of the Yeast Saccharomyces: Metabolism and Gene Expression* (Strathern, J.N., Jones, E.W., and Broach, J.R., eds.), pp. 335–360, Cold Spring Harbor Laboratory, Cold Spring Harbor, NY.

Carswell, E.A., Old, L.J., Kassel, R.L., et al. (1975) *Proc. Natl. Acad. Sci. USA* 72, 3666–3670.

Couderc, R., and Baratti, J. (1980) *Agric. Biol. Chem.* 44, 2279–2289.

Cregg, J.M., and Madden, K.R. (1987) in *Biological Research on Industrial Yeasts*, vol. 2 (Steward, G.G., Russel, I., Klein, R.D., and Hiebsch, R.R., eds.), pp. 1–18, CRC Press, Inc., Boca Raton, FL.

Cregg, J.M., and Madden, K.R. (1988) *Dev. Ind. Microbiol.* 29, 33–41.

Cregg, J.M., and Madden, K.R. (1989) *Mol. Gen. Genet.* 219, 320–323.
Cregg, J.M., Barringer, K.J., Hessler, A.Y., and Madden, K.R. (1985) *Mol. Cell. Biol.* 5, 3376–3385.
Cregg, J.M., Tschopp, J.F., Stillman, C.A., et al. (1987) *Bio/Technology* 5, 479–485.
Cregg, J.M., Madden, K.R., Barringer, K.J., Thill, G.R., and Stillman, C.A. (1989a) *Mol. Cell. Biol.* 9, 1316–1323.
Cregg, J.M., Digan, M.E., Tschopp, J.F., et al. (1989b) in *Genetics and Molecular Biology of Industrial Microorganisms* (Hershberger, C.L., Queener, S.W., and Hegeman, G., eds.), pp. 343–352, American Society for Microbiology, Washington, DC.
Digan, M.E., Tschopp, J.F., Grinna, L.S., et al. (1988) *Dev. Ind. Microbiol.* 29, 59–65.
Digan, M.E., Lair, S.V., Brierley, R.A., et al. (1989) *Bio/Technology* 7, 160–165.
Ellis, S.B., Brust, P.F., Koutz, P.J., et al. (1985) *Mol. Cell. Biol.* 5, 1111–1121.
Grinna, L.S., and Tschopp, J.F. (1989) *Yeast* 5, 107–115.
Hagenson, M.J., Holden, K.A., Parker, K.A., et al. (1989) *Enzyme Microb. Technol.* 11, 650–656.
Hallewell, R.A., Mills, R., Olsen, P.T., et al. (1987) *Bio/Technology* 5, 363–366.
Harder, W., and Veenhuis, M. (1989) in *The Yeasts,* vol. 3 (Rose, A.H., and Harrison, J.S., eds.), pp. 289–316, Academic Press, San Diego, CA.
Hinnen, A., Hicks, J.B., and Fink, G.R. (1978) *Proc. Natl. Acad. Sci. USA* 75, 1929–1933.
Hitzeman, R.A., Chen, C.Y., Hagie, F.E., et al. (1983) *Nucleic Acids Res.* 11, 2745–2763.
Ito, H., Murata, K., and Kimura, A. (1983) *Agric. Biol. Chem.* 48, 341–347.
Jackson, K.W., Malke, H., Gerlach, D., Ferretti, J.J., and Tang, J. (1986) *Biochemistry* 25, 108–114.
Jolles, P., and Jolles, J. (1984) *Mol. Cell. Biochem.* 63, 165–189.
Kingsman, S.M., Kingsman, A.J., and Mellor, J. (1987) *TIBTECH* 5, 53–57.
Kniskern, P.J., Hagopian, A., Montgomery, D.L., et al. (1986) *Gene* 46, 135–141.
Koutz, P., Davis, G.R., Stillman, C., et al. (1989) *Yeast* 5, 167–177.
MacKay, V.L. (1987) in *Biological Research on Industrial Yeasts*, vol. 2 (Steward, G.G., Russel, I., Klein, R.D., and Hiebsch, R.R., eds.), pp. 27–36, CRC Press, Inc., Boca Raton, FL.
Malke, H., Gerlach, D., Hohler, W., and Ferretti, J.J. (1984) *Mol. Gen. Genet.* 196, 360–363.
Malke, H., Roe, B., and Ferretti, J.J. (1985) *Gene* 34, 357–362.
Martin, M. (1982) in *Streptokinase in Chronic Arterial Disease* (Martin, M., and Auel, H., eds.), p. 1, CRC Press, Inc., Boca Raton, FL.
Moonen, P., Mermod, J.J., Ernst, J.F., Hirschi, M., and Delamarter, J.F. (1987) *Proc. Natl. Acad. Sci. USA* 84, 4428–4431.
Pennica, D., Nedwin, G.E., Hayflick, J.S., et al. (1984) *Nature* 312, 724–729.
Perlman, D., Halvorson, H.O., and Cannon, L.E. (1982) *Proc. Natl. Acad. Sci. USA* 79, 781–785.
Reddy, V.A., Johnson, R.S., Biemann, K., Williams, R.S., and Ziegler, F.D. (1988) *J. Biol. Chem.* 263, 6978–6985.
Rothstein, R.J. (1983) *Methods Enzymol.* 101, 202–210.
Schekman, R., and Novick, P. (1982) in *The Molecular Biology of the Yeast Saccharomyces: Metabolism and Gene Expression* (Strathern, J.N., Jones, E.W.,

and Broach, J.R., eds.), pp. 363-393, Cold Spring Harbor Laboratory, Cold Spring Harbor, NY.
Schultz, L.D., Tanner, K.J., Hoffman, E.A., et al. (1987) *Gene* 54, 113-123.
Shuster, J.R. (1987) in *Biological Research on Industrial Yeast*, vol. 2 (Steward, G.G., Russel, L., Klein, R.D., and Hiebsch, R.R., eds.), pp. 19-25, CRC Press, Inc., Boca Raton, FL.
Siegel, R.S., and Brierley, R.A. (1989) *Biotech. Bioeng.* 34, 403-404.
Sreekrishna, K., Rica, H., Potenz, B., et al. (1988) *J. Basic Microbiol.* 28, 265-278.
Sreekrishna, K., Nelles, L., Potenz, R., et al. (1989) *Biochemistry* 28, 4117-4125.
Tiollais, P., Charnay, P., and Vyas, G.N. (1981) *Science* 213, 406-411.
Trimble, R.B., and Maley, F. (1977) *J. Biol. Chem.* 252, 4409-4412.
Tschopp, J.F., Brust, P.F., Cregg, J.M., Stillman, C.A., and Gingeras, T.R. (1987a) *Nucleic Acids Res.* 15, 3859-3876.
Tschopp, J.F., Sverlow, G., Kosson, R., Craig, W., and Grinna, L. (1987b) *Bio/Technology* 5, 1305-1308.
Van Arsdell, J.N., Kwok, S., Schweickart, V.L., et al. (1987) *Bio/Technology* 5, 60-64.
Veenhuis, M., and Harder, W. (1987) in *Peroxisomes in Biology and Medicine* (Fahimi, H.D., and Sies, H., eds.), pp. 437-460, Springer-Verlag KG, Berlin.
Veenhuis, M., Van Dijken, J.P., and Harder, W. (1983) *Adv. Microb. Physiol.* 24, 1-82.
Wegner, E.H. (1983) U.S. patent 4,414,329.
Yamada, M., Furutani, Y., Notake, M., et al. (1985) *J. Biotech.* 3, 141-153.
Ziegler, F.D., Maley, F., and Trimble, R.B. (1988) *J. Biol. Chem.* 263, 6986-6992.

PART V

Ecology

CHAPTER 14

Ecology of Methylotrophic Bacteria

R.S. Hanson
E.V. Wattenberg

Kaserer (1906) and Sohngen (1906) first reported the existence of methane-oxidizing bacteria. The ubiquity of these organisms and their importance to the conservation of organic material was generally recognized by 1920 (Aiyer 1920). Later, Hutton and Zobell (1949) succeeded in isolating pure cultures of methane-oxidizing bacteria from several sources including gas field soils, beach sand, and mud from marine and freshwater samples.

Until the studies of Whittenbury et al. (1970), the only bacteria known to oxidize methane were the obligate methanotrophs (bacteria that grow aerobically on methane as their sole carbon and energy source), *Methanomonas methanica*, later renamed *Methylomonas methanica* (Whittenbury and Krieg 1984), and *Methylococcus capsulatus* (Foster and Davis 1966). In 1970, Whittenbury et al. (1970) reported the isolation of over 100 strains of methane-utilizing bacteria and classified them into five genera. Although Whittenbury et al. (1970) considered this classification premature, subsequent investigators regarded this grouping as justifiable not only for Whittenbury's original isolates but also for more recently described methylo-

---

Preparation of this manuscript and previously unreported research results were funded in part by a grant to R.S.H. from the National Science Foundation (NSF/BR-8903833-01).

trophs (Gal'chenko et al. 1978 and 1986; Gal'chenko and Andreev 1984; Hanson et al. 1990; Trotsenko 1983). Current classification of methanotrophs—by phospholipid fatty acid composition, DNA-DNA homology, biochemical characterization, fine structure, resting stage, 5S and 16S ribosomal RNA sequence analysis, and protein profile from SDS polyacrylamide gel electrophoresis—support the division of the methanotrophs into the five original genera (Gal'chenko et al. 1986; Gal'chenko and Andreev 1984; Hanson et al. 1990).

The five genera fall into two major groups as described in detail elsewhere in this volume. Roughly, type I methanotrophs including *Methylococcus, Methylomonas*, and *Methylobacter*, contain stacks of disc-shaped membrane structures which are distributed throughout the cell. By contrast, type II methanotrophs *Methylosinus* and *Methylocystis*, have paired intracytoplasmic membranes that are found primarily at the periphery of the cell. Furthermore, type I organisms assimilate formaldehyde through the ribulose monophosphate (RMP) pathway, whereas type II organisms assimilate formaldehyde through the serine pathway (Whittenbury and Dalton 1981; Whittenbury and Krieg 1984). Bacteria in each group share several other unique characteristics (Whittenbury and Krieg 1984). *Methylococcus capsulatus* has been separated into a third group (type X) because its DNA has a higher G&C content and it also contains ribulose bisphosphate carboxylase (Whittenbury and Krieg, 1984).

14.1 ROLE OF METHANE IN THE BIOSPHERE

Methanotrophs form the major biological sink for methane in the biosphere (Bouwman 1989; Cicerone and Oremland 1988). Probably the best-characterized example of their link in the methane cycle takes place in stratified, eutrophic, freshwater lakes (Abramochinka et al. 1987; Anthony 1982; Hanson 1980; Harrits and Hanson 1980; Rudd and Hamilton 1978; Rudd and Taylor 1980). The upper layer of the lake, the epilimnion, receives sunlight, is well-oxygenated, and harbors photosynthetic phytoplankton. These organisms can fix carbon dioxide for the production of oxygen and the synthesis of organic material. Among the recipients and utilizers of this organic material are the methanogens, anaerobic producers of methane that grow in the sediment in the lower layer of the lake, the hypolimnion. Methanotrophs reside in the metalimnion, an intermediate zone that is characterized by a steep temperature, oxygen, and methane gradient. The methane released by the anaerobic methanogens diffuses up the metalimnion. There the aerobic methanotrophs oxidize the methane and produce cell material and carbon dioxide which help sustain the phytoplankton and grazers in the food chain. Understanding the role of methanotrophs in recycling methane has become increasingly important since methane has been identified as a significant greenhouse gas.

Methane, like the other greenhouse gases in the earth's atmosphere including CO_2, CO, and NO_2, has the dual property of allowing the sun's energy to penetrate the earth's atmosphere while blocking the escape of the earth's infrared radiation. By trapping the earth's radiation, greenhouse gases may contribute to global warming (Schneider 1989). Methane is coming into prominence as a greenhouse gas because it absorbs infrared radiation more efficiently than CO_2, the most abundant greenhouse gas, and its concentration in the atmosphere is increasing rapidly (Bouwman 1989; Graedel and Crutzen 1989; Pearce 1989). In addition, methane can effect the concentration of CO_2, ozone, hydroxyl radicals, and CO in the troposphere and the concentration of chlorine in the stratosphere (Cicerone and Oremland 1988; Johnston 1984). Finally, atmospheric methane has a relatively long half-life of 9.6 years (Cicerone and Oremland 1988).

The rise in atmospheric methane roughly correlates in time with the industrial revolution and the growth in human population (Cicerone and Oremland 1988; Graedel and Crutzen 1989). This trend points to a likely role of humankind in influencing methane levels in the atmosphere. The concentration of methane appears to have remained close to 700 ppb from the end of the last ice age (approximately 10,000 years ago), up until 300 years ago (Cicerone and Oremland 1988; Graedel and Crutzen 1989). At that time, methane levels began to rise, accelerating rapidly over the last century. The current methane concentration has been estimated at 1,700 ppb and appears to be increasing at the alarming rate of 1% per year (Cicerone and Oremland 1988; Graedel and Crutzen 1989).

Humans could influence the release of methane through many avenues. For example, out of the 400–600 $\times$ 10^{12} g of methane released per year, tapping natural gas and coal mining yield an estimated 30–90 $\times$ 10^{12} g of methane per year (Bouwman 1989; Cicerone and Oremland 1988; Pearman and Fraser 1988). Microbial production of methane from landfills produces another 30–70 $\times$ 10^{12} g methane per year (Bouwman 1989; Cicerone and Oremland 1988; Pearman and Fraser 1988). Yet many models indicate that the majority of the methane released per year comes from agricultural sources (Bouwman 1989; Cicerone and Oremland 1988; Pearman and Fraser 1988; Seiler 1983). Together, rice paddies, livestock, and the burning of tropical forests and grasses to create more farmland have been estimated to produce almost 400 $\times$ 10^{12} g methane per year (Bouwman 1989; Cicerone and Oremland 1988; Pearman and Fraser 1988; Seiler 1983).

Cattle and other ruminants harbor methanogenic bacteria. These microbes contain cellulolytic enzymes for the primary degradation of the polymers ingested by the animals. As a byproduct of their metabolism these bacteria produce methane. A 500-kg domestic cow may release 200 liters of methane per day (Cicerone and Oremland 1988). Even termites, which have a similar relationship with methanogens, have been implicated as a potentially significant source of atmospheric methane (Cicerone and Oremland 1988; Pearce 1989; Seiler 1983).

The cultivation of rice contributes to atmospheric methane by amplifying a natural mechanism for bypassing the major biological methane sink. Due to their vascular system, under some conditions rice and other plants can essentially siphon the methane from the soil (Pearce 1989). Consequently, the major fraction of methane produced in rice paddies is released to the air instead of being recycled by methanotrophs.

Although methanotrophs are an important methane sink in soil and water, the major methane sink in the atmosphere is oxidation by hydroxyl radicals in the troposphere (Cicerone and Oremland 1988). By-products of the oxidation of methane by hydroxyl radicals include CO_2 and CO. Therefore, an increase in atmospheric methane may indirectly contribute to the greenhouse effect by increasing the concentration of other greenhouse gases. The concentration of atmospheric methane may also be influenced by the concentration of CO, a product of automobile exhaust as well as of methane oxidation. Because hydroxyl radicals are also a sink for CO, increases in CO reduce the OH available for methane oxidation. Conversely, an increase in methane can reduce the concentration of hydroxyl radicals available for the removal of CO and other chemicals.

Calculations of the contribution to atmospheric methane from various sources and sinks are based largely on estimates (Cicerone and Oremland 1988). The grave potential of methane as a greenhouse gas warrants further study in order to pinpoint the sources of increased methane and to develop the means to control them.

14.2 ECOLOGICAL IMPORTANCE OF SOLUBLE METHANE MONOOXYGENASE IN METHANOTROPHIC BACTERIA

Methanotrophs employ a monooxygenase to catalyze the first reaction in the oxidation of methane to methanol. This methane monooxygenase has received growing interest as a potentially valuable tool for industrial scale biotransformation and ecologically important biodegradation reactions. Both a membrane-bound methane monooxygenase (mMMO) and a soluble methane monooxygenase (sMMO) have been identified (Dalton and Higgins 1987). Differences between the enzymes with respect to structure, sensitivity to inhibitors, and substrate specificity suggest that the soluble enzyme does not result from a simple modification of the particulate enzyme (Burrows et al. 1984; Dalton et al. 1983; Dalton and Stirling 1987; Scott et al. 1981). The particulate enzyme has eluded purification largely because enzyme activity is lost upon solubilization of the cell membrane (Burrows et al. 1984). In contrast, sMMO has been purified and characterized from both *Methylococcus capsulatus* (Bath) (Colby and Dalton 1978; Green and Dalton 1986; Woodland and Dalton 1984) and *Methylosinus trichosporium* OB3b (Fox et al. 1989; Fox and Lipscomb 1988).

14.2 Ecological Importance of Soluble Methane Monooxygenase

The soluble enzyme isolated from *Methylosinus trichosporium* OB3b has three components: Component A, a hydroxylase of mol wt 240 kDa, is a non-heme iron protein composed of three subunits in an alpha$_2$-beta$_2$-gamma$_2$ structure (Fox et al. 1989). The molecular weights for alpha, beta, and gamma are 54 kDa, 43 kDa, and 23 kDa, respectively. Component B, a putative regulatory protein, has a molecular weight of 15.8 kDa (Fox et al. 1989). Finally, a reductase called component C, is an iron-sulfur flavoprotein of mol wt 39 kDa (Fox et al. 1989). The structure of the sMMO isolated from *Methylococcus capsulatus* (Bath) resembles that of the *Methylosinus trichosporium* OB3b enzyme, although the molecular weights of the subunits differ slightly (Woodland and Dalton 1984).

The sMMO has a wider substrate specificity than the pMMO (Burrows et al. 1984). The soluble enzyme can catalyze the incorporation of oxygen into a broad range of substrates including alkanes, alkenes, aromatic hydrocarbons, alicyclic hydrocarbons, halogenated alkanes, and chlorinated aromatic hydrocarbons (Burrows et al. 1984; Dalton and Higgins 1987; Haber et al. 1983; Higgins et al. 1980). This enzyme may be useful on an industrial scale both for the production of chemicals and the breakdown of industrial pollutants. For example, sMMO might aid in the production of alcohols from alkanes and in the formation of specific epoxides for subsequent construction of polymers (Large and Bamforth 1988). In addition, sMMO can catalyze the breakdown of trichloroethylene (TCE), a major industrial pollutant (Oldenhuis et al. 1989; Tsien et al. 1989).

In *Methylosinus trichosporium* OB3b, the expression of sMMO correlates with the rapid oxidation of halogenated, low-molecular-weight hydrocarbons (Oldenhuis et al. 1989; Tsien et al. 1989). Cells containing sMMO can oxidize TCE and other halogenated, low-molecular-weight hydrocarbons at rates exceeding 10 mmol per h per g dry weight, while cells containing only the particulate enzyme do not significantly oxidize TCE.

Methylosinus trichosporium OB3b cells do not express sMMO constitutively. Instead, Western blot analysis shows that the level of sMMO rises as the concentration of copper in the media decreases. The appearance of rapid TCE oxidation follows the same trend. Although the rate of methane oxidation remains relatively constant from 0.00 to 1.00 μM copper, significant TCE degradation only appears when the copper concentration is reduced to 0.15 μM per gram dry cell weight (Tsien et al. 1989).

The mechanism by which copper regulates the expression of sMMO remains unknown (Dalton and Higgins 1987; Stanley et al. 1983). The expression of sMMO and rapid TCE oxidation has only been observed in the pure cultures of few methanotrophs, *Methylosinus trichosporium* OB3b, closely related type II methanotrophs, and *Methylococcus capsulatus* (Bath) (Green and Dalton 1989; Oldenhuis et al. 1989; Tsien et al. 1990). The development of immunological probes for sMMO or a simple colorimetric assay for the enzyme (Brusseau et al. 1990) should help determine whether methanotrophs in the environment contain the genes for sMMO and also under what conditions such genes may be expressed.

14.3 METHODS FOR ESTIMATION OF POPULATIONS OF METHANE-UTILIZING BACTERIA IN SOIL AND WATER SAMPLES

Many methods have been developed for the identification and quantitation of methylotrophs in the environment. These methods exploit the unique characteristics of these organisms which range from their ability to grow on methane to their specific nucleic acid sequences. A thorough study of methylotrophs in the environment may require a combination of different approaches.

One method for estimating the populations of methanotrophic bacteria in natural samples is by measuring the rate of conversion of $^{14}C-CH_4$ to $^{14}CO_2$ and ^{14}C-cell material (Flett et al. 1976; Harrits and Hanson 1980; Patt et al. 1974 and 1976; Rudd and Hamilton 1978; Rudd and Taylor 1980). In pure cultures, the specific rate of methane oxidation is approximately 500 nmoles/mg dry weight per minute. When 50 μM ^{14}C-methane is added to soil and water samples, a crude estimate of the biomass of methylotrophs can be calculated by dividing the observed rate of methane oxidation by the assumed specific rate. This technique is valuable for surveying large numbers of environmental samples for methanotrophs (Harrits and Hanson 1980; Rudd et al. 1974) and for comparing the metabolic activities of methanotrophs under different environmental conditions.

Viable cell counts have also been used to determine the presence of methanotrophic bacteria. Unfortunately, this method can yield misleading results due to the poor colony-forming efficiencies of methanotrophs in pure culture (Hanson 1980). Only a small fraction of the total bacterial population of soils and waters observed by direct microscopic counts can be cultured (Bone and Balkwill 1986). Plating soil and water samples involves many variables which can influence the viability of the cells. For example, some methanotrophs require non-methane-utilizing consorts to provide growth factors, remove inhibitory metabolic products, or otherwise stimulate their growth (Anthony 1982; Hanson 1980; Lamb and Garver 1980). These bacteria may not form colonies when diluted and spread on the surface of solid media. The choice of media, incubation temperature, and even impurities present in agar or methane can also create selective pressure. Given our current knowledge, it is nearly impossible to anticipate the range of growth conditions required to cultivate all of the methanotrophs in a sample (Hanson 1980; Whittenbury and Dalton 1981).

To reduce the background of colonies of non-methane-utilizing microbes, often seen during attempts to enumerate and purify methanotrophs. Gal'chenko (1975) has substituted purified silica gel for agar as a solidifying agent. Others have spread or collected microbes from water samples onto sterile filters and then placed them on agar or porous supports soaked in a mineral salts medium (Hanson 1980). Alternatively, those colonies that assimilate carbon from methane can be distinguished from others by radioautography following incubation of the samples under a ^{14}C-methane atmosphere. These techniques have been reviewed by Hanson (1980), Heyer et al. (1977), Gal'chenko et al. (1978 and 1986) and Whittenbury and Dalton (1981).

Techniques to enumerate and identify bacteria without culturing them have been reviewed by Bahlool and Schmidt (1980) and Holden and Tiedje (1988). These include the use of antibodies, enzymatic assays, phospholipid analysis, and nucleic acid probes. Rigorous application of these techniques requires good recoveries of cells and cell components from the samples. The methods for efficient recoveries and problems associated with the procedures have been described by Bahlool and Schmidt (1980) and Holden and Tiedje (1988).

Antibodies represent a potentially effective means to identify methanotrophs without culturing them. To accurately determine the number and identity of methanotrophs in a sample, the antisera must cross-react with all methanotrophs of interest, yet not detect non-methane utilizers. At present, there is insufficient information on the taxonomy of methylotrophs, the serotypes within species, and the strains that dominate in the environment to ensure preparation of an adequate collection of antisera (Hanson et al. 1990; Whittenbury and Krieg 1984). Still, antisera can be prepared against those bacteria available in culture collections or isolated from the environment by conventional means. The specificity and efficiency of the antibodies for detecting methanotrophs could be assessed by cross-checking a given sample with a battery of other detection methods.

Abramochinka (1987), Gal'chenko et al. (1988) and Reed and Dugan (1978) used indirect fluorescent-antibody-membrane (FA) filter-staining techniques to determine the populations of individual species or groups of methanotrophic bacteria in fresh water and in samples from the Black Sea. Unfortunately, control experiments revealed that the recovery of cells added to water and soil samples was not quantitative (Reed and Dugan 1978). Although 93–98% of *Methylosinus trichosporium* added to the lake waters was detected by FA procedures, the recovery of *Methylomonas methanica* varied from 1.8–29.2%. The recovery rate when either organism was added to sediment samples was approximately 10%. Attempts to estimate numbers of individual cells in samples was also complicated by the presence of cell aggregates and rosettes.

Antibodies can also be used in Western blot analysis to determine the presence of enzymes unique to methylotrophs. For example, methanol dehydrogenases can be divided into five groups based on their isoelectric points and electrophoretic mobilities in polyacrylamide gels (Anthony 1982; Hanson et al. 1990). There are two and perhaps three groups of methanol dehydrogenases that are distinguishable by cross-reactivity with antibodies that correspond to these groupings (Anthony 1980; Hanson et al. 1990). sMMO can also be detected by Western blot analysis (Tsien et al. 1989).

Phospholipid analysis may provide a useful approach for measuring changes in community structure and populations of methanotrophs in bioreactors and natural samples (Nichols et al. 1986 and 1987). Nichols et al. (1987) compared the phospholipid ester-linked normal and lipopolysaccharide (LPS) hydroxy fatty acids of different bacteria and observed that the monounsaturated fatty phospholipid acid 18:1 Δ 10C and 18:1 Δ 11C were uniquely present in type II methanotrophic bacteria. The 18:1 Δ 10C fatty acid constituted 50% of the phospholipid fatty acids (PLFAs) of *Methylosinus trichosporium* OB3b while 18:1 Δ 11C was the dominant component (84–89%) of the PLFAs of *Methylobacterium organophilum*. When natural gas and air were passed through a soil column, an increase of the 18:1 Δ 10C PLFA to 16–28% of the total fatty acids was observed. Phospholipid analysis is not yet practical since it requires expensive instrumentation and the LPS hydroxy fatty acid and PLFA profiles of a large number of bacteria remain unknown.

Physiological groups and species can also be identified with nucleic acid probes (Chumakov 1987; Fox et al. 1980). Fluorescent signature probes can be used to distinguish and enumerate a few methylotrophic cells from among several thousand nonmethylotrophic bacteria in a sample fixed on a microscope slide (DeLong et al. 1989; Giovannoni et al. 1988; Tsien et al. 1990). Signature probes can be derived from the 16S rRNA sequences of selected organisms. Tsuji et al. (1990) have recently sequenced the 16S rRNAs of over 15 strains of methylotrophs. Phylogenetic analysis of the sequences provided the information to develop two probes that are specific for either type I or type II methanotrophs. Because the 16S rRNA sequences from all the Proteobacteria are not yet available, there is no guarantee that these signature probe sequences are unique only to methylotrophs. Fortunately, the specificity of fluorescent-labelled probes can be rapidly tested by the methods of DeLong et al. (1989) and Giovannoni et al. (1988). The sensitivity of this technique depends on the amount of 16S rRNA per cell and on the fluorescent yield of hybridized probe molecules. Image enhancement or increasing the number of fluorescent dye molecules per probe molecule might serve to amplify the sensitivity of the fluorescent probes (Giovannoni et al. 1988; Porter 1988).

A type II signature probe sequence (5'-CCCTGAGTTATTCCGAAC-3') was selected from sequences that appear in the 16S rRNAs of all type II methylotrophs examined. This probe hybridizes to the RNA extracted from type II methanotrophs but not to RNA extracted from other Proteobacteria or type I methanotrophs (Tsien et al. 1990). When this signature probe was labelled with a fluorescent dye and incubated with cells fixed on a microscope slide as described by DeLong et al. (1989), it hybridized only to cells of type II methanotrophs (Figure 14–1). Likewise, a signature probe

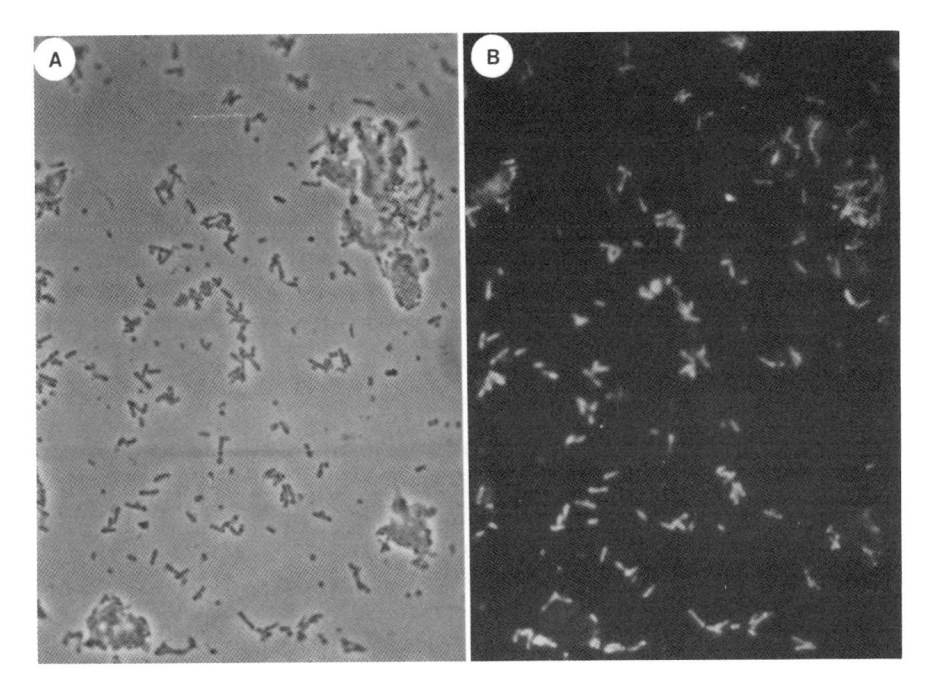

14.3 Estimation of Populations of Methane-Utilizing Bacteria

FIGURE 14-1 The use of a type-specific, fluorescently labelled, oligodeoxynucleotide probe complementary to a 16S rRNA sequence to differentiate type I and type II methanotrophs. Formaldehyde-fixed cells of *Methylosinus trichosporium* OB3b and *Methylomonas gracilis* were hybridized with a X-rhodamine-labelled probe, which was designed for the recognition of type II methanotrophs. (A) Phase contrast microscopy shows rod-shaped *Methylosinus trichosporium* OB3b cells and short-rod-shaped cells of *Methylomonas gracilis*. (B) Epifluorescence microscopy shows the binding of the X-rhodamine-labelled probe to cells of *Methylosinus trichosporium* OB3b, which is a type II methanotroph.

created for type I methanotrophs (5'-GGTCCGAAGATCCCCCGCTT-3') specifically hybridizes to RNAs extracted from type I methanotrophs. The accumulation of 16S rRNA sequence data from other methylotrophs will facilitate the development of more genus and species specific probes.

The sensitivity of the 16S rRNA hybridization reaction illustrated in Figure 14-1 depends on the total population of bacteria in the sample, the fraction represented by the species of interest and the amount of ribosomal RNA per cell. In addition, the hybridization conditions, including incubation temperature and the ionic strength of the buffers, can significantly influence the specificity and sensitivity of methods that employ signature probes (Olsen et al. 1986).

It is possible to hybridize signature probes to RNA isolated from environmental samples and bonded to membrane filters by slot-blotting procedures as a means of tracking phylogenetically coherent groups of microorganisms in their natural setting without culturing (Stahl et al. 1988). This method permits detection of specific groups of organisms in a sample. However, the amount of 16S rRNA per cell varies with the growth rate, the amount of RNA per methylotrophic bacterial cell in environmental samples is not known, making it difficult to estimate population sizes by this method.

Hanson et al. (1990) have proposed using nucleic acid probes that are homologous to highly conserved genes to measure the population sizes of methylotrophic bacteria in the environment. For example, the methanol dehydrogenase (MDH) genes of two methane and methanol-utilizing bacteria have been cloned (Bastien et al. 1988; Harms et al. 1988; Machlin et al. 1987; Nunn and Lidstrom 1986), and two have been sequenced (Harms et al. 1988; Machlin and Hanson 1988). Methanol dehydrogenase is encoded by a single copy gene that is highly conserved in Gram-negative methylotrophs. The average genome content per cell is estimated to be between 1.5 and 2.0 in slowly growing cells. Therefore, the number of MDH genes in a DNA sample provides an estimate of the number of methylotrophs (methane plus methanol oxidizers) in the sample. In addition, restriction-fragment-polymorphism analysis of the MDH genes may aid in the identification of different species of methylotrophs (Hanson et al. 1990). Probes derived from the genes encoding the sMMO may help distinguish methanotrophs from nonmethane utilizers.

Assays for enzymes which are unique to different methylotrophs could be useful for estimating populations of each group of methylotrophs in the biomass collected from various sources. Hexulose-6-phosphate synthetase and hexulose-6-phosphate isomerase in an extract indicate the presence of type I methylotrophs. Type II methylotrophs and, to a lesser extent, some type I methylotrophs contain hydroxypyruvate reductase and serine-glyoxylate amino transferase (Anthony 1982). Finally, the activity of the dye-linked methanol dehydrogenase reflects the relative abundance of Gram-negative methylotrophs (Anthony 1982; Hanson 1980).

14.4 METHANE PRODUCTION AND OXIDATION IN AQUATIC ECOSYSTEMS

This section of the chapter describes some of the many studies of methane production and consumption in different types of aquatic environments, along with the factors believed to control the activities of methanotrophic bacteria. Reviews by Heyer (1977), Hanson (1980), Rudd and Taylor (1980), and Seiler (1984) describe other studies.

Methane-oxidizing bacteria have been detected in most of the freshwater, marine, and terrestrial environments examined (Gal'chenko et al. 1988; Hanson 1980; Heyer et al. 1984; Whittenbury 1970). Heyer (1977) concluded that the distribution and population size of methane-utilizing bacteria is related to the methane concentration and rate of methane production in each habitat. Natural sources of methanotrophs include meadows and woods, water and soil from ponds and rivers, mud from calcareous swamps, and the sediments and oxic-anoxic interfaces of freshwater lakes (Heyer 1977). Methanotrophs also thrive in environments such as the soil from oil fields and sewage sludge. Acid peat bogs, salt lakes and sediments from the Baltic Sea also appear to contain significant populations of methanotrophic bacteria (Heyer et al. 1984; Iverson et al. 1987). The acidic soils of coniferous forests and heath failed to produce methanotrophs in enrichment cultures. Marine environments, as well as flowing waters, contain only low numbers of these microbes (Heyer 1977). Methanotrophs isolated from fresh waters and soils are remarkably similar in serotype to those found in marine environments and acidic peat bogs except that methanotrophs isolated from seawater require NaCl for growth (Seiburth et al. 1987; Gal'chenko et al. 1988).

Eutrophic lakes are a large source of methane and methanotrophs. In all of the lakes studied, most of the methane was oxidized as it diffused to a point in the water column where oxygen was available. Little methane escaped to the atmosphere, especially in deep, permanently stratified lakes like Lake Kivu (Central Africa) in which both thermogenic and biogenic methane are available (Jannasch 1975). Several studies (Cappenberg 1972; Flett et al. 1976; Hanson 1980; Jannasch 1975; Saralov et al. 1985) demonstrated that methanotrophic bacteria or methane-oxidizing activities are highest where methane and oxygen are simultaneously present at the oxic-anoxic interface.

Flett et al. (1976), Harrits and Hanson (1980), Patt et al. (1974), and Rudd and Hamilton (1978) found that methane oxidation is confined to the thermocline during summer stratification in some eutrophic lakes in North America. The locations of methane-oxidation activities were determined by measuring the conversion of $^{14}CH_4$ to $^{14}CO_2$ and radioactive biomass. The maximum rates of methane oxidation occurred in the thermocline where the dissolved oxygen was 1 ppm or less. The rate of methane production in Lake Mendota (Wisconsin) (Harrits and Hanson 1980) was estimated to be 67.5 mmoles C per m^2 per day, compared to a primary production rate of 250 mmol C per m^2 per day during the summer of 1977. Approximately 45% of the methane produced was oxidized and converted to biomass, 47% accumulated in the hypolimnion, and only 8% was released to the atmosphere during the summer months. 36-60% of the total carbon input of lakes in Canada and Michigan was converted to methane. In Lake 227 (Canada), 36% of this methane was recycled by oxidation (Flett et al. 1976). The methane production rate in Lake 227 was about one-sixth of that in Lake Mendota. Approximately 11% of the methane was oxidized, 5% was released to the atmosphere, and the remainder was stored in the hypolimnion until fall turnover.

During fall turnover prior to freeze-up, the whole-lake rate of methane oxidation increased due to distribution of oxygen throughout the water column. At this time, methane oxidation occurred throughout the water columns of both Lake Mendota and Lake 227. Approximately 90% of the methane evasion from Lake 227 occurred during and shortly after overturn (Rudd and Hamilton 1978). After winter stratification, the water column in Lake Mendota was depleted of methane so that methanotrophic activity was confined to the methane-containing sediment surfaces.

Mono Lake (California) is a meromictic lake in an area of active volcanism (Oremland et al. 1987). The lake is hypersaline, alkaline (pH 9.8), and rich in carbonate (0.5 M) and sulfate (110 mM). The lake has been stratified year-round since 1983 because of a large influx of water which decreased the salinity of the surface waters. Although anaerobic methane oxidation occurs in the sulfate-rich bottom waters, most of the methane produced in the lake diffuses up to the oxic/anoxic boundary (oxycline) for oxidation.

In Lake Steckin (FRG) an oligotrophic lake, the methane oxidation rate ranged between 475 and 960 μmol per h per m² in the upper sediment layers. The number of methanotrophic bacteria in littoral and profundal sediments varied between 8×10^3 and 6×10^5 per g. In the interstitial waters of sediments, the concentration of methane was 90 to 560 μmol per liter. Little methane was found in the water, suggesting nearly complete oxidation of methane in the upper layers of the sediment.

Harris et al. (1982) measured methane fluxes in the Great Dismal Swamp (Virginia). This peat bog is one of the largest swamp forests in the United States. The depth of the swamp varies with changes in the seasonal rainfall. During the dry months (June-January, 1980), when the peat horizon was not covered with water, methane was consumed from the atmosphere and from the anaerobic areas below the surface where methanogenesis occurred. When the peat was saturated with water, the swamp produced methane at rates as high as 20 g C per m² per day. Thus, wetlands and swamps may be a significant sink for atmospheric methane during periods of drought but a source of methane during rainy seasons (Harris et al. 1985).

14.5 DISTRIBUTION OF METHANE-OXIDIZING BACTERIA IN FRESHWATER ENVIRONMENTS

Abramochinkina et al. (1987) and Saralov et al. (1985) employed most probable number procedures to determine the numbers of methane-oxidizing bacteria in the water and bottom sediments of freshwater lakes. Species were identified by indirect immunofluorescence. In Lake Dalgoe, a eutrophic lake in the Southern Moscow Region (USSR), oxygen was present in the upper 3.5–4 m of a 4.6-m water column. The sediments were anoxic. Methane was rapidly oxidized at the oxic/anoxic interface at 2.5–3.5 m where the oxygen concentration was 37 μM or less. The average rate of methane oxidation was 78 micromoles per m² per hour. Methanotrophs comprised

14.5 Distribution of Methane-Oxidizing Bacteria in Freshwater Environments

11.9% of the microbial population in sediment samples. *Methylomonas methanica* accounted for 50% of the methanotrophs, while *Methylosinus trichosporium* and *Methylocystis methanolicus* accounted for 12.2 and 10.5%, respectively. Other methanotrophs detected included *Methylocystic echinoides, M. chroococcum*, lesser numbers of other type II methanotrophs, and *Methylococcus capsulatus*. The maximum number of methanotrophs (5×10^8 per g) were found at 18–20 cm below the surface of the bottom silty layer, which is covered by shallow oxygenated water.

Flett et al. (1976), Rudd (1974), and Rudd and Hamilton (1978) proposed that methanotrophic bacteria are confined to a narrow zone within the chemocline of Lake 227 during the summer because of their sensitivity to oxygen concentrations above 1.0 mg per liter in the epilimnion. An oxygen-sensitive, nitrogen-fixation process appears to cause this temporal oxygen intolerance. Sansone and Martens (1978) also observed that methane oxidation was sensitive to oxygen above 1.9 mg per liter when dissolved inorganic nitrogen (DIN) concentrations were below 14 μM in samples taken from a marine environment (Cape Lookout Bight, North Carolina). After fall turnover and during the winter when high dissolved inorganic nitrogen concentrations (>20 μM $NO_3^- + NH_4^+$) are distributed throughout the water column, methane oxidation occurs in the presence of higher concentrations of dissolved oxygen. This trend indicates that the methanotrophs are not inherently oxygen sensitive.

Nitrogen fixation coincident with methane oxidation has been observed in several freshwater lakes. In these lakes, one might expect the type II methanotrophs to be dominant because only these bacteria and *Methylococcus capsulatus* are known to fix nitrogen (Whittenbury and Krieg 1984). Therefore, type II methanotrophs would be expected to be the dominant methane oxidizers in oligotrophic environments. When Saralov et al. (1985) examined nitrogen fixation under microaerophilic conditions in the absence of cyanobacteria in two acidic, polyhumic lakes (pH 4.4–5.4), they detected the type II methanotrophs, *Methylosinus trichosporium* and *Methylosinus sporium*.

In a study of nine lakes and streams in the USSR, Saralov et al. (1985) found low numbers of methanotrophs (400–800/ml) in flowing waters but up to 4×10^6 per ml in metalimnion samples of eutrophic stratified lakes. Of the small numbers of methanotrophs present in an oligotrophic lake (Lake Svente), the type II methane utilizers *Methylosinus trichosporium, M. sporium*, and *Methylocystis parvus* dominated. In metalimnion samples from a eutrophic lake, *Methylomonas methanica* and *M. rubrum* were dominant although type II methanotrophs were also present.

Reed and Dugan (1978) used a fluorescent antibody-membrane staining technique to study the autecology of two methanotrophs, *Methylomonas methanica* and *Methylosinus trichosporium*, in Cleveland Harbor (Lake Erie, USA). The maximum populations of *Methylomonas methanica* (2.8×10^4 cells/g) and *Methylosinus trichosporium* (approximately 5×10^5 cells/g) were found at the oxic/anoxic interface at the sediment surface or within 1 m of the surface. The relative numbers of *Methylomonas methanica* decreased dramatically as the water cooled in November.

Heyer and Suckow (1985) described methane oxidation in three peat bog lakes of pH 3.8–5.3. The highest rate of methane production observed was 405 μmol per h per m². Nearly all of the methane produced in the anaerobic layers of the peat was recycled. The rate of methane oxidation in the upper layers of the peat mire was 90–429 μmol per h per m², when measured in closed serum vials under in situ conditions. The upper layer of the mire, where the most rapid methane oxidation occurred, contained the highest numbers of methanotrophic bacteria, estimated at $3-5.5 \times 10^3$ cells/ml by the MPN method. Enrichment cultures yielded type II methanotrophs, *Methylosinus sporium, M. trichosporium*, and *Methylocystis* species.

14.6 METHANOTROPHIC BACTERIA IN SEAWATER

Twenty-three of 32 enrichments inoculated with Sargasso Sea samples resulted in growth of methanotrophs (Sieburth et al. 1987) while none of the 72 samples from the Atlantic Ocean Gulf Stream yielded positive enrichments. An accurate assessment of the number and diversity of species of methanotrophs could not be determined due to the restricted range of enrichment conditions. The bacterium isolated from these enrichments was a type I obligate methylotroph named *Methylomonas pelagica*. Growth of this bacterium depends on NaCl or seawater but is inhibited by ambient sunlight. Except for these properties, this bacterium resembles other *Methylomonas* strains. Antibodies prepared against killed cells of *Methylomonas pelagia* cross-reacted with terrestrial species of *Methylomonas methanica*. It was proposed that this bacterium, which is present in the mixed oxygenated waters of the Sargasso Sea, grows on the methane produced during fermentation in the organic particulates that accumulate at the thermocline (Sieburth et al. 1987).

Gal'chenko et al. (1988) employed indirect immunofluorescence using antibodies prepared against 14 methanotrophs isolated from soils and fresh waters, and end-point dilution assays to determine the numbers and species of methanotrophs in sediments and waters of the Black Sea. Using immunofluorescence, the highest numbers of methanotrophic bacteria in sediments (2.3×10^5 to 1.3×10^6 cells/ml) were found at a depth of 20–25 cm. These numbers are three orders of magnitude larger than those obtained with viable counting procedures, demonstrating that antibodies can detect methanotrophs more efficiently than methods that rely on the recovery of viable cells. In bottom sediments covered by oxygenated water, 65–70% of the methane-utilizing bacteria detected by immunofluorescence were type I methanotrophs of the genera *Methylobacter* and *Methylomonas*. In water above the sediments, *Methylobacter* and the type II methylotrophs of the genus *Methylocystis* were most abundant. Their numbers were greatest (approximately 10^5 cells/ml) at 150–200 m, the zone between aerobic and anaerobic waters.

A methane-consuming mussel (family Mytilidae) lives close to hydrocarbon seeps in the Gulf of Mexico where the methane levels are 50–200 µmol per liter. Methane oxidation is confined to the gills of the animals and may be performed by the intracellular bacteria found there (Childress et al. 1986). These intracellular symbionts may supply the mussel with carbon. Methane consumption was measured at 1.36 (± 0.23) µmol per h per g dry gill tissue, or 0.74–0.90 µmol per h per g of whole animal. Transmission electron microscopy of gill tissue sections revealed numerous intracellular coccoid bacteria with stacked intracytoplasmic membranes resembling those found in type I methanotrophs.

14.7 METHANE UTILIZATION IN SOILS

Methane oxidation in soils was first recognized in 1906 (Sohngen 1906). A recent study showed that soils not previously exposed to exogenous methane exhibit methane oxidation at rates of 0.34–74 µmol per h per g dry soil. Exposing the soil to 20 kPa of methane for six days to enrich for methanotrophs increased the oxidation rate to 10–13 µmol methane per h per g dry soil (Megraw and Knowles 1989). When Megraw and Knowles (1989) conducted laboratory studies on cultivated humisol preparations, they observed that 0.27 µmol CO_2 was produced per mol methane oxidized. The remaining 73% of the methane carbon was incorporated into cell material. This ratio was similar to that observed with pure cultures of type II methanotrophs (Megraw and Knowles 1989; Whittenbury et al. 1970), but higher than that observed in freshwater systems by Harrits and Hanson (1980) and Rudd and Hamilton (1978). In freshwater systems, only 25–30% of the methane oxidized was assimilated into cell material. In marine systems, 98% of the methane oxidized was recovered as CO_2 (Megraw and Knowles 1989).

14.8 ARE SOILS AND AQUATIC ENVIRONMENTS SINKS FOR ATMOSPHERIC METHANE?

The affinities for methane exhibited by methanotrophs are probably to be too low to enable these organisms to cause significant oxidation of atmospheric methane (Megraw and Knowles 1989). Megraw and Knowles (1989) and Conrad (1984) calculated that the amount of methane in the aqueous phase of soils in equilibrium with the atmosphere would be 2.5 nM at 20°C. The K_m values for methane vary from 26 µM for impure cultures (Harrison

1973) to 66 µM and 160 µM for pure cultures of *Methylosinus trichosporium* OB3b and *Methylococcus capsulatus* (Bath) respectively (Colby and Dalton 1978). In fresh water, the values for methane oxidation are 5 µM (Rudd and Hamilton 1978); in sediments, 10 µM (Lidstrom and Sommers 1964); and in soils, 37 µM (Megraw and Knowles 1989). These K_m values suggest that methanotrophs in water and soils do not derive most of their methane from the atmosphere but instead probably live on the methane produced by methanogens (Megraw and Knowles 1989).

Exceptions to these observations may exist. Some tropical and tundra soils are capable of consuming methane from the atmosphere (Seiler 1983). In the open ocean, low rates of methane consumption were reported when the dissolved methane concentrations were 5–10 nM (Ward et al. 1987), and the consumption of atmospheric methane by dry soils of the Great Dismal Swamp was measured by Harris et al. (1982).

14.9 FORTUITOUS METABOLISM OF METHANOTROPHS AND ITS ENVIRONMENTAL SIGNIFICANCE

Methanotrophs can oxidize a wide variety of compounds besides one-carbon molecules (Bedard and Knowles 1989; Higgins et al. 1981). The obligate methanotrophic bacteria obtain no energy from the metabolism of these compounds since they can only use one-carbon (C_1) compounds as sole carbon sources (Higgins et al. 1981).

Many studies suggest that ammonia oxidation is carried out by methanotrophs in freshwater lakes (Bedard and Knowles 1989; Hanson 1980; Harrits and Hanson 1980; Knowles and Lean 1987). It has also been proposed that under some conditions, methanotrophs essentially perform all of the ammonia oxidation in some humisols (Bedard and Knowles 1989). In support of this suggestion, Megraw and Knowles (1989) isolated a mixed culture from humisols that exhibits methane-dependent oxidation of ammonia.

Methanotrophic bacteria can oxidize ammonia to nitrite, and conversely, other ammonia oxidizers can oxidize methane to CO_2 (Ward et al. 1987). The similarities in the metabolic abilities of these two groups of bacteria have recently been reviewed by Bedard and Knowles (1989). Both groups of organisms contain monooxygenases with broad substrate specificities (Bedard and Knowles 1989; Dalton and Higgins 1987; Dalton and Stirling 1987; Haber et al. 1983; Higgins et al. 1981; Ward 1987). All methanotrophic bacteria examined contain a hydroxylamine oxidoreductase (Bedard and Knowles 1989; E. Topp and R. Hanson, unpublished observations). Both groups of bacteria can also oxidize carbon monoxide, bromomethane, ethylene, propylene, cyclohexane, benzene, and phenol. In addition, methanotrophs and ammonia oxidizers both contain intracytoplasmic membranes (Bedard and Knowles 1989).

Biological oxidation provides an important sink for atmospheric CO (Meyer 1985). Carboxydobacteria, known to utilize CO as a sole carbon and energy source, exhibit half-saturation constants for this substrate that are considerably higher than the concentrations of CO dissolved in the interstitial water of soils from the atmosphere (Bedard and Knowles 1989). Curiously, both methanotrophic bacteria and ammonia-oxidizing bacteria have much higher affinities for CO than carboxydobacteria (Bedard and Knowles 1989; Dalton and Stirling, 1987).

Several halogenated low-molecular-weight hydrocarbons have become significant environmental pollutants. Since many of these compounds are toxic and potentially carcinogenic, their presence in soil and groundwater poses a serious health risk (Miller and Guengerich 1983), and their removal challenges current water treatment technologies (Sayre 1988). Trichloroethylene (TCE) is one of the most widespread and persistent of this group of chemicals. Several studies revealed that a wide variety of aerobic bacteria (Little et al. 1988) that contain mono- and dioxygenases (Nelson et al. 1988; Wackett et al. 1989) can metabolize TCE. These organisms include ammonia-oxidizing bacteria (Arciero et al. 1989), toluene-oxidizing bacteria, and cultures of methanotrophic bacteria (Wackett et al. 1989). TCE oxidation is stimulated in soils and aquifers exposed to methane suggesting that the use of methanotrophs for the bioremediation of contaminated waters and soils is a practical objective (Wilson 1988; Wilson and Wilson 1985).

14.10 ANAEROBIC METHANE OXIDATION

Reeburgh (1976, 1980, 1982, and 1983) has reviewed the anaerobic metabolism of methane and other hydrocarbons. His own measurements of methane distribution and $\delta\ ^{13}CH_4$ values indicate that methane is converted to CO_2 in the absence of oxygen (Reeburgh and Heggie 1977). This study suggests that anaerobic methane oxidation requires sulfate and is inhibited by oxygen. When Kosiur and Warford (1979) measured the conversion of ^{14}C-acetate and ^{14}C-lactate to ^{14}C-CH_4 and ^{14}C-CO_2 in Santa Barbara Basin sediments, they also concluded that methane can be oxidized to CO_2 under anaerobic conditions. The rate of oxidation in marine sediments was estimated to be 68–232 μmol per liter per year (Kosiur and Warford 1979; Reeburgh and Heggie 1977).

Panganiban et al. (1979) showed that radioactive methane is converted to CO_2 in samples harvested from the anaerobic sediment surface of the eutrophic Lake Mendota (Wisconsin). In these samples, oxygen completely inhibited the oxidation of methane. Furthermore, methane carbon was not assimilated into cell material. Anaerobic enrichment cultures from Lake Mendota sediments exhibited methane-dependent sulfate reduction and sulfate-dependent methane oxidation (Hanson 1980). The enriched cultures required acetate or lactate as carbon sources, but were dependent on methane as a source of energy. The bacteria from these cultures remain to be characterized.

In a careful study of methane oxidation in Big Soda Lake (Nevada) an alkaline, saline lake, Iversen et al. (1987) observed that methane oxidation rates were low (1.3 nmol per liter per day) in the oxic waters of the aerobic mixolimnion and in the water column above. Rates of oxidation of ^{14}C-CH$_4$ to ^{14}C-CO$_2$ increased in parallel to the methane concentration profile deeper in the water column. Rates were highest (49–85 μmol per liter per day) in the anoxic monimolimnion (bottom layer). Residual methane in the monimolimnion of Big Soda Lake was enriched in ^{13}C, an indication of anaerobic oxidation (Iverson et al. 1987).

The concept that sulfate is the terminal electron acceptor in anaerobic methane oxidation is favored but unproven. The studies described by Panganiban et al. (1979) and Hanson (1980) indicate that sulfate reduction is dependent on methane oxidation. Furthermore, the zones of sulfate depletion and methane oxidation coincide in marine sediments (Iverson and Jorgenson 1985; Reeburgh and Heggie 1977) and in the anaerobic water columns of a sulfate-rich lake (Iverson et al. 1987).

Several other studies also support the hypothesis that anaerobic oxidation of methane occurs in both freshwater and marine environments (Iverson and Jorgenson 1985; Lidstrom 1983; Martens and Brenner 1977; Reeburgh 1976). The investigation of the microbiology and biochemical mechanisms of anaerobic methane oxidation remains a fruitful and significant area of research.

14.11 ECOLOGY AND DIVERSITY OF BACTERIA THAT GROW ON C$_1$ SUBSTRATES OTHER THAN METHANE

Methanol is formed by reactions as diverse as the chemical oxidation of methane in the trophosphere (Ehalt 1974) and the demethylation of plant lignins and pectins (Anthony 1982). Formate is an end-product of fermentations of organic compounds and a byproduct of industrial processes. The decomposition of fish produces methylamines from carnatine and lecithin as does the degradation of some pesticides (Anthony 1982) and the reduction of trimethylamine-N-oxide, an osmoregulatory compound found in marine fish and invertebrates (Anthony 1982). Chlorinated methanes, which are used as industrial solvents, are produced during treatment of water with chlorine. Dimethylsulfide arises from the cleavage of dimethyl-β-propiothetin, a product of algal sulfur metabolism (Andreae 1980) and is transported into the atmosphere. Bacteria that utilize all of these compounds have been isolated (Anthony 1982; Hanson et al. 1990; Lidstrom 1990).

The largest group of bacteria that utilize methanol and methylamines are the pink-pigmented facultative methylotrophs (PPFMs). They resemble *Methylobacterium* sp. strain AM1 and *Methylobacterium organophilum* strain XX. The PPFMs use the serine pathway for formaldehyde assimilation. They grow on a wide range of heterotrophic carbon and energy sources as well as on C_1 compounds (Green et al. 1988; Hanson et al. 1990; Jenkins and Jones 1987), although only one species utilizes methane (Patt et al. 1974).

Some methanol- and methylamine-utilizing bacteria are obligate methylotrophs and cannot use compounds with carbon-carbon bonds as sources of energy or carbon. Nearly all obligate methanol-utilizing bacteria employ the ribulose monophosphate (RuMP) pathway for formaldehyde fixation and resemble the well-characterized species *Methylophilus methylotrophus* (Anthony 1982). Some methanol-utilizing bacteria are also capable of denitrification (Anthony 1982; Harder and Attwood 1978).

Restricted facultative methylotrophs that grow on a more limited range of multicarbon compounds than the PPFMs belong to the genera *Bacillus* and *Hyphomicrobium*. Members of the genus *Hyphomicrobium* are serine pathway methylotrophs, capable of utilizing nitrate as a terminal electron acceptor (Lidstrom 1990) for anaerobic respiration. They are very versatile in terms of their metabolic abilities (Harder and Attwood 1978). Gram-positive methylotrophs include the *Bacillus* species mentioned above, and *Arthrobacter, Nocardia*, and *Mycobacterium* species. All of these Gram-positive methylotrophs employ the RuMP pathway for formaldehyde fixation (Lidstrom 1990).

Lidstrom (1990) has listed the properties of several other bacteria that utilize methanol and methylamines as carbon or energy sources. This list includes those facultative phototrophic bacteria that employ methanol as a photoreductant and belong to the genus *Rhodopseudomonas*, and also *Paracoccus denitrificans*, a Gram-negative bacterium that can grow as an aerobic heterotroph, utilize nitrate as a terminal electron acceptor, and can act as a chemolithotroph employing hydrogen or thiosulfate as energy sources.

Three mycelial fungi and several yeast isolates belonging to the genera *Hansenula, Candida, Pichia, Torulopsis, Kloechera*, and *Saccharomyces* can also utilize methanol as a carbon and energy source (Anthony 1982). The fungi and yeasts differ from bacteria both in the properties of their catabolic enzymes and the pathways used for formaldehyde assimilation.

The bacteria that utilize methanol, methylamines, and methylated sulfur species have been isolated from plant materials, soils, freshwater, marine environments, and air (Anthony 1982; Hanson 1980; Kuono and Ozaki 1974; Lidstrom 1990). The PPFMs appear to be the most common methanol-utilizing bacteria found in soils (Kuono and Ozaki 1974) and on plant surfaces (Corpe 1985). Kuono and Ozaki (1974) have shown that over one-fourth of the methylotrophs isolated from soil, compost, wastewater, and water samples were PPFM-like bacteria. Corpe (1985) found over 500

PPFM-colony-forming units per cm² on a variety of plant leaf surfaces. These organisms apparently obtain nutrients from the demethylation of plant polymers. Corpe (1985) also found much smaller numbers of PPFMs on glass, wood, plastics, brick, concrete, and, very occasionally, in air. Except for those resembling *Hyphomicrobium* species, PPFMs and other serine-pathway methylotrophs have not been isolated from ocean waters (Strand and Lidstrom 1984).

Unfortunately, although several groups have isolated and characterized new strains for biotechnological purposes, the taxonomy of the bacteria that grow on methanol, methylamines, and methylated sulfur species, except for the genus *Hyphomicrobium* (Harder and Attwood 1978), remains in a confused state (Green et al. 1988; Lidstrom 1990; also see Chapter 1). The inability to identify specific strains and the paucity of ecological studies have also resulted in limited knowledge about their distribution or roles in nature.

REFERENCES

Abramochinka, F.N., Bezrukova, L.V., Koshelev, A.V., Gal'chenko, V.F., and Ivanov, M.V. (1987) *Microbiol.* 56, 375–382.
Aiyer, P.A.S. (1920) *Mem. Dept. Agr. India. Chem. Ser.* 5, 173–180.
Andreae, M.O. (1980) *Limnol. Oceanogr.* 25, 1054–1063.
Anthony, C. (1982) *The Biochemistry of Methylotrophs,* Academic Press, London.
Arciero, D., Vannelli, T., Logan, M., and Hooper, A.B. (1989). *Biochem. Res. Comm.* 159, 640–643.
Bahlool, B.B., and Schmidt, E.L. (1980) *Adv. Microbial. Ecol.* 4, 203–241.
Bastien, C., Machlin, S., Zhang, Y., Donaldson, K., and Hanson, R.S. (1988) *Appl. Environ. Microbiol.* 55 (12), 3124–3130.
Bedard, C., and Knowles, R. (1989) *Microbiol. Rev.* 53, 68–84.
Bone, T.L., and Balkwill, D.L. (1986). *Appl. Env. Microbiol.* 51, 462–468.
Bouwman, A.F. (1989) *Netherlands J. Agricultural Science* 37, 13–19.
Brusseau, G.A., Tsien, H.C., Hanson, R.S., and Wackett, L.P. (1990) *Biodegradation* 1, 19–29.
Burrows, K.J., Cornish, A., Scott, D., and Higgins, I.J. (1984) *J. Gen. Micro.* 130, 3327–3333.
Cappenberg, T.E. (1972) *Hydrobiologia.* 40, 471–485.
Childress, J.J., Fisher, C.R., Brooks, J.M., et al. (1986) *Science* 233 (4770), 1306–1308.
Chumakov, K.M. (1987) *Sov. Sci. Rev. D. Physicochem. Biol.* 7, 51–94.
Cicerone, R.J., and Oremland, R.S. (1988) *Global Biogeochem. Cycles* 2 (4), 299–327.
Colby, J.D., and Dalton, H. (1978) *Biochem. J.* 165, 395–402.
Conrad, R. (1984) in *Current Perspectives in Microbial Ecology* (Klug, I.J. and Reddy, C.A., eds.), pp. 461–467, American Society for Microbiology, Washington, DC.
Corpe, W.A. (1985) *J. Micro. Methods.* 3(3–4), 215–233.

Dalton, H., and Higgins, I.J. (1987) in *Microbial Growth on C_1 Compounds* (Van Verseveld, H.W., and Duine, J.A., eds.), pp. 89–94, Martinus Nijhoff Publishers, Dordrecht, The Netherlands.
Dalton, H., and Stirling, D.I. (1987) *Philos. Trans. Roy. Soc. London, Ser. B.* 297, 481–496.
Dalton, H.S., Stanley, H., Prior, S.D., and Leak, D.J. (1983) *Biotech. Lett.* 5, 487.
DeLong, E.F., Wickham, G.S., and Pace, N.R. (1989) *Science* 243, 1360–1363.
Ehalt, D.H. (1974) *Tellus.* 26, 58–70.
Flett, R.J., Schindler, D.W., Hamilton, R.D., and Campbell, N.E.R. (1976) *Canad. J. Fish Aquatic Sci.* 37, 494–505.
Foster, J.W., and Davis, R.H. (1966) *J. Bacteriol.* 91, 1924–1931.
Fox, B.G., and Lipscomb, J.D. (1988) *Biochem. Biophys. Res. Comm.* 154, 165–170.
Fox, B.G., Froland, W.A., Dege, J., and Lipscomb, J.D. (1989) *J. Biol. Chem.* 264, 10023–10033.
Fox, G.E., Stackebrandt, E., Hespell, R.B., et al. (1980) *Science* 209, 457–463.
Gal'chenko, V.F. (1975) *Appl. Biochem. Microbiol.* (in Russian) 11, 447–450.
Gal'chenko, V.F., and Andreev, L.V. (1984) in *Microbial Growth on C_1 Compounds* (Crawford, R.L., and Hanson, R.S., eds.), pp. 269–281, American Society for Microbiology, Washington, DC.
Gal'chenko, V.F., Shiskina, V.N., Suzina, N.E., and Trotsenko, Y.A (1978) *Microbiology (USSR)* 46, 723–728.
Gal'chenko, V.F., Andreev, L.V., and Trotsenko, Yu.A. (1986) in *Taxonomy and identification of obligate methanotrophic bacteria.* Academy of Sciences, USSR. Puschino, USSR (in Russian).
Gal'chenko, V.F., Abramochkina, F.N., Bezrukova, L.V., Sokolov, E.N., and Ivanov, M.V. (1988) *Microbiology (USSR)* 47, 248–253.
Giovannoni, S.J., DeLong, E.F., Olsen, G.J., and Pace, N.R. (1988) *J. Bacteriol.* 170, 720–726.
Graedel, T.E., and Crutzen, P.J. (1989) *Sci. Am.* 260(9), 58–68.
Green, J., and Dalton, H. (1986) *Biochem. J.* 236(1), 155–162C.
Green, J., and Dalton, H. (1989) *J. Biol. Chem.* 264, 17698–17703.
Green, P.N., Bousfield, I.J., and Hood, D. (1988) *Int. J. Syst. Bact.* 38, 124–127.
Haber, C.L., Allen, L.N., and Hanson, R.S. (1983) *Science* 221, 1147–1151.
Hanson, R.S. (1980) *Adv. Appl. Microbiol.* 26, 3–39.
Hanson, R.S., Netrusov, A.I., and Tsuji, K. (1990) in *The Procaryotes*, 2nd ed., (Balowes, A., Trüper, H.G., Dworkin, M., Harder, W., and Schleifer, K.H., eds.), Springer-Verlag, New York (in press).
Harder, W., and Attwood, M. (1978). *Adv. Microbial. Physiol.* 17, 303–359.
Harms, N., DeVries, G.E., Maurer, K., Hoogendak, J., and Stouthamer, A.H. (1988) *J. Bacteriol.* 169, 3969–3975.
Harris, R.C., Sebucher, D.I., and Day, F.P. (1982) *Nature* 247, 673–674.
Harris, R.C., Gorham, E., Sebacher, D.I., Bartlett, K.B., and Flebbe, P.A. (1985) 315(6021), 652–654.
Harrison, D.E.F. (1973) *J. Appl. Bacteriol.* 36, 301–308.
Harrits, S., and Hanson, R.S. (1980) *Limnol. Oceanogr.* 25, 412–421.
Heyer, J. (1977) in *Microbial Growth on C_1 Compounds*, (Skryabin, G.K., Ivanov, M.V., Kondratjeva, E.N., et al., eds.), pp. 19–21, USSR Academy of Sciences, Puschino, USSR.

Heyer, J., and Suckow, R. (1985) *Limnologica* (Berlin) 6, 247-266.
Heyer, J., Malashenko, Y., Berger, U., and Budkova, E. (1984) *Zatschrift fr allegmaine Microbiologie* 10, 725-744.
Higgins, I.J., Best, D.J., and Hammond, R.C. (1980) *Nature* 286, 561-564.
Higgins, I.J., Best, D.J., Hammond, R.C., and Scott, D. (1981) *Microbial. Rev.* 45, 556-590.
Holden, W.E., and Tiedje, J.M. (1988) *Ecology* 69, 561-568.
Hutton, W.E., and Zobell, C.E. (1949) *J. Bacteriol.* 58, 463-473.
Iverson, N., and Jorgenson, B.B. (1985) *Limnol. Oceanogr.* 30, 944-955.
Iverson, N., Oremland, R.S., and Klug, M. (1987) *Limnol. Oceanogr.* 32, 804-814.
Jannasch, H.W. (1975) *Limnol. Oceanogr.* 20, 860-864.
Jenkins, O., and Jones, D. (1987) *J. Gen. Microbiol.* 131, 335-344.
Johnston, H.S. (1984) *Ann. Rev. Phys. Chem.* 35, 481-505.
Kaserer, J. (1906) *Bacteriol. Abt.* 15, 573-576.
Knowles, R., and Lean, D.R.S. (1987) *Can. J. Fish. Aquatic Sci.* 44, 743-749.
Kosiur, D.R., and Warford, A.L. (1979) *Estuarine Coastal Marine Sci.* 8, 379-385.
Kuono, K., and Ozaki, A. (1974) in *Microbiol Growth on C_1 Compounds*, (Terui, G., ed.), pp. 11-21, Society of Fermentation Technology, Osaka, Japan.
Lamb, S.C., and Garver, J.C. (1980) *Biotechnol. Bioeng.* 22, 2097-2118.
Large, P.J., and Bamforth, C.W. (1988) in *Methylotrophy and Biotechnology*, pp. 180-188, Longman Scientific and Technical, Essex, England.
Lidstrom, M.E. (1983) *Limnol. Oceanogr.* 28, 1247-1251.
Lidstrom, M. (1990) in *The Procaryotes*. (Balows, A., Trüper, H.G., Dworkin, M., Harder, W., and Schleifer, K.H., eds.), Springer-Verlag, New York (in press).
Lidstrom, M.L., and Sommers, L. (1984) *Appl. Env. Microbiol.* 47, 1255-1260.
Little, C.D., Palumbo, A.V., Herbes, S.E., et al. (1988) *Appl. Env. Microbiol.* 54, 951-956.
Machlin, S.M., and Hanson, R.S. (1988) *J. Bacteriol.* 170, 4739-4747.
Machlin, S.M., Tam, P.E., Bastien, C.A., and Hanson, R.S. (1987) *J. Bacteriol.* 170, 141-148.
Martens, C.S., and Brenner, R.A. (1977) *Limnol. Oceanogr.* 22, 10-14.
Megraw, S.R., and Knowles, R. (1989) 62, 359-366.
Meyer, O. (1985) in *Microbial Gas Metabolism; Mechanistic, Metabolic, and Biotechnical Aspects* (Pool, R.K., and Dow, C.S., ed.), pp. 131-151, Academic Press, Inc., London.
Miller, R.E., and Guengerich, F.P. (1983) *Cancer Res.* 43, 1145-1152.
Nelson, M.J.K., Montgomery, S.O., and Pritchard, P.H. (1988) *Appl. Env. Microbiol.* 54, 604-606.
Nichols, P.D., Smith, G.A., Antworth, C.P., Hanson, R.S., and White, D.C. (1986) *FEMS Microbiol. Lett.* 31 (6), 327-336.
Nichols, P.D., Hensen, J.M., Anthworth, C.P., et al. (1987) *Environ. Toxic Chemistry* 6, 89-97.
Nunn, D.N., and Lidstrom, M.E. (1986). *J. Bacteriol.* 166 (2), 581-590.
Oldenhuis, R., Vink, R.L.J.M., Janssen, D.B., and Witholt, B. (1989) *Appl. Env. Microbiol.* 55, 2819-2826.
Olsen, G.J., Lane, D.J., Giovannoni, S.J., Pace, N.R., and Stahl, D.A. (1986) *Ann. Rev. Microbiol.* 40, 337-365.
Oremland, R.S., Miller, L.G., and Whitticar, M.J. (1987) *Geochimica et Cosmochimica Acta.* 51, 2915-2929.

Panganiban, A.T., Patt, T.E., Hart, W., and Hanson, R.S. (1979) *Appl. Env. Microbiol.* 37, 303–309.
Patt, T.E., Cole, G.C., Bland, J., and Hanson, R.S. (1974) *J. Bacteriol.* 120, 955–964.
Patt, T.E., Cole, G.C., and Hanson, R.S. (1976) *Int. J. Syst. Bacteriol.* 26, 226–229.
Pearce, F. (1989) *New Scientist* 122 (1663), 37–41.
Pearman, G.I., and Fraser, P.J. (1988) *Nature* 332, 489–490.
Porter, K. (1988) *Ecology* 69, 558–560.
Reeburgh, W.S. (1976) *Earth Planet Sci. Lett.* 28, 337–344.
Reeburgh, W.S. (1980) *Earth Planet Sci. Lett.* 47, 345–352.
Reeburgh, W.S. (1982) in *The Dynamic Environment of the Ocean Floor* (Fanning, K., and Manheim, F.T., eds.), pp. 203–217, Heath, Lexington, MA.
Reeburgh, W.S. (1983) *Ann. Rev. Earth Planet Sci.* 11, 269–298.
Reeburgh, W.S., and Heggie, D.T. (1977) *Limnol. Oceanogr.* 22, 1–9.
Reed, W.N., and Dugan, P.R. (1978) *J. Gen. Microbiol.* 113, 1389.
Rudd, J.W.M., and Hamilton, R.D. (1978) *Limnol. Oceanogr.* 23, 337–348.
Rudd, J.W., and Taylor, C.D. (1980) *Adv. Aquat. Microbiol.* 2, 77–150.
Rudd, J.W.M., Hamilton, R.D., and Campbell, N.E.R. (1974) *Limnol. Oceanogr.* 19, 519–524.
Sansone, F.J., and Martens, C.S. (1978) *Limnol. Oceanogr.* 23, 349–355.
Saralov, A.I., Krylova, I.N., Saralova, E.E., and Kuznetsov, S.I. (1985) *Microbiology (USSR)* 53, 695–700.
Sayre, I.M. (1988) *J. Am. Water Works Assoc.* 80, 53–60.
Schneider, S.H. (1989) *Science* 243, 771–781.
Scott, D., Brannan, J., and Higgins, I.J. (1981) *J. Gen. Microbiol.* 125, 63–72.
Seiler, W. (1983) in *Current Prospectives in Microbial Ecology* (Klug, M.J., and Reddy, C.A., ed.), pp. 468–477, American Society for Microbiology, Washington, DC.
Sieburth, J., Johnson, P.W., Eberhardt, M.A., et al. (1987) *Current Microbiol.* 14, 285–293.
Sohngen, N.L. (1906) *Zent. Bakt. Parasitenk II.* 15, 513–517.
Stahl, D.A., Flesher, B., Mansfield, H., and Montgomery, L. (1988) *Appl. Env. Microbiol.* 54, 1079–1084.
Stainthorpe, A.C., Murrell, J.C., Salmond, G.P.C., Dalton, H., and Lees, V. (1989) *Arch. Microbiol.* 152, 154–159.
Stanley, S.H., Prior, S.D., Leak, D.J., and Dalton, H. (1983) *Biotech. Lett.* 5, 487–492.
Strand, S.E., and Lidstrom, M.E. (1984) *FEMS Microbiol. Lett.* 21, 247–251.
Trotsenko, Y.A. (1983) *Acta Biotechnol.* 3, 269–277.
Tsien, H.C., Brusseau, G.A., Hanson, R.S., and Wackett, L.P. (1989) *Appl. Env. Microbiol.* 55, 3155–3161.
Tsien, H.C., Bratina, B.J., Tsuji, K., and Hanson, R.S. (1990) *Appl. Env. Microbiol.* 56, 2858–2865.
Tsuji, K., Tsien, H.C., Hanson, R.S., et al. (1990) *J. Gen. Microbiol.* 136, 1–10.
Wackett, L.P., Brusseau, G.A., Householder, S.R., and Hanson, R.S. (1989) *Appl. Env. Microbiol.* 55 (11), 2960–2964.
Ward, B.B., Kirkpatrick, K.A., Novelli, P.C., and Scranton, M.I. (1987) *Nature* 327, 226–228.

Whittenbury, R., and Dalton, H. (1981) in *The Prokaryotes* (Starr, M., Stolp, H., Truper, H.G., Balowes, A., and Schlegel, H.G., ed.) pp. 894–902, Springer-Verlag, KG, Berlin, FRG.

Whittenbury, R., and Krieg, N.R. (1984) in *Bergey's Manual of Systematic Bacteriology, vol. 1* (Krieg, N.R., and Holt, J.G., eds.), pp. 256–262, Williams and Wilkins, Baltimore, MD.

Whittenbury, R., Phillips, K.C., and Wilkinson, J.F. (1970) *J. Gen. Microbiol.* 61, 205–218.

Wilkinson, T.G., Topiwala, H.H., and Hamer, G. (1974) *Biotechnol. Bioengin.* 16, 41–59.

Wilson, J.T. (1988) *Biotech.* 2, 75–77.

Wilson, J.T., and Wilson, B.H. (1985) *Appl. Env. Microbiol.* 49, 242–243.

Woodland, M.P., and Dalton, H. (1984) *Anal. Biochem.* 139, 459–465.

Zobell, C.E. (1946) *Bacteriological Reviews* 10, 1–49.

INDEX

Acetaldehyde, 242, 255
Acetaldehyde dehydrogenase, 140
Acetobacter aceti, 296, 299
Acetobacter methanolicus, 17, 108
Acetone, 182–183
Acetylene, 136–138
N-Acetylglutamate kinase, 291
Acetylornithinase, 291
N-Acetylornithine glutamate
    transacetylase, 291
Achromobacter, 16
Achromobacter parvulus, 120
Acinetobacter, 16, 55
Aconitase, 102
Acrolein, 255
Active transport, 128–132
Aerobic growth
    balances and yields for, 151–152,
        162–164
    product formation, 157–167
Air-lift Loop Reactor, 215–218
Alcaligenes, 16
Alcohol assay, 242–243

Alcohol dehydrogenase, 238–239, 296,
    299
Alcohol oxidase, 70–71, 122, 184,
    186, 235, 239–244, 254–255,
    306, 311
    commercial production and uses,
        193–196
Alcohol oxidase promoter, 197
Alcohol oxidoreductases, 238–243
Aldehyde, commercial production of,
    254–255
Aldehyde dehydrogenase, 134–135,
    142, 144, 244
Aldehyde dismutase, 235
Aldehyde oxidoreductases, 243–245
Aldehyde reductase, 264
Aldolase, 81–82, 84–85, 90, 96
Aldose reductase, 263
Alkylhalidase, 237
Allose, 266
Alteromonas thalasso-methanolica, 14
Amicyanine, 135
Amidase, 235–237

Index

Amine dehydrogenase, 129
Amine oxidase, 70–71, 129–130, 134, 140–141, 179
Amine oxidoreductases, 245–246
Amino acid biosynthesis, regulation and genetics in bacteria, 289–294
D-Amino acid oxidase, 70
Amino acids
  commercial production of, 264–265
  synthesis from RuMP pathway intermediates, 91–92
D-Amino-acylase, 237
5-Aminolevulinic acid, 161
Ammonia-oxidizing bacteria, 340
AMP deaminase, 260
Amycolatopsis, 140
Anabaena flos-aquae, 64
Anaerobic growth, 167–169
  kinetics, balances, and yields for, 152–155
  methane oxidation, 341–342
Ancalomicrobium, 64
Anthranilate synthase, 293
AOX promoter, 306–318
Aphanenomenon flos-aquae, 62
Appendages, bacterial, 55–58
Aquatic ecosystem, 326, 338–340
  identification and quantitation of methylotrophs in, 330–334
  methane in, 334–336
  methane-oxidizing bacteria in, 336–338
Arginine biosynthesis, 291–292
Aromatic amino acid biosynthesis, 292–294
Arthrobacter, 16, 108, 119, 129–131, 134, 140–143, 178–180, 237, 245, 259, 343
Aspartate kinase, 290–291
Assimilation reactions, regulation of, 139
ATP, commercial production of, 259–261
ATPase, 51
Auxotrophy, in methylotrophic bacteria, 281–283

Bacillus, 16, 115, 118, 138, 140, 177, 182, 238, 343
Bacillus brevis, 17
Bacillus methanicus. See *Methylomonas methanica*

Benzaldehyde, 242
Benzene, 340
Benzylamine, 134
Biocatalysis, 235
Bioenergetic concepts, 150–157
Biogas, as SCP substrate, 210, 225–226
Biomass. *See also* Growth yield; Single cell protein
  weight fraction of carbon in, 150
Bioremediation, 157–160, 341
Biosensors, 235, 239, 243, 246
Blastobacter aminooxidans, 55
Bleach, 241
Brevibacterium, 16, 140
Bromomethane, 340
2,3-Butanediol dehydrogenase, 238

Calvin cycle. *See* Ribulose bisphosphate pathway
Candida, 26, 184, 240, 266, 343
Candida boidinii, 26–27, 29–34, 71, 161, 185–186, 190, 240, 254–255, 259–265
Candida cariosilignicola, 27, 29, 32
Candida entomophila, 27, 29, 30, 33
Candida glucophila, 29
Candida maris, 27, 29, 32
Candida methanolica, 124
Candida methanolophaga, 27, 29
Candida methanolovescens, 32
Candida methanosorbosa, 27, 29, 33
Candida methylica, 27, 29, 33
Candida molischiana. See Hansenula capsulata
Candida nanaspora, 29
Candida nemodendra, 27, 29, 32
Candida nitratophila, 27, 29, 32
Candida ooitensis, 27, 29
Candida ovalis, 27, 29
Candida pignaliae, 27, 29, 32
Candida pinus, 27, 29, 32
Candida sonorensis, 27, 29–30, 32–33
Candida succiphila, 27, 29, 32
Candida utilis, 191, 238–239
Carbohydrate, synthesis from products of RuMP pathway, 88–91
Carbon assimilation, 79–109. *See also specific pathways*
Carbon monoxide dehydrogenase, 55
Carbon monoxide oxidase, 235
Carbon monoxide oxidoreductases, 245

Carbon monoxide-utilizing bacteria, 55, 158, 168, 340–341
Carbon tetrachloride, 159
Carbon-sulfur bond-attacking oxidoreductases, 246
Carboxylic ester hydrolases, 236
β-Carotene, 6
Carotenoid, 6
Catalase, 70–71, 123, 194–195, 235, 240–242, 254
Cell wall
  bacterial, 40–43
  mannans of methanol-utilizing yeast, 27, 30, 33–34
Chemical mutagenesis, 281
Chemicals, commercial production of, 254–257
Chlorinated aliphatic hydrocarbons, 157–160, 169
Chloromethane, 158, 163
Chorismate mutase, 293
Circulation time, in fermentor, 212–214
Citrate synthase, 102
Citric acid production, 266
Cloning vector
  bacterial, 289
  yeast, 307–312
Clostridium thermoaceticum, 247
Codon usage, in bacteria, 277–281
Coenzyme Q$_{10}$. *See* Ubiquinone
Complementation mapping, in bacteria, 284–286
Conjugation, bacterial, 283–284, 289
Consensus sequence, in bacterial promoters, 275–277
Copper, 47, 53, 132, 135, 329
Corynebacterium, 16
cou mutants, 282
Crystalloid inclusion, 60–62
Cyanate lyase, 237
Cyanide hydratase, 237
Cyanide lyase, 237
Cyclohexane, 340
Cyst, bacterial, 68–69
Cytochrome *c*, 135–136, 294–295
  commercial production of, 257–258
  interaction with methanol dehydrogenase, 116–117
Cytochrome *c* peroxidase, 195, 241–242
Cytoplasmic inclusions, bacterial, 58–68

Cytoplasmic membranes, bacterial, 6–10, 43–44, 326

dcm genes, 276
Dehalogenases, 237–238
3-Deoxy-D-arabinoheptulose-7-phosphate synthase, 292–293
Detergent manufacture, 241–242
Diaphorase, 235
Dichloromethane, 158, 163
Dichloromethane dehalogenase, 237
Dihydroxyacetone, commercial production of, 196–197
Dihydroxyacetone cycle
  distribution and occurrence of, 108
  reactions of, 98–101
  in yeast, 80, 98–101
Dihydroxyacetone kinase, 196, 262
Dihydroxyacetone reductase, 239, 261–263
Dihydroxyacetone synthase, 70, 98–101, 122–123, 196
Dimethyl disulfide, 160, 163
Dimethyl formamide, 159
Dimethyl sulfate, 159
Dimethyl sulfide, 159–160, 163
Dimethyl sulfoxide, 159, 246
Dimethylamine, 159, 168
Dimethylformamidase, 235–237
Dipodascus, 29
DNA-DNA hybridization, in methane-oxidizing bacteria, 9
Down-stream processing, in SCP production, 222–224

Ecology
  of methanol-utilizing yeast, 34
  of methylotrophic bacteria, 325–344
Ecosystem misuse, 205–206
Electron transport, 135–136
  in methylamine-grown methylotrophs, 117
Energy consumption, in fermentor, 219–221
Energy regularity, 150
Energy transduction, 118
Enolase, 89, 97, 102
Enzymes
  of industrial potential, 233–247
  profile of methanol-utilizing yeast, 31
Epilimnion, 326, 337

Index

Epoxides, commercial production of, 256–257
Esterase, 235
Ethanol dehydrogenase, 177
Ethylene, 340
Eubacterium acidaminophilum, 246
Eubacterium limosum, 178
Eutrophic lake, 335–337, 341–342
Exospore, bacterial, 68–69
Extracellular polymer, 40–41

Facultative methylotroph, 4, 16–17, 174
   aromatic amino acid synthesis in, 292–294
   mixed substrate utilization by, 177–180
   RuMP pathway in, 90
Fatty acids, of methylotrophic bacteria, 7–10, 15, 332
FCCP, 131
Feedback repression, 139
Feedstuffs, 207–208, 227–230
Fermentor
   bubble behavior in loops, 214–215
   circulation time in, 212–214
   design principles, 211–222
   energy consumption and gas utilization in, 219–221
   loop, 214–217
   mixing in, 211–212
   operation of aerated loops, 221
   for SCP production, 211–222
   for yeast culture, 310–312
Fimbrae, 55
Fish fry feed, 227
Fish meal, 207, 226–228
Flagella, 57–58
Flavobacterium, 16, 256
Fluorescent-antibody-membrane filter-staining technique, 331
F_{420}-NADP-oxidoreductase, 259
Formaldehyde
   assimilation via serine pathway, 101–107
   commercial production of, 254–255
   growth on, 158, 161–162, 165–166
   heat and free energy of combustion, 150
   oxidation of, 131–135
      by dissimilatory DHA cycle, 98–100
      by dissimilatory RuMP cycle, 91–93
      to formate, 112, 118–119, 123–124
   three-carbon compounds from, 88
   toxic levels of, 119, 123–124, 134
   transport of, 131
Formaldehyde dehydrogenase, 51, 118–119, 123, 134, 137, 186, 235, 243–244, 254, 260
Formaldehyde dismutase, 244, 256
Formaldehyde reductase, 123
Formamide, 159
Formate
   commercial production of, 256
   growth on, 158, 161–162, 165–166, 168
   heat and free energy of combustion, 150
   oxidation to carbon dioxide, 112, 119–120, 124
Formate dehydrogenase, 113, 119–120, 123, 137, 186, 244–245, 254, 259–260
   commercial production and uses, 195
   NAD^+-dependent, 51, 119–120, 124, 134, 176
   $NADP^+$-dependent, 119–120
Formate oxidoreductases, 243–245
S-Formylglutathione, 120
Fossil fuel, 327
Free energy of combustion, 150
Freshwater ecosystems, 326, 334–336, 341–342
Fructose 1,6-bisphosphatase, 81–82, 85, 99–100, 139, 187
Fructose 1,6-bisphosphate aldolase, 96, 99–100, 140, 142, 144
Fumarase, 102

Galactomyces, 29
β-Galactosidase, *E. coli*, expression in yeast, 310, 312–314
Gaprin, 228
Gas utilization, in fermentor, 219–221
Gas vacuole, 62–64
GC content, of methanol-utilizing yeast, 27, 30, 33–34
Gene cloning, in bacteria, 289
Genetic exchange systems, bacterial, 283–284
Genetics, bacterial, 273–301

Glucose-6-phosphate dehydrogenase, 89, 92–95, 100, 112, 120–121, 134, 140, 300
Glucose phosphate isomerase, 89, 92, 94, 100
Glutamate production, 264
Glyceraldehyde phosphate dehydrogenase, 81–82, 84, 90, 99
Glycerate kinase, 102, 105, 139, 296, 300
Glycerol
  commercial production of, 196–197
  production in yeast, 261–263
Glycerol dehydrogenase, 261–262
Glycerol kinase, 196, 261, 263
Glycerol-3-phosphate dehydrogenase, 262–263
Glycogen storage, 63, 67–68
Glycosylation, of heterologous proteins in yeast, 318–319
gpd genes, 300
Greenhouse gases, 327–328
Growth yield, 118, 161, 165–167
  aerobic, 151–152, 162–164
  anaerobic, 152–155, 167–169
  statistical estimation of parameters, 155–157

1-Haloalkane dehalogenase, 237
Hansenula, 26, 184, 240, 343
Hansenula capsulata, 26–28, 30–34
Hansenula glucozyma, 27–28, 31–32
Hansenula henricii, 27–28, 32
Hansenula ofunaensis, 27–28, 33, 261–262
Hansenula philodendra, 27–28, 32
Hansenula polymorpha, 27–28, 31–35, 69–71, 122, 158, 161, 184–191, 193–196, 238, 240–243, 256, 261, 265
Hansenula wickerhamii, 27–28, 32
Heat of combustion, 150
Heat of fermentation, 152
Heavy metal, in polyphosphate body, 67
Hepatitis B surface antigen, expression in yeast, 310, 313–315
Heterologous gene expression, in yeast, 197, 305–320
Hexose phosphate isomerase, 100
3-Hexulose phosphate isomerase, 89–90, 92, 94, 112

Hexulose phosphate pathway, 7
Hexulose phosphate synthase, 89–90, 112, 120, 179
3-Hexulose phosphate synthase, 92–94
Hexulose-6-phosphate isomerase, 140–144, 334
Hexulose-6-phosphate synthase, 130–131, 139–144, 334
Hfr-like mapping, 285–288
Holdfast, 40–41
Homoisocitrate lyase, 107
Homoserine dehydrogenase, 290–291
Homoserine kinase, 290–291
Hydrocarbon seep, 339
Hydrogen cyanide, 159
Hydrogen peroxide generation, 241–242
Hydrogen sulfide, 160
Hydrolase, industrial potential of, 236–238
L-α-Hydroxyacid oxidase, 70
Hydroxylamine oxidoreductase, 340
S-Hydroxymethyl glutathione, 123
Hydroxypyruvate reductase, 102, 104, 139, 334
Hyphomicrobia, 103
Hyphomicrobium, 16, 40, 54–55, 65, 109, 116, 118, 129, 158–160, 163, 176–177, 236–239, 244, 246, 266, 343
Hyphomicrobium facilis, 282
Hyphomicrobium vulgare, 58–59
Hypolimnion, 326, 335

Iditol, 263
Iditol dehydrogenase, 263
Immunological assay, for methylotrophic bacteria, 331–332
Intracellular transport, of heterologous proteins in yeast, 319
Invertase, *S. cerevisiae*, expression in *Pichia*, 310, 313–314, 317–319
Isocitrate lyase, 102–103, 106–107, 109, 140, 143, 179
Isopropanol, 182–183

Jet Loop Reactor, 215–219, 221, 223–224

2-Keto 3-deoxy 6-phosphogluconate aldolase, 88–89, 91, 95, 108
β-Ketoester reductase, 235, 245
Ketone reductase, 238
Kloeckera, 26, 185–187, 190, 239, 343

Landfill, 327
Leucine production, 265
Lipid cyst, 68–69
Lipids, synthesis from RuMP pathway intermediates, 91–92
Loop fermentor, 214–217
Lyase, industrial potential of, 236–238
Lysozyme, bovine, expression in yeast, 311, 313–314, 318–319

Maintenance requirement, 160–161, 164–169
Malate dehydrogenase, 102, 105
Malate synthase, 179
Malate thiokinase, 102, 105–106
Malic enzyme, 119
Malnutrition, 205–206
Malyl-CoA lyase, 102–103, 106, 139, 296, 299–300
Malyl-CoA synthetase. *See* Malate thiokinase
Mapping
 in bacteria, 284–285
 complementation, 285–286
 conjugation, 284
 Hfr-like, 285–288
Marine ecosystems, 334–336, 338–339, 341–342
Marker exchange, 275, 283
mcl genes, 299–300
mdh genes, 334
Melanine, 6
Membranous sheet, 58–60
Meromictic lake, 336
Mesosome, 44
Metachromatic body. *See* Polyphosphate body
Metalimnion, 326, 337
Methane
 anaerobic oxidation of, 341–342
 in aquatic ecosystems, 334–336
 atmospheric, sinks for, 339–340
 in biosphere, 326–328
 growth on, 158, 160–162, 166
 heat and free energy of combustion, 150

oxidation of, 112–114, 132–133
 as SCP substrate, 210
 in soils, 339–340
Methane monooxygenase, 51, 53, 113–114, 132–133, 136–138, 235, 238, 246, 256–257, 294
 ecological importance of, 328–330
Methane oxidation genes, 294
Methanethiol, 159–160, 163
Methanethiol oxidase, 246
Methane-utilizing bacteria, 112–114
 in aquatic ecosystems, 336–338
 estimation in environmental samples, 330–334
 taxonomy of, 4–21
 ultrastructure of, 44–51
Methane-utilizing yeast, 34
Methanobacterium formicicum, 168–169
Methanobrevibacter arboriphilus, 168
Methanococcus vannielii, 67
Methanol
 growth on, 158, 161–162, 164–169
 heat and free energy of combustion, 150
 as industrial feedstock, 253
 mixed culture growing on, 180–181
 oxidation of, 133
 oxidation to formaldehyde, 112, 114–116, 122–123
 as SCP substrate, 210, 254
 transport of, 128
Methanol dehydrogenase, 51, 53, 113–118, 128, 131–138, 235, 239, 258, 294–301, 331–334
 dye-linked, 334
 interaction with cytochrome *c*, 116–117
Methanol oxidase, 31–33, 122–123
Methanol oxidation genes, 294–299
Methanolomonas glucoseoxidans, 19
Methanol-regulated genes, in yeast, 307–308
Methanolsarcina, 168
Methanol-utilizing bacteria, 343
 obligate, 13–16
 restricted facultative, 13–16
 ultrastructure of, 51–54
Methanol-utilizing yeast
 applications of, 34–35
 identification and nomenclature of, 26–30
Methanomonas, 6
Methanomonas margaritae, 47, 52

Methanomonas methanica. See
  Methylomonas methanica
Methanomonas methanooxidans, 45
Methanomonas methylovora, 19, 264
Methanosarcina, 62–63
Methanosarcina barkeri, 62, 159, 168
Methanosarcina vacuolata, 64
Methazotroph, 174
Methionine production, 265
Methyl ketone, commercial
  production of, 257
Methyl sulfate, 159
Methylamine, 135, 163, 168, 177–179
  oxidation of, 117, 133–134
  transport of, 129–130
Methylamine dehydrogenase, 117,
  129, 135, 245
Methylamine oxidase, 245
Methylamine-utilizing bacteria, 54–
  55, 343
Methylated sulfur compounds, 343–
  344
Methylene chloride, 182
N^5, N^{10}-Methylene tetrahydrofolate
  dehydrogenase, 119
Methylglutamate dehydrogenase, 129
O-Methyl-L-homoserine, 265
Methyl-D-mannose, 266
Methylobacillaceae, 15–16, 21
Methylobacillus, 13–15, 18–19, 21
Methylobacillus flagellatum, 121,
  128–129, 135, 274–275, 277,
  280–282, 284–293, 300
Methylobacillus glycogenes, 13
Methylobacter, 11, 45, 68–69, 326,
  338–339
Methylobacterium, 15–21, 40, 64,
  276–277, 295
Methylobacterium ethanolicum, 16,
  47
Methylobacterium extorquens, 4, 17,
  109, 115–118, 128–129, 138–
  139, 276–279, 282, 284, 295–
  300
Methylobacterium fujisawaense, 17
Methylobacterium hypolimneticum,
  16, 47
Methylobacterium mesophilicum, 17
Methylobacterium organophilum, 8,
  16, 41, 47, 53, 58, 109, 139,
  178, 239, 256, 276–277, 283,
  295–300, 332
Methylobacterium radiotolerans, 17
Methylobacterium rhodesianum, 17
Methylobacterium rhodium, 17, 55,
  58
Methylobacterium zatmanii, 17
Methylococcaceae, 5, 9, 20–21
Methylococceae, 10
Methylococcus, 5–6, 10–11, 18–20, 45,
  57, 68–69, 162, 326
Methylococcus capsulatus, 11–12, 44–
  48, 52–53, 60, 94, 114, 119–
  120, 132–133, 136, 138, 158,
  162, 175, 226, 256, 283, 294,
  325–329, 337, 340
Methylococcus gracilis, 8, 11
Methylococcus luteus, 6
Methylococcus mobilis, 6, 13, 47, 57,
  61–62, 67
Methylococcus thermophilus, 6, 8, 12,
  131, 136
Methylocystis, 6, 10–12, 18–19, 45,
  57, 68, 326, 338–339
Methylocystis chroococcum, 337
Methylocystis echinoides, 7, 13, 48,
  52–57, 337
Methylocystis methanolicus, 337
Methylocystis parvus, 8, 13, 69, 337
Methylomicrobium, 17
Methylomonadaceae, 5
Methylomonas, 5–6, 10–11, 18, 45–
  46, 60–61, 68–69, 128, 132,
  134–135, 139, 175–176, 257–
  258, 276, 295–297, 326, 338
Methylomonas albus, 8, 13, 19, 136,
  276, 295–297
Methylomonas aminofaciens, 265
Methylomonas clara, 14, 275–276
Methylomonas gracilis, 333
Methylomonas methanica, 4–6, 8–12,
  48, 51–54, 281, 325, 331, 337–
  338
Methylomonas methanolica, 52, 161–
  162
Methylomonas methanolophila, 265
Methylomonas methanolovorans, 167
Methylomonas methanoxidans, 54
Methylomonas pelagica, 6, 13, 338
Methylomonas rubra, 6–8, 11, 131,
  136
Methylomonas rubrum, 337
Methylomonas thalassica, 14
Methylophaga, 14–15, 18, 21
Methylophaga marina, 14
Methylophaga thalassica, 14
Methylophilus, 14–15, 18–19, 21
Methylophilus methanolovorus, 19, 66

Methylophilus methylotrophus, 14, 19, 94, 108, 115–116, 121, 131, 135–138, 210, 213, 274, 282–286, 291–292, 295, 301
Methylophilus viscogenes, 285–286
Methylosineae, 9–10
Methylosinus, 5–6, 10–12, 18–19, 45, 47, 57, 68–69, 283, 326
Methylosinus sporium, 8, 12–13, 69, 337–338
Methylosinus trichosporium, 8, 12–13, 40, 47–52, 68, 109, 113–114, 132, 136, 138, 238, 246, 256, 293, 328–333, 337–340
Methylosporovibrio methanica, 295–296
Methylotrophic bacteria, relation to other prokaryotes, 17–20
Methylovarius, 5–6, 9–11, 18–19
Methylovarius luteus, 8, 11–12
Methylovarius ucrainicus, 8, 11–12
Methylovarius vinelandii, 11–12
Methylovarius whittenburyi, 11–12
Methylphosphonate, 236
Microbody, 69–71
Microcystis aquaticus, 64
Mixed culture, 173–175
　growth on methanol, 180–181
　growth on mixed substrates, 182–183
Mixed substrate utilization, 174
　applied aspects of, 191–197
　by mixed bacterial cultures, 180–183
　by pure bacterial culture, 175–180
　by yeast, 184–197
Mixing, in fermentor, 211–212
mmo genes, 294
Molecular genetics, bacterial, 273–301
Monod model, 153
Monomethylamine, 159
Monomethylsulfatase, 235–236
Monooxygenases, 246–247
Moraxella, 120
mox genes, 276–279, 294–299
Mussel, methane-consuming, 339
Mutagenesis
　in bacteria, 281–283
　chemical, 281
　site-directed, 275, 283
　transposon, 281–283
Mycobacterium, 16, 44, 140, 343
Mycoprotein, 209

NAD(P)H, commercial production of, 259
NADH dehydrogenase, 51, 134
NADH oxidase, 136
Natural gas
　as industrial feedstock, 253
　as SCP substrate, 211, 224–226
Nitrogen fixation, 337
Nitrogen source, 177–178
Nocardia, 115, 134–135, 140–145, 178, 239, 243–244, 292, 343
Nocardia corallina, 256
Nostoc muscorum, 64
Nucleic acid probe, for methylotrophic bacteria, 332–334

Obligate methylotroph, 4, 13–16, 20–21, 174
　aromatic amino acid synthesis in, 292–294
　mixed substrate utilization by, 175–176
　RuMP pathway in, 89
Oligotrophic lake, 336–337
One-carbon compounds
　regulation of assimilation of, 139
　transport into cells, 128–132
ORF 2, 297–298
Ornithine transcarbamylase, 291
Oxic/anoxic boundary, 335–338
Oxidation pathways
　cyclic, 120–122
　bacterial, 111–122
　regulation of, 128–138
　in yeast, 122–124
Oxidoreductases, industrial potential of, 238–247

Paracoccus denitrificans, 53, 107, 129, 131, 133, 135–136, 167, 277, 283, 296–299, 343
Pathogenic bacteria, 181
Peat bogs, 335–336
Pentose phosphate epimerase, 81–82, 86, 89–90, 96, 99–100
Pentose phosphate isomerase, 81–82, 87, 89–90, 96, 99
Peptostreptococcus productus, 158, 168
Peroxidase, 235
Peroxisome, 69–71, 122–123, 260, 306
pgi genes, 300

Phenol, 182, 340
Phenylalanine hydroxylase, 293
Phenylalanine production, 265
Phosphoenolpyruvate carboxykinase, 92, 97
Phosphoenolpyruvate carboxylase, 92, 97, 102, 105, 139, 296, 299–300
Phosphoenolpyruvate synthetase, 89, 92, 97
6-Phosphofructokinase, 90, 96, 144, 187
Phosphoglucoisomerase, 112, 120, 300
6-Phosphogluconate dehydrase, 121
6-Phosphogluconate dehydrogenase, 89, 91–92, 95, 100, 112, 120–121, 134, 140
3-Phosphoglycerate dehydrogenase, 292
Phosphoglycerate kinase, 81–82, 84, 90, 99
Phosphoglycerate mutase, 89, 97, 102, 105
Phosphohexulosisomerase, 120
Phospholipids, of methylotrophic bacteria, 9, 332
Phosphoriboisomerase, 112
Phosphoribulokinase, 81–82, 86–87, 139
Phosphoric ester hydrolases, 236
Phosphoserine phosphatase, 292
Pichia, 26, 184, 240, 343
Pichia angusta. See Hansenula polymorpha
Pichia capsulata. See Hansenula capsulata
Pichia cellobiosa. See Pichia methanolica
Pichia finlandica. See Hansenula wickerhamii
Pichia glucozyma. See Hansenula glucozyma
Pichia henricii. See Hansenula henricii
Pichia kodamae, 27–28, 33
Pichia methanolica, 27–28, 30–34
Pichia methylovora, 27–28
Pichia minuta, 27–28, 31–32
Pichia naganishii, 27–28, 32
Pichia pastoris, 27–28, 30–34, 69, 122, 161, 193, 197, 240, 243, 255

heterologous gene expression in, 305–320
Pichia philodendra. See Hansenula philodendra
Pichia pini. See Pichia pinus
Pichia pinus, 26–28, 32, 191
Pichia trehalophila, 27–28, 32
Pili, 55
Pink-pigmented facultative methylotrophs (PPFM), 21, 53, 58, 64, 67, 103, 109, 133, 266, 343–344
Plant seed meal, 226–228
Plasmid
   bacterial, 274–275, 284–285
   prime, 284–285
   suicide, 282, 287
Plasmid IncP1, 284
Plasmid pAO804, 308–310
Plasmid pAS8-121, 287
Plasmid pMD100, 282
Plasmid pMO75, 282
Plasmid pRK2013, 282
Plasmid pULB113, 285
Plasmid R68.45, 284
Polar body, 60–61
Polar organelle, 60–61
Poly-β-hydroxyalkanoate granule, 64–65
Poly-β-hydroxybutyrate granule, 64–65
Poly-β-hydroxybutyric acid, commercial production of, 265–266
Poly-β-hydroxyvalerate granule, 64
Polyol, production in yeast, 263–264
Polyphosphate body, 65–67
Polyphosphate overplus phenomenon, 66
Post-translational modification, of heterologous proteins in yeast, 306, 317–319
ppc genes, 299–300
PPFM. See Pink-pigmented facultative methylotrophs
PQQ. See Pyrroloquinoline quinone
pqq genes, 296–297
Prephenate dehydratase, 293
Prephenate dehydrogenase, 293
Pressure cycle fermentor, 213, 219–220
Prime plasmid, 284–285
Prodiginine, 6
Product formation

aerobic, 151–152, 157–167
anaerobic, 152–155, 167–169
Promoter
  bacterial, 275–277
  foreign, expression in bacteria, 277
  yeast, 306–308. *See also AOX promoter*
Propeller Loop Reactor, 216–217
Propionaldehyde, 255
Propionaldehyde dehydrogenase, 140
Propylene, 340
Propylene oxide, 256
Prostheca, 58–59
Prosthecomicrobium pneumaticum, 64
Protaminobacter, 16
Protaminobacter ruber. *See Pseudomonas extorquens*
Protein deficiency, 205–206
Protein factors, modulation of C_1 enzymes by, 136–138
Protein profile, of methane-oxidizing bacteria, 11–13
Proteobacteria
  alpha group of, 20
  beta group of, 19
  gamma group of, 19
Protomonas, 4, 17
Protomonas extorquens, 259
Protonophore, 131–132
Pruteen, 210, 228
Pseudomonas, 16, 109, 119–120, 129, 135, 158, 161–165, 176, 178, 181, 237–238, 264, 266, 292–293
Pseudomonas aminovorans, 109
Pseudomonas carboxydovorans, 55
Pseudomonas extorquens, 52, 265–266
Pseudomonas methanica, 4
Pseudomonas methylotropha, 52, 162
Pseudomonas oxalaticus, 107
Pseudomonas putida, 243, 256
Pseudomonas rhodos. *See Methylobacterium rhodium*
Pseudomonas rosea, 162–163
Pyrroloquinoline quinone (PQQ), 114–116, 135, 239
  commercial production of, 258–259
Pyruvate carboxylase, 92, 97
Pyruvate dehydrogenase, 103, 119, 176
Pyruvate kinase, 92, 97–98, 103, 119
Pyruvate phosphate dikinase, 92, 98

Quinone, of methanol-utilizing yeast, 27, 30, 33–34

recA gene, 274–275, 277, 280
RecA protein, 274–275
Recombination, in bacteria, 274–275
Respiratory chain, 135–136
  commercial production of coenzymes, 257–261
Resting cells, bacterial, 10, 68–69
Restricted facultative methylotroph, 4, 13–16, 21, 343
Restricted methylotroph, mixed substrate utilization by, 176–177
Retroconjugation, 284
Rhodococcus erythropolis, 135, 243
Rhodomicrobium, 54–55
Rhodomicrobium vannielii, 69
Rhodopseudomonas, 343
Rhodopseudomonas acidophila, 107
Rhodopseudomonas gelatinosa, 55
Rhodopseudomonas sphaeroides, 106
Rhodospirillum rubrum, 65
Rhodotorula glutinis, 34
Rhodotorula rubra, 34
Ribose 5-phosphate isomerase, 187
Ribosomal RNA, bacterial, 9, 332–334
Ribulose bisphosphate (RuBP) pathway
  distribution and occurrence of, 107–108
  reactions of, 80–87
  regulation of, 139
Ribulose monophosphate (RuMP) pathway, 7, 80, 87–98, 119, 326, 343
  assimilatory, 113
  dissimilatory, 120–122
  distribution and occurrence of, 108
  regulation of, 139–145
Ribulose-1,5-bisphosphate carboxylase, 81–83, 139
Ribulose-1,5-bisphosphate oxygenase, 139
Ribulose 5-phosphate epimerase, 187
Rice paddy, 327–328
Rod-shaped inclusion, 60–62
RuBP pathway. *See* Ribulose bisphosphate pathway
Ruminants, 327–328
RuMP pathway. *See* Ribulose monophosphate pathway

Saccharomyces, 343

Index

Saccharomyces cerevisiae, 66, 213, 306
Scale-up, of yeast culture, 310–312
Secondary alcohol dehydrogenase, 257
Secretion, of heterologous proteins in yeast, 306, 316–319
Sediments, 335–338, 340–342
Sedoheptulose bisphosphatase, 81, 85, 88, 90, 96, 108
Sedoheptulose bisphosphate aldolase, 99
Sedoheptulose bisphosphate phosphatase, 99
Serine biosynthesis, 264–265, 291–292
Serine dehydratase, 264
Serine glyoxylate aminotransferase, 139, 334
Serine pathway, 7, 80, 119, 264, 326, 343
  distribution and occurrence of, 108–109
  genetics of, 299–300
  reactions of, 101–107
  regulation of, 139
Serine transhydroxymethylase, 102, 104, 139, 264
Serine-glycine aminotransferase, 102, 104
Shikimate pathway, 292
Signal sequence, 319
Signature probe, 332–334
Single cell protein, 25
  drying of, 224
  feedstock for, 253
  as feedstuff, 207–208
  in human diet, 206–207
  marketable products and markets, 226–230
  production of
    autotrophic, 208–209
    costs of, 211
    down-stream processing, 222–224
    fermentor design for, 211–222
    methylotrophic, 209–210
    non-C_1 heterotrophic, 209
    production requirements, 210–211
    substrates for, 224–226
  public opinion about, 229–230
  quality of, 227–228
  sterility of, 223–224
  substrates for, 160–161, 180–181
  use of, 205–210
  water removal from, 222–223
  yeast, 34–35, 183, 192–193, 306
Site-directed mutagenesis, 275, 283
Soil
  identification and quantitation of methylotrophs in, 330–334
  methane in, 339–340
  methylotrophs in, 335
Sorbitol, commercial production of, 196, 263
Sorbitol dehydrogenase, 196
SOS genes, 274
Spinae, 55–57
Sporobolomyces gracilis, 34
Sporobolomyces roseus, 34
Star-forming bacteria, 55
Stemphylium loti, 237
Storage polymer, 64–68
Streptokinase, expression in yeast, 313–315
Succinate dehydrogenase, 51, 102–103
Sugar/methanol mixture, 184–191
Suicide plasmid, 282, 287
Sulfate ester hydrolases, 236
Surface layer, bacterial, 40–55
Swamps, 335–336
Symbiont, 339

Taxonomic distance value, 17–18
Taxonomy
  of methylotrophic bacteria, 3–20, 326
  of methylotrophic yeast, 25–35
Terminal oxidase, 135–137
Thiobacillus, 159–160
Thiobacillus novellus, 107
Thiobacillus thioparus, 159–160, 163
Thiobacillus versutus, 107
thr genes, 290–291
Threonine
  biosynthesis of, 289–291
  transport of, 289–290
Threonine synthetase, 291
Torulopsis, 26, 240, 343
Transaldolase, 82, 87–89, 96, 99, 141
Transduction, bacterial, 283
Transformation
  bacterial, 283
  in yeast, 306, 308–310
Transketolase, 81–82, 85–86, 89–90, 96, 99
Transport, 128–132

of amino acids, 289–290
of formaldehyde, 131
of methanol, 128
of methylamine, 129–130
Transposon mutagenesis, 281–283
Trichloroethylene, 329, 341
Trichloromethane, 158
Trimethyl ammonium salts, 159
Trimethyl phosphate, 236
Trimethyl sulfonium salts, 159
Trimethylamine, 159, 168, 175
Trimethylamine N-oxide, 159
Triokinase, 99–101
Triose phosphate isomerase, 81–82, 84, 90, 99
trp genes, 293–294
Tryptophan production, 265
Tryptophan synthase, 293–294
Tubules, 60–62
Tumor necrosis factor, expression in yeast, 310, 313–316
Tyrosine production, 265

Ubiquinone
   commercial production of, 259
   of methanol-utilizing bacteria, 15
Ultrastructure
   of methylotrophic bacteria, 40–69
   of methylotrophic yeast, 69–71

Valine production, 265
Viable cell count, 330

Vibrio succinogenes, 158
Vitamin B_{12} production, 266
Volutin body. *See* Polyphosphate body

Wastewater treatment, 180, 182–183, 234
Western blot, 331–332

Xanthobacter autotrophicus, 237, 283
Xanthomonas maltophilia, 51
Xenobiotics, 234
Xylitol, 264
Xylitol dehydrogenase, 264
Xylose isomerase, 264
Xylulose monophosphate pathway, 186

Yeast
   cells with specific catalytic activity, 195–196
   dihydroxyacetone cycle in, 80, 98–101
   expression of heterologous proteins in, 197, 305–320
   glycerol production in, 261–263
   mixed substrate utilization by, 184–197
   oxidation pathways in, 122–124
   polyol production in, 263–264
   single cell protein, 192–193
   taxonomy of, 25–35